W9-CSF-274

Laser Probes
for Combustion Chemistry

Laser Probes
for Combustion Chemistry

David R. Crosley, EDITOR

SRI International

Based on a symposium
sponsored by the Division
of Physical Chemistry at
the 178th Meeting of the
American Chemical Society,
Washington, D.C.,
September 9–14, 1979.

ACS SYMPOSIUM SERIES 134

AMERICAN CHEMICAL SOCIETY
WASHINGTON, D. C.   1980

6417-0056 ✓   sep/ae chem

Library of Congress CIP Data

Laser probes for combustion chemistry.
(ACS symposium series; 134 ISSN 0097–6156)

Includes bibliographies and index.

1. Combustion—Congresses. 2. Laser spectroscopy—Congresses.
I. Crosley, David R., 1941–   . II. American Chemical Society. Division of Physical Chemistry. III. Series: American Chemical Society. ACS symposium series; 134.

QD516.L23   541.3'61   80–17137
ISBN 0–8412–0570–1   ACSMC8   134 1–495 1980

Copyright © 1980

American Chemical Society

All Rights Reserved. The appearance of the code at the bottom of the first page of each article in this volume indicates the copyright owner's consent that reprographic copies of the article may be made for personal or internal use or for the personal or internal use of specific clients. This consent is given on the condition, however, that the copier pay the stated per copy fee through the Copyright Clearance Center, Inc. for copying beyond that permitted by Sections 107 or 108 of the U.S. Copyright Law. This consent does not extend to copying or transmission by any means—graphic or electronic—for any other purpose, such as for general distribution, for advertising or promotional purposes, for creating new collective works, for resale, or for information storage and retrieval systems.

The citation of trade names and/or names of manufacturers in this publication is not to be construed as an endorsement or as approval by ACS of the commercial products or services referenced herein; nor should the mere reference herein to any drawing, specification, chemical process, or other data be regarded as a license or as a conveyance of any right or permission, to the holder, reader, or any other person or corporation, to manufacture, reproduce, use, or sell any patented invention or copyrighted work that may in any way be related thereto.

PRINTED IN THE UNITED STATES OF AMERICA

QD 516
L 23
CHEM

ACS Symposium Series

M. Joan Comstock, *Series Editor*

Advisory Board

David L. Allara

Kenneth B. Bischoff

Donald G. Crosby

Donald D. Dollberg

Robert E. Feeney

Jack Halpern

Brian M. Harney

Robert A. Hofstader

W. Jeffrey Howe

James D. Idol, Jr.

James P. Lodge

Leon Petrakis

F. Sherwood Rowland

Alan C. Sartorelli

Raymond B. Seymour

Gunter Zweig

02181

FOREWORD

The ACS SYMPOSIUM SERIES was founded in 1974 to provide a medium for publishing symposia quickly in book form. The format of the Series parallels that of the continuing ADVANCES IN CHEMISTRY SERIES except that in order to save time the papers are not typeset but are reproduced as they are submitted by the authors in camera-ready form. Papers are reviewed under the supervision of the Editors with the assistance of the Series Advisory Board and are selected to maintain the integrity of the symposia; however, verbatim reproductions of previously published papers are not accepted. Both reviews and reports of research are acceptable since symposia may embrace both types of presentation.

CONTENTS

Preface xi

OVERVIEWS

1. Lasers, Chemistry, and Combustion 3
   David R. Crosley

2. Laser Probes for Combustion Applications 19
   J. R. McDonald

LASER-INDUCED FLUORESCENCE: MOLECULES

3. Laser-Induced Fluorescence Spectroscopy in Flames 61
   John W. Daily

4. Laser Probes of Premixed Laminar Methane–Air Flames and
   Comparison with Theory 85
   James H. Bechtel

5. Laser-Induced Fluorescence: A Powerful Tool for the Study of
   Flame Chemistry 103
   C. H. Muller, III, Keith Schofield, and Martin Steinberg

6. Laser-Induced Fluorescence Spectroscopy Applied to the Hydroxyl
   Radical in Flames 131
   M. J. Cottereau and D. Stepowski

7. A Multilevel Model of Response to Laser-Fluorescence Excitation
   in the Hydroxyl Radical 137
   Anthony J. Kotlar, Alan Gelb, and David R. Crosley

8. Saturated-Fluorescence Measurements of the Hydroxyl Radical ... 145
   Robert P. Lucht, D. W. Sweeney, and N. M. Laurendeau

9. Nitric Oxide Detection in Flames by Laser Fluorescence 153
   Daniel R. Grieser and Russell H. Barnes

10. Laser-Induced Fluorescence of Polycyclic Aromatic Hydro-
    carbons in a Flame 159
    Donald S. Coe and Jeffrey I. Steinfeld

11. Flow Visualization in Supersonic Flows 167
    N. L. Rapagnani and Steven J. Davis

LASER-INDUCED FLUORESCENCE: ATOMS

12. What Really Does Happen to Electronically Excited Atoms
    in Flames? 175
    Kermit C. Smyth, Peter K. Schenck, and W. Gary Mallard

13. **Collisional Ionization of Sodium Atoms Excited by One- and Two-Photon Absorption in a Hydrogen–Oxygen–Argon Flame** **183**
C. A. Van Dijk and C. Th. J. Alkemade

14. **On Saturated Fluorescence of Alkali Metals in Flames** **189**
C. H. Muller, III, Martin Steinberg, and Keith Schofield

15. **Saturation Broadening in Flames and Plasmas as Obtained by Fluorescence Excitation Profiles** **195**
N. Omenetto, J. Bower, J. Bradshaw, S. Nikdel, and J. D. Winefordner

16. **Determination of Flame and Plasma Temperatures and Density Profiles by Means of Laser-Excited Fluorescence** **199**
J. Bradshaw, S. Nikdel, R. Reeves, J. Bower, N. Omenetto, and J. D. Winefordner

SPONTANEOUS RAMAN SCATTERING

17. **Raman-Scattering Measurements of Combustion Properties** **207**
Marshall Lapp

18. **Temperature from Rotational and Vibrational Raman Scattering: Effects of Vibrational–Rotational Interactions and Other Corrections** **231**
Michael C. Drake, C. Asawaroengchai, and Gerd M. Rosenblatt

19. **Temperature–Velocity Correlation Measurements for Turbulent Diffusion Flames from Vibrational Raman-Scattering Data** **239**
S. Warshaw, Marshall Lapp, C. M. Penney, and Michael C. Drake

20. **Observations of Fast Turbulent Mixing in Gases Using a Continuous-Wave Laser** **247**
C. M. Penney, S. Warshaw, Marshall Lapp, and Michael C. Drake

21. **A Nd:YAG Laser Multipass Cell for Pulsed Raman-Scattering Diagnostics** **255**
Domenic A. Santavicca

22. **Time-Resolved Raman Spectroscopy in a Stratified-Charge Engine** .. **259**
J. Ray Smith

COHERENT RAMAN SPECTROSCOPY

23. **Spatially Precise Laser Diagnostics for Practical Combustor Probing** **271**
Alan C. Eckbreth

24. **CARS Measurements in Simulated Practical Combustion Environments** **303**
Gary L. Switzer, William M. Roquemore, Royce P. Bradley, Paul W. Schreiber, and Won B. Roh

25. **Update on CARS Diagnostics of Reactive Media at ONERA** **311**
M. Péalat, B. Attal, S. Druet, and J. P. Taran

26. **The Application of Single-Pulse Nonlinear Raman Techniques to a Liquid Photolytic Reaction** **319**
William G. Von Holle and Roy A. McWilliams

MODELLING AND KINETICS

27. **Detailed Modelling of Combustion: A Noninterfering Diagnostic Tool** 311
    Elaine S. Oran, Jay P. Boris, and M. J. Fritts

28. **Rate of Methane Oxidation Controlled by Free Radicals** 357
    John R. Creighton

29. **The Detailed Modelling of Premixed, Laminar, Steady-State Flames. Results for Ozone** 365
    Joseph M. Heimerl and T. P. Coffee

30. **On the Rate of the $O + N_2$ Reaction** 375
    Daniel J. Seery and M. F. Zabielski

31. **Reactions of $C_2(X^1\Sigma_g^+)$ and $(a^3\Pi_i)$ Produced by Multiphoton UV Excimer Laser Photolysis** 381
    Louise R. Pasternack, J. R. McDonald, and V. M. Donnelly

32. **Pulsed-Laser Studies of the Kinetics of $C_2O(\tilde{A}^3\Pi_i$ and $\tilde{X}^3\Sigma^-)$** 389
    V. M. Donnelly, William M. Pitts, and A. P. Baronavski

33. **Kinetics of CH Radical Reactions Important to Hydrocarbon Combustion Systems** 397
    J. E. Butler, J. W. Fleming, L. P. Goss, and M. C. Lin

34. **Carbon Monoxide Laser Resonance Absorption Studies of $O(^3P)$ + 1-Alkynes and Methylene Radical Reactions** 403
    W. M. Shaub and M. C. Lin

OTHER DIAGNOSTIC TECHNIQUES

35. **Absorption Spectroscopy of Combustion Gases Using a Tunable IR Diode Laser** 413
    R. K. Hanson, P. L. Varghese, S. M. Schoenung, and P. K. Falcone

36. **Multiangular Absorption Measurements in a Methane Diffusion Jet** 427
    Robert J. Santoro, H. G. Semerjian, P. J. Emmerman, R. Goulard, and R. Shabahang

37. **Temperature Measurement in Turbulent Flames Via Rayleigh Scattering** 435
    Robert W. Dibble, R. E. Hollenbach, and G. D. Rambach

38. **Droplet-Size Measurements in Reacting Flows by Laser Interferometry** 443
    Umberto Ghezzi, Aldo Coghe, and Fausto Gamma

39. **Continuous-Wave Intracavity Dye Laser Spectroscopy: Dependence of Enhancement on Pumping Power** 451
    Stephen J. Harris

40. **The Use of Photoacoustic Spectroscopy to Characterize and Monitor Soot in Combustion Processes** 457
    D. K. Killinger, J. Moore, and S. M. Japar

Index 463

PREFACE

Emission spectroscopy and, to a lesser degree, absorption spectroscopy have provided considerable information on and insight into the chemistry occurring during the process of combustion. In particular, many of the transient free-radical molecules important in the chain reactions were identified and characterized through their emission spectra in flames. Now, new laser spectroscopic techniques offer the promise of obtaining more detailed and precise information, especially for the ground electronic states of many of the molecules involved in combustion.

Over the past few years, several of these laser-based spectroscopic methods have been demonstrated and developed as probes of combustion systems. Taken as a group, they generally offer excellent spatial and temporal resolution, high sensitivity, and species selectivity. Each has categories of particular species and conditions for which it is best suited, and the several methods complement one another well. Although the techniques remain in various stages of development, useful results on both laboratory flames and practical combustors now are beginning to emerge. The profiles of species concentrations and temperatures produced can be compared with detailed models of the combustion chemistry, whose design and manipulation into realistic simulations have been made possible by mathematical and computational techniques evolving over roughly the same time period. The understanding of combustion chemistry that will result from this combination of laser probes and computer models will be important in efforts to design clean and efficient means of combustion of fuel in use now and those anticipated for the future.

This symposium is an attempt to capture an expression of the state of the art in this fascinating, fast-moving, and important field. Fall 1979 was a significant time for laser-combustion diagnostics, for there are just now appearing—as can be seen in this volume—papers describing the use of laser-based techniques to provide chemical information, in contrast to most past publications that have dealt solely or largely with feasibility demonstrations and/or technique development. The chapters in this symposium form an excellent and quite complete representation of the present capabilities of the laser methods and of the most active areas of research on current problems associated with their development. They range in general tone from preliminary reports of research in new areas of endeavor to descriptions of relatively complete pieces of work. It should be noted that the symposium focused on what may be termed laser spectroscopic probes, that is, techniques that are species selective and hence serve to furnish data directly relevant to the combustion chemistry. Thus some important laser-based flame probe methods such as laser Doppler velocimetry and laser schlieren and holographic techniques are not emphasized here.

One aspect evident in assembling these chapters was the menagerie of disciplines represented by the authors. Included among them are physical chemists, analytical chemists, physicists, mathematicians, mechanical engineers, and aerospace engineers, plus two whose formal degrees are in applied science. In addition these chapters are divided almost equally by institutional origin, among universities, government laboratories, and private (profit and nonprofit) corporations. Clearly the development, and application of, these methods has been and is a multidisciplinary effort. Nonetheless, it is appropriate that the symposium was held under the sponsorship of the Physical Chemistry Division of the American Chemical Society. The field of physical chemistry encompasses the spectroscopy, chemical kinetics, collisional energy transfer, gas dynamics, and thermodynamics that form the basic subfields underlying the development of laser probes for combustion and the understanding of combustion chemistry.

I hope that this book will serve to describe the excitement of current research in this field and the promise of a close coupling between the laser diagnostics and the modelling and chemical kinetics studies. Thanks are extended to the many contributors and to the publishers, and especially to Judy Turner for her assistance in organizing the symposium and assembling this volume.

SRI International
333 Ravenswood Avenue
Menlo Park, CA 94025

April, 1980

DAVID R. CROSLEY

OVERVIEWS

Lasers, Chemistry, and Combustion

DAVID R. CROSLEY

Molecular Physics Laboratory, SRI International, Menlo Park, CA 94025

The development of a detailed, microscopically based understanding of the chemistry of combustion represents a considerable
challenge. This is because so many chemical and physical processes interact together to produce the phenomenon of combustion.
What we know as a flame involves a large chemical reaction network
through which energy is produced, resulting in steep gradients of
both molecular concentrations and temperature; but the reaction
rates and mechanistic paths of that network are sensitively dependent on those parameters. Thus, the chemistry is woven inex-
tricably with heat transfer and diffusion of reactive species
through the gas flow.

All of these aspects must be considered together with care
and attention given to the proper selection of the molecular input
parameters (rate constants and transport coefficients) as well as
to detailed, unambiguous verification of the results. This is no
small task, and we have at hand such a picture for only the simplest of flames: O_3 decomposition, H_2/O_2, H_2/F_2 and, to a lesser
degree, CH_4/O_2. But if the goal of a microscopically based description can be accomplished for a larger class of combustion
systems, then we shall have a powerful tool--a picture of combustion with an independent foundation, having the predictive ability, quantitative and qualitative, which comes with understanding
a phenomenon instead of merely parametrizing it.

Recently, there have appeared two important new technological
advances, which make possible this kind of detailed, fundamental
approach. Laser methods enable us to measure species concentrations and temperatures for a wide variety of molecules over a
range of conditions, and to do so with high spatial resolution.
Large scale computational techniques permit the formulation of
realistic simulation models of the combustion system, including

0-8412-0570-1/80/47-134-003$05.00/0
© 1980 American Chemical Society

the chemistry and the fluid dynamics, and can incorporate sensitivity analysis to indicate which species and rate constants merit special attention. A coupling of these tools will provide a microscopically-based description of combustion with the required detailed experimental verification. Additionally, reliable reaction rate measurements are now available for many combustion reactions of interest. In fact, those same developments in laser techniques used for flame probing have also made possible new reaction rate measurements, particularly for transient species, and have greatly improved the quality and acquisition of fundamental spectroscopic data for such molecules.

Laser Combustion Probes

There are several ways in which lasers have been used to probe flames. Most important, from the standpoint of understanding chemical phenomena in combustion, are laser spectroscopic probe methods which provide concentrations and temperatures of particular molecular species. There are several of these, each with specific relative advantages and disadvantages. They all share some common features, however. A key one results from the fact that laser beams can be focussed down to very small diameters, of the order of 0.05 mm or less with ease. Because in flames there can be significant concentration differences occurring over distances of the order of 0.1 mm, this degree of spatial resolution is necessary to obtain meaningful data. The use of pulsed lasers provide temporal resolution as well, which is important when turbulent, detonative or other time dependent phenomena are present. In general, the laser forms a nonintrusive probe which does not perturb the gas flow or chemical reactions, although the use of an intense laser can create a significant excited state population in the species it is pumping, and cause complications due to differences in reactivity between excited and ground states (1).

Schematic energy level diagrams for the most widely used probe methods are shown in Fig. 1. In each case, light of a characteristic frequency is scattered, emitted, and/or absorbed by the molecule, so that a measurement of that frequency serves to identify the molecule probed. The intensity of scattered or emitted radiation can be related to the concentration of the molecule responsible. From measurements on different internal quantum states (vibrational and/or rotational) of the system, a population distribution can be obtained. If that degree of freedom is in thermal equilibrium within the flame, a temperature can be deduced; if not, the population distribution itself is then of direct interest.

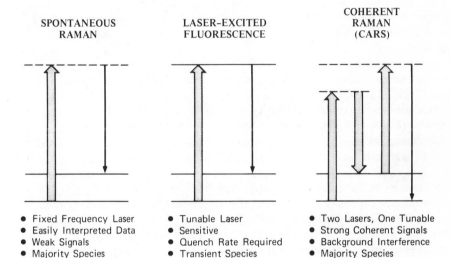

Figure 1.   Schematic energy-level diagrams for the three most widely used spectroscopic laser combustion probes: (– – –), virtual states; (———), real states. Thick arrows represent laser photons and thin arrows indicate scattered or emitted photons.

Raman scattering (2), now often termed spontaneous Raman scattering (SRS), is a well-established and relatively simple technique requiring only a single fixed frequency laser. Upon pumping by the laser to a virtual state, indicated by a dashed line in the figure, the molecule scatters a second photon, returning to a different level. Importantly, SRS is a method for which the signals are linear in laser intensity and molecular concentration, and is unaffected by collisions, so that the relationship between measured intensities and concentrations is straightforward. However, the scattering cross section is very small, resulting in low signal levels. Hence, SRS is generally limited to those species which are the major constituents of the flame (fuel, oxidant, principal exhaust gases, and in air flames, N_2) and appears useful only for flames which are not highly luminous or sooting.

Laser-induced fluorescence (LIF) depends on the absorption of a photon to a real molecular state, and is therefore a much more sensitive technique, capable of detection of sub-part-per-billion concentrations. Thus, this is the most suitable for measurement of those minor species which are the transient intermediates in the reaction network. Here a tunable laser is required, as well as an electronic absorption system falling in an appropriate wavelength region; serendipitously, many of the important transient species have band systems which are suitably located for application of LIF probing. The ability to sensitively detect transitions originating from electronically as well as vibrationally excited levels of a number of molecules offers the possibility of inquiring into the participation of non-equilibrium chemistry in combustion processes.

A quantitative interpretation of the LIF signals requires knowledge of collisional quenching rates, since at pressures of the order of an atmosphere most of the molecules excited by the laser do not radiate but are removed from the excited state by quenching. There are several ways of dealing with this problem. One attractive solution is to increase the laser intensity to the point of optical saturation of the transition, so that stimulated emission competes with quenching as an excited state removal route. Even here a properly founded analysis would appear to require knowledge of rates of energy transfer among internal levels (3,4,5). However, the indications (6) are such that relative measurements--profiles of species concentrations through the flame--should not suffer greatly from this problem.

There exists a family of nonlinear Raman techniques, of which

coherent anti-Stokes Raman scattering (CARS) appears the most
generally promising as a combustion diagnostic (7). The CARS
process may be viewed in the following way. A strong pump laser
connects the ground and a virtual state, whence a tunable probe
laser returns the molecule to an excited level, and a second pump
laser photon elevates it to a second virtual state. This mixing
process results in the scattering of a fourth photon whose energy
is that of the difference between this higher virtual level and
the ground state. Most important, that photon is scattered as a
coherent beam of high intensity and small divergence, yielding
high signal levels as well as a high degree of discrimination
against background flame luminosity. Thus, CARS is suitable for
use in highly luminous systems such as sooting flames.

The resonant CARS signals occur when the energy difference
between the pump and probe lasers equals that of some level spacing in the molecule, but there also appears a nonresonant (but
coherent) background signal at all wavelengths. This has largely
limited CARS to probes of majority species, although recent developments (8) involving polarization of the beams have successfully
reduced this background and lowered the detectability limits.
Additionally, the use of three incoming laser beams instead of
two, in various geometrical configurations, have relaxed the line-
of-sight restrictions on earlier CARS measurements in gases,
providing spatial resolution comparable to SRS and LIF (7).

The nature of the Raman methods makes it possible, using an
optical multichannel analyzer and, for CARS, a broad-band probe
laser, to obtain multispecies, multilevel information from a
single laser pulse. This is not in general possible for LIF using
a single laser, since here the laser wavelength must be tuned to
the absorption line of a particular species.

In Fig. 2 are shown in a schematic fashion, experimental setups for three other techniques also used as combustion diagnostics. Laser absorption is not as sensitive as LIF but is applicable to a wider class of molecules, especially using lasers in
the infrared regions (9). It is a line-of-sight technique, that
is, an average is necessarily made along the entire path of the
laser beam as it traverses the flame. This can be a restriction
if there exist inhomogenieties of concentration, although developments in tomographic techniques (10) are attacking this problem.
An important advantage here is that the absorption spectra may be
analyzed in a simple, straightforward way without requiring assumptions on lineshapes or collisional rates.

Optogalvanic spectroscopy (11,12) is the detection of changes
in the degree of ionization of the flame upon irradiation with a

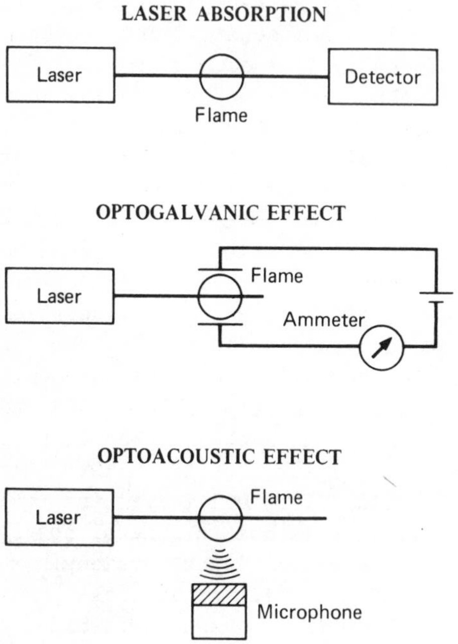

Figure 2.   Schematics of experimental setups for three laser spectroscopic probe methods

laser, and is extremely sensitive. Optoacoustic spectroscopy
(13,14) depends on the detection of sound waves following absorption and subsequent collisional degradation of laser energy. If
the laser is pulsed, a measurement of the arrival time of the
sound pulses at the microphone permits a spatially resolved determination of the speed of sound within the flame (13), which is
an important parameter in fluid dynamics treatments.

Not illustrated is the use of multiphoton excitation of
fluorescence (12,15), thus far demonstrated in flame systems only
for excitation of atoms. It affords the means to excite otherwise inaccessible states and offers other potential advantages in
spatial resolution and for optically thick flames, in spite of
inherently low signal levels.

It is to be emphasized that these laser techniques should
not be viewed as competitive, but rather as complementary. The
combination of LIF, eminently suitable for transient molecules,
but not majority species (due to a lack of suitably located absorption bands), and CARS, useful for the major constituents but
not the minor ones (because of signal strength limitations), is
exemplary in this respect. Properly designed laser probe experiments on flames will involve selection and application of those
techniques best suited for the desired measurements.

From a spectroscopist's point of view, the laser probe techniques offer high resolution spectra with ease. In addition, the
hot flames can provide significant populations in levels otherwise accessible only with difficulty. In Fig. 3 is exhibited an
LIF excitation scan, in which a spectrometer with fixed wavelength and bandpass detects the fluorescence as the laser is tuned
across a series of absorption lines (16). The molecule here is
NH, present in an ammonia/oxygen flame. The large, off-scale
peaks are excitations from the ground vibrational level of the
molecule, while the smaller marked lines are those of the densely
packed head of the (1,1) band. This excitation, originating from
an excited vibrational level, was not detectable in the low pressure, room temperature investigation (17) of this system. Such a
symbiotic relationship--using flames to advance the spectroscopy
and the laser spectroscopic techniques to probe flames--is characteristic of areas such as this where basic research and applications interdiffuse.

Modelling and Kinetics

Among these several laser techniques is thus found the capability to measure nearly every microscopic observable of interest
in a combustion system. Coupled with other more conventional

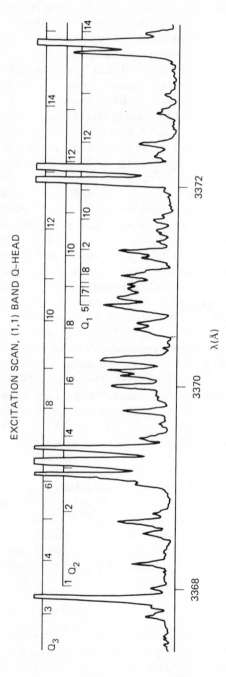

Figure 3. Excitation scan (0.3-cm^{-1} laser bandwidth) through the Q-head of the (1,1) band of the $A^3\Pi_i - X^3\Sigma^-$ system of NH in an atmospheric pressure NH_3–O_2 flame. The off-scale peaks are (0,0) band P-branches.

probes, a quite complete experimental characterization can be
gained. But the proper use of such data demands it be collated
within a theoretical framework--a model--which permits intercon-
necting pieces to be discerned and generalizations to be made.
Only in that way can we develop an understanding of the chemistry
of combustion so that we can later predict behavior beyond the
range of conditions covered, and parameters fitted, in previous
experiments.

One can readily write down the fluid dynamics equations describing the conservation of mass, energy and momentum throughout
the chemically reacting gas flow which constitutes the flame, with
explicit inclusion of the chemical transformations between molecules. In practice, due principally to the complex interplay and
feedback between energy released in chemical reactions, the molecular motion and energy transport, and the temperature dependent
reaction rates, obtaining a confident solution to these equations
is most difficult and challenging.

Two kinds of problems arise. The first arises from the
coupling of differential equations over vastly disparate characteristic time scales, a problem yielding to solution through the
application of mathematical and computational ingenuity (<u>18</u>,<u>19</u>).

The second is concerned with the need to have a complete and
sensible chemical mechanism, valid over a wide range of temperature. Even a relatively simple combustion system will involve
dozens of reactions, so that a well established reaction rate data
base is essential. It is equivalently essential that the results
be verified by comparison with detailed experimental data--such as
that provided by laser probes. For example, in a study of the
ozone decomposition flame (<u>20</u>), it was found that certain alternative but wrong choices of key input parameters were not discern-
ible if flame speed were used as the sole predicted result for
verification; however, these choices did produce considerable
differences in the profiles of the transient oxygen atom concentration and the temperature.

In general, it appears likely that a fit to the intermediate
species profiles should provide the most sensitive means of model
verification, even when the result of direct interest is a bulk
parameter such as flame speed or rate of energy release. Laser
probe methods can be also extremely useful in the absence of full
verification, however. Even a semiquantitative measurement of
some species in a flame can constitute the clue to the inclusion
within the model of an entire subnetwork of chemical reactions.
Coupled closely to model predictions, the laser probe results can
pinpoint species and reactions which merit special attention, that

is, those to which the overall results are sensitive. It is in
such an interactive way that the capabilities of both the laser
probes and the computer simulation models can lead to real advances in our understanding of combustion chemistry.

Laser-Induced Fluorescence in OH

As previously mentioned, LIF is the method of choice for
detection of the transient intermediate species which form the
keys to the chemistry of combustion.

Table 1

Combustion Intermediates

Detected by Laser-Induced Fluorescence

OH

CH, CN, C_2

NH, NH_2

NO, NO_2, HNO

S_2, SH, SO, SO_2

CS, CS_2

CH_2O, CH_3O

C_2O, HCN

A listing of the combustion intermediates which have been observed using LIF, in flames and/or in flow systems, is given in
Table 1. The level of activity in this field can be gauged by the
fact that there have been four new entries on this list in the
past six months.

OH, at the top of the list, is by far the most popular molecule for investigation by LIF, for several reasons. Ubiquitous
in combustion processes, it is a key participant in many reaction
networks and mechanisms. Its spectroscopic data base is unusually
well characterized, both in terms of line positions and intensity
relationships. It has a small enough number of energy levels to
be computationally tractable, so that it is beginning to serve as
a testing ground for models of optical saturation in molecules
(4,5). And finally, its absorption bands lie in a convenient
wavelength region--the range covered by frequency doubling the
efficient and stable rhodamine dyes.

For many of the other species, further research (in some
cases a considerable amount) needs to be performed in order to
establish a firm spectroscopic data base for the quantitative
analysis of LIF data. It should be noted that the prospects for
unambiguous LIF detection of some other species, particularly

larger molecules within a complex mixture, are not as bright due
to spectroscopic limitations such as less well-defined absorption
structure or the existence of predissociation.

Some of our recent studies of LIF on OH in flames demonstrate
the close connection between current work in other areas of physical chemistry--in this case, state-to-state collisional energy
transfer--and the development of diagnostic tools for combustion.
In these experiments, measurements are made of the collisional
redistribution of excited state population following laser excitation of OH to individual levels, in an atmospheric pressure flame.

At such a pressure, the excited molecules undergo many collisions before radiating; the questions addressed here concern
both a description--in terms of state-dependent energy transfer
rates--and a diagnostic exploitation of those collisions. In the
experiments, OH in the partially burnt gases of a methane-air
flame is excited to individual rotational levels of the $v'=0$
vibrational level of the $A^2\Sigma^+$ state, and measurements of the
resulting fluorescence dispersed through a monochromator provide
populations of individual levels.

In the first series of experiments (21), measurements are
made of the rotational distribution within $v'=0$. Figure 4 shows
a portion of each of the spectra obtained upon pumping two different rotational levels, $N'=1$ and $N'=10$. The obvious qualitative
differences demonstrate that thermalization does not occur prior
to emission. Rather, the observed distributions reflect a competition between rotational relaxation and quenching, and require
detailed state-dependent collision rates for a description. Such
a description, with explicit inclusion of the temperature dependence of the rates as a parameter, has been used to obtain the
temperature within the flame from spectra such as these (3).
Additionally, on a less detailed basis, the current results show
that the ratio of the rotational relaxation rate to the quenching
rate decreases with increasing rotation.

A further interesting result concerns spin component conservation. Each rotational level of this $^2\Sigma$ state has two spin components, one (termed F_1) with the spin and rotational angular
momentum vectors parallel, and the other (F_2) having them antipar-
allel. Excitation of an F_1 component results in transfer primarily to F_1 components of other rotational levels, and similarly
for F_2 excitation. In a collision, the OH would as soon exchange
several hundred cm^{-1} of energy as flip its spin around at no
energy cost. Similar results were also observed in the fluorescence scans of the NH molecule (16).

Such collisional selection rules are at the heart of current

research in molecular collision dynamics in systems significantly
simpler than flames. Yet they not only appear in the flames, but
have decided implications for diagnostic measurements as well.
From Fig. 4 it can be seen that, for finite bandpass detection,
one will obtain different fluorescent intensities per emitting
molecule depending on the level pumped. This can produce systematic errors in both the determination of absolute concentrations
and the use of excitation scans to obtain ground state rotational
temperatures (21). Also, the lack of a thermal distribution imposes restrictions on models of and data analysis in optical
saturation techniques.

If the OH undergoes rotational energy transfer, it will
undergo vibrational energy transfer as well (22). Figure 5 shows
the emission from the $v'=1$ level following excitation of the $N'=4$
rotational level in $v'=0$. This results from molecules which have
been collisionally transferred upwards some 3000 cm^{-1}. Also shown
in Fig. 5, on the same intensity scale, is a small portion of the
emission from $v'=0$. From a ratio of these intensities, we find
that the $v'=1$ population N_1 is about 3.5% of that in $v'=0$, N_0.

This ratio can be used to determine the temperature in the
following way. A steady state balance (ignoring the radiative
decay rate) is applied to N_1:

$$V\exp(-\Delta E/kT)N_0 = (V + Q)N_1$$

Here, Q is the collisional quenching rate and V is the downward
$(1 \rightarrow 0)$ collisional vibrational transfer rate. The upward transfer
rate is then given by detailed balancing averaging over rotational
levels--an assumption briefly discussed below, but whose applicability receives support from the fact that the same ratio N_1/N_0
was obtained upon pumping each of three different rotational
levels in $v'=0$. An estimate of the ratio V/Q is taken from previous low pressure studies (23).

The result for the temperature is $1460 \pm 50^\circ$K. This compares
favorably with the ground state OH rotational temperature of
1350°K, obtained from an excitation scan through the R-branch
region of the (0,0) band.

A promising possibility for this method would involve the use
of two photomultipliers, simultaneously measuring emission from
$v'=1$ and $v'=0$ on a single laser shot. This would provide a time-
resolved measurement of the temperature, the concentration of the
key reactive species OH, and their correlation. Such data is of
importance in studying time-dependent phenomena such as reactive
turbulence or detonations.

The experiment poses an intriguing fundamental question:

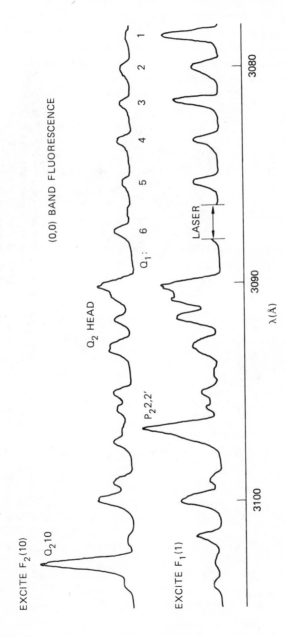

Figure 4.   Portions of the rotationally resolved fluorescent emission in the (0,0) band following excitation of individual rotational levels within $v' = 0$ of the $A^2\Sigma^+$ state of OH, in an atmospheric pressure CH_4–air flame.   Top: excitation into $N' = 10$, $J' = 21/2$; bottom: excitation into $N' = 1$, $J' = 3/2$.

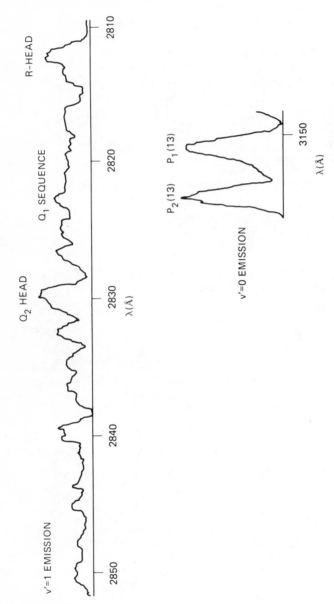

Figure 5. Fluorescence scans of emission following excitation of $N' = 4$, $J' = 9/2$ of the $v' = 0$ level in OH in a CH_4–air flame. Top: (1,0) band fluorescence, emitted by molecules collisionally transferred upwards to $v' = 1$; bottom: two rotational lines in the (0,0) band, emitted by molecules in the $N' = 12$ level of $v' = 0$. Both scans are on the same intensity scale.

how does one deal with detailed balancing for a system not totally
at thermal equilibrium? The temperature obtained is presumably
the translational temperature of the flame gases. However, the
rotational distribution results described earlier demonstrate
that, for the rotational degree of freedom in $v'= 0$, a thermal
distribution does not obtain. In addition, downward vibrational
transfer rates have earlier (23) been shown to have a rotational
level dependence. Nonetheless, the results obtained for pumping
the three different rotational levels--including one whose rotational energy was greater than the vibrational energy difference--
were in good agreement. The answers to these questions and subsequent full development of the method as a diagnostic, will require both further experiments and a careful consideration of
detailed, state-to-state molecular dynamics.

Summary

Laser-based spectroscopic probes promise a wealth of detailed
data--concentrations and temperatures of specific individual molecules under high spatial resolution--necessary to understand the
chemistry of combustion. Of the probe techniques, the methods of
spontaneous and coherent Raman scattering for major species, and
laser-induced fluorescence for minor species, form attractive
complements. Computational developments now permit realistic and
detailed simulation models of combustion systems; advances in
combustion will result from a combination of these laser probes
and computer models. Finally, the close coupling between current
research in other areas of physical chemistry and the development
of laser diagnostics is illustrated by recent LIF experiments on
OH in flames.

Literature Cited

(Papers given as numbers are those in the current symposium
volume).

1. Muller, C.H., Steinberg, M., and Schofield, K. Paper 14.
2. Lapp, M., Paper 17.
3. Daily, J.W., Paper 3.
4. Kotlar, A.J., Gelb, A., and Crosley, D.R., Paper 7.
5. Lucht, R.P., Sweeney, D.W., and Laurendeau, N. Paper 8.
6. Cottereau, M.J., and Stepowski, D., Paper 6.
7. Eckbreth, A.C., Paper 23.
8. Rahn, L.A., Zych, L.J., and Mattern, P.L., Opt. Comm. 30,
     249 (1979).
9. Hanson, R.K., Varghese, P.L., Schoenung, S.M., and Falcone,
     P.K., Paper 35.

10. Santoro, R.J., Emmerman, P.J., Goulard, R., Semerjian, H.G.,
     and Shabahang, R., Paper 36.
11. Smyth, K.C., Schenck, P.K., and Mallard, W.G., Paper 12.
12. Van Dijk, C.A., and Alkemade, C. th. J., Paper 13.
13. Allen, J.E., Anderson, W.R., and Crosley, D.R., Optics
     Letters $\underline{1}$, 118 (1977).
14. Killinger, D.K., Moore, J., and Japar, S.M., Paper 40.
15. Allen, J.E., Anderson, W.R., Crosley, D.R., and Fansler,
     T.D., Seventeenth Symposium (Internatioral) on Combustion,
     1979, p. 797.
16. Smith. G.P., and Crosley, D.R., to be published.
17. Anderson, W.R., and Crosley, D.R., Chem. Phys. Lett. $\underline{62}$,
     275 (1979).
18. Oran, E.S., Boris, J.P., and Fritts, M.J., Paper 27.
19. American Chemical Society Symposium on Reaction Mechanisms,
     Models and Computers, published as J. Phys. Chem. $\underline{81}$, #25
     (1977).
20. Heimerl, J.M., and Coffee, T.P., Paper 29.
21. Smith, G.P., Crosley, D.R., and Davis, L.W., Eastern
     Sectional Meeting, Atlanta, GA, November 1979; Smith, G.P.,
     and Crosley, D.R., to be published.
22. Crosley, D.R., and Smith, G.P., Applied Optics, in press
     (1980).
23. Lengel, R.K., and Crosley, D.R., J. Chem. Phys. $\underline{68}$, 5309
     (1978).

RECEIVED February 1, 1980.

Laser Probes for Combustion Applications

J. R. McDONALD

Chemistry Division, Code 6110, Naval Research Laboratory, Washington, D.C. 20375

Within the past few years lasers have found an ever increasing role as diagnostic instruments to probe difficult environments such as in remote atmospheric sensing and measurements in hostile environments, e.g. arcs, plasmas, discharges, and combustion sources. Laser applications as a combustion diagnostic is the subject of the discussion of this paper. The pioneering efforts in this area began with the use of lasers to carry out spontaneous Raman scattering measurements to determine temperature and major constituent concentrations. This is now a mature field, and the capability and limitations of spontaneous Raman measurements are well documented. In the following talk Marshal Lapp will discuss the application of this technique; it will therefore not be further discussed now. Moreover, we will not discuss other versions of Raman spectroscopy such as resonant, near resonant, and time resolved Raman spectroscopy. Each of these techniques also has potential application in combustion study. The subject matter of this paper will be limited to two specific topics; (a) Coherent Anti-Stokes Raman Spectroscopy and (b) Saturated Laser-Induced Fluorescence Spectroscopy. This limitation of subject matter also excludes such techniques as laser absorption, low power laser-induced spontaneous emission, optoacoustic and optogalvanic spectroscopy and laser light scattering techniques involving particulate and refractive index gradient scattering. Many of these areas are the subject of invited papers and poster session presentations in this symposium.

As I have indicated, this presentation will be divided into two parts. In the first part we will discuss the development of Coherent Anti-Stokes Raman Spectroscopy, the problems inherent in applications to combustion sources, recent developments which address operational problems, and the state-of-the-art today. This will be followed by a similar discussion involving the use of saturated laser-induced fluorescence spectroscopy as a combustion diagnostic.

This chapter not subject to U.S. copyright.
Published 1980 American Chemical Society

Coherent Anti-Stokes Raman Spectroscopy

The Generation of Signals. Several good reviews of CARS
have recently appeared (1, 2, 3). I refer you to these references for an in-depth treatment of the subject.

In CARS two incident laser beams at frequencies ω_1 and ω_2
(often referred to as pump and probe beams) incident upon a sample interact through the third order nonlinear susceptibility to
generate a coherent beam at a frequency, ω_3, as depicted in Fig. 1
The CARS signal, as shown, is at the frequency $\omega_3 = 2\,\omega_1 - \omega_2$. When
the difference $(\omega_1 - \omega_2)$ coincides with a Raman active resonance
the magnitude of the radiation at ω_3 can become very large. For
efficient signal generation the incident laser beams must be
aligned such that the three wave mixing process is properly phase
matched as noted at the top of Fig. 1. For gases which are essentially dispersionless, phase matching occurs for a collinear
overlaping beam configuration. Although this configuration is
easy to setup it also creates some problems, i.e. long active
CARS beam overlap volumes and signal generation in optical components. We will return to this subject.

In a CARS experiment the signal intensity, I_3, at the frequency, ω_3, is given in Eq. (1)

$$I_3 = \left[\frac{4\pi^2 \omega_3}{c^2} \right]^2 (I_1^2\, I_2\, |\chi|^2\, z^2), \tag{1}$$

where I_1 and I_2 are the intensities of the pump and probe beams
at ω_1 and ω_2, c is the speed of light, χ is the third-order
nonlinear susceptibility and z is the phase matched interaction
distance. The third-order susceptibility can be written as the
sum of a resonant and a nonresonant term:

$$\chi = \sum_j (\chi' + i\chi'')_j + \chi^{nr}, \tag{2}$$

The resonant susceptibility associated with a homogeneously
broadened Raman transition, j, is,

$$(\chi' + i\chi'')_j = K_j \frac{\Gamma_j}{2\Delta\omega_j - i\Gamma_j}, \tag{3}$$

where Γ_j is the Raman linewidth for transition j, $\Delta\omega \equiv -(\omega_1 - \omega_2)$
is the detuning frequency. K is the modulus of the suscepti-
bility, and is given as,

a) GENERAL

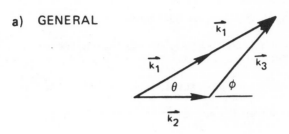

b) COLLINEAR

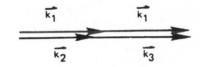

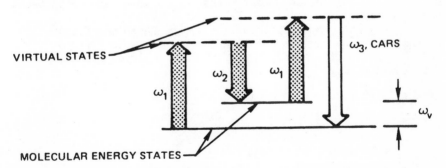

United Technologies Research Center

Figure 1.   Phase-matching diagrams for CARS signal generation are shown in a and b. The energy diagram at the bottom shows the energy matching scheme for signal generation (9).

$$K_j = \frac{4\pi c^4}{h\,\omega_2^4}\,N\Delta_j g_j\,\left(\frac{\delta\sigma}{\delta\Omega}\right)_j\,\frac{1}{r_j} \tag{4}$$

$N \equiv$ the total species number density;

$\Delta_j \equiv$ the normalized population difference between the transition levels;

$g_j \equiv$ the linestrength factor; and

$\left(\frac{\delta\sigma}{\delta\Omega}\right)_j \equiv$ the Raman crossection for transition j.

Complications arise because of the cross-terms in the square of the third order susceptibility;

$$|\chi|^2 = \chi_2'^2 + 2\chi_2'\chi^{nr} + \chi^{nr^2} + \tag{5}$$

$$\chi_1''^2 + 2\,\chi_1''\,\chi_2'' + \chi_2''^2,$$

the term $2\,\chi_2'\,\chi^{nr}$ may be either positive or negative exhibiting either constructive or destructive interference effects. This effect gives the characteristic CARS spectrum which under high sensitivity and resolution demonstrates strong interference patterns near resonances.

In summary, then, Coherent Antistokes Raman spectra are characterized by several complicating factors which are not characteristic of spectroscopy as usually employed for analytical measurements, i.e.

$$I_3 \propto I_1^2\,I_2$$

$$I_3 \propto N^2$$

$$I_3 \propto z^2$$

I_3 contains nonresonant background contributions and

I_3 must be deconvoluted from the laser line shapes; the Raman resonance line shapes; the detector slit function; and the polarization properties of the laser and signal fields.

In spite of these complicating features several major laboratories have undertaken major and expensive experimental projects to develop CARS as an analytical tool for combustion probing. This is because CARS offers the means for analytically probing combustion phenomena which cannot easily be done by other means. Some of the advantages and potential advantages of CARS are:

(a) It can be a point source probe;

(b) It involves a noninterfering measurement;

(c) It can be made in short experiments on a time scale faster than turbulently changing phenomena;

(d) It can be used in highly luminous and particulate laden environments to give temperature and major species number density measurements.

Experimental Setups. The experimental configurations used for CARS measurements have common features in most laboratories now. Figure 2 shows one such arrangement which is typical. A pulsed laser, in most cases a Q-switched Nd:YAG, is used to generate the ω_1 frequency at 532 nm. In a few cases experimenters use ruby or other solid state pulsed lasers. Typically the 532 nm beam is split and the second beam is used to pump a dye laser – dye laser amplifier leg. This produces the ω_2 probe frequency. The dye laser is tunable by choice of dye (for broad band applications) or by dispersive elements in the oscillator (for narrow band generation). The ω_1 and ω_2 beams are then optically recombined spatially and temporally in the flame zone to be probed. The ω_3 antistokes frequency is then separated from the ω_1 and ω_2 beams by dispersing elements, shown in Figure 2 as prisms, and is then further analyzed with a monochromator. The detection may be photoelectric using a gated photomultiplier or may by use of an OMA if one wishes to detect the whole spectrum on each shot. The latter technique offers powerful advantages in that it allows complete spectra to be sampled in a single 20 ns experiment. I will return to this point.

Results With Colinear Beams. To give one a feeling for the potential of the technique Figure 3 shows the CARS signal of H_2 in an N_2 bath as a function of concentration ($\underline{4}$). We see that the signal demonstrates the N^2 concentration dependence between about 10^2 and 5×10^5 ppm H_2 in N_2. This is a dynamic range of $\sim$ 5000.

Figure 4 shows the CARS spectrum of D_2 gas scanned using a narrow band probe (ω_2) laser within the bright discharge region of an electrical discharge lamp. One can see the Q-branch band heads from both the v"=0 and v"=1 levels. These spectra can be used to determine both the rotational and vibrational population distributions of D_2 within the discharge.

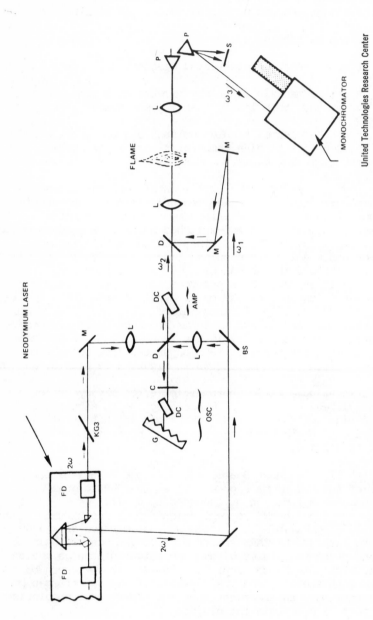

Figure 2. A typical experimental arrangement for measuring CARS spectra. FD denotes frequency doubler, KG3 is a 1.06-μm absorbing filter, L is for lens, g is a grating, DC is a dye cell, D is a dichroic beam splitter, m is a mirror, and P is a prism (9).

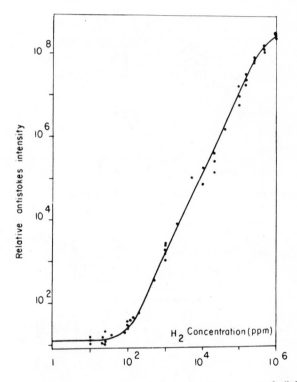

Applied Physics Letters

Figure 3.   Plot of CARS signal vs. H_2 concentration in N_2 gas (4).   CARS spectrum of discharged gas (D_2 48 torr).

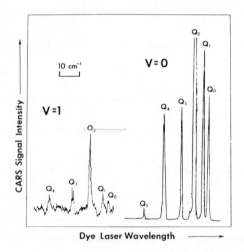

Figure 4.   Plot of the CARS signal for D₂ gas within the bright region of an electric discharge lamp

With a similar experimental setup Figure 5 shows the spectrum of N_2 probed within the homogeneous region of a flat flame burner in an experiment done a couple of years ago at NRL. As we mentioned in the introduction, considerable data analysis must be carried out to extract a temperature from spectra such as shown in Figure 5. It turns out that the only way to conveniently extract this information is to carry out detailed computer modelling of the spectra with appropriate spectroscopic and instrumental operating parameters. Theoretical CARS analytical programs have been developed by several groups to reduce data such as shown in Figure 5. The reader is refered to articles by Shaub, et al. for a description of these procedures and for programs to carry out the modelling. Figure 6 shows modelled CARS spectra for N_2 in a flame as a function of the bulk gas temperature assuming rotational and vibrational equilibrium for N_2 in the flame (7).

Another species present in high concentration in combustors is water. Both experimental and theoretical H_2O spectra for a premixed CH_4-air flame are shown in Figure 7 (8). Because of the large spontaneous Raman cross section for H_2O there is little or no interference from background signals. Water, however, is difficult to use as a probe because of the extreme complexity of the spectrum as compared to typical diatomics. The computer match up generated by Hall et al. in Figure 7 is excellent and gives a good fit to the independently measured temperature.

Single Pulse Measurements. For CARS to be a useful probe for flame dynamics in turbulent media it is necessary to obtain information in a time short compared to the transient behavior taking place in the combustor. Any type of scanning proceedure to extract spectral information is suitable only for homogeneous laboratory flames. To extract instantaneous information a technique called broad band CARS has been developed (9). This is shown schematically in Figure 8. In this technique the narrow band ω_1 pulse is mixed with a broad band ω_2 beam. This technique produces an ω_3 spectrum for all resonances $(2\omega_1-\omega_2)$ within the spectral width of the ω_2 beam. The efficiency of such a process is obviously much lower than for a narrow band ω_2 beam because of the nonresonant frequencies contained within the ω_2 bandwidth. For many applications this is not a practical limitation, however. To detect the entire CARS spectrum for a given single pulse measurement requires polychromatic detectors. The technique typically employed is to use a device called an optical multichannel analyzer in place of the single frequency photomultiplier detector at the exit slit of the monochromator. Such high sensitivity devices exist now from several commercial vendors.

Figures 9 and 10 show such single pulse measurements made in our laboratory for methane and methane in a mixed gas system (10). Since this early work broad band CARS has been developed and applied to more and more complex systems including single pulse

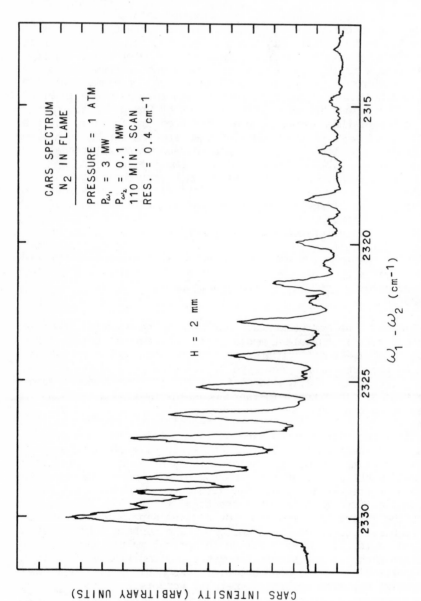

Figure 5. Plot of the CARS spectrum of N_2 gas in the combustion zone of a homogeneous flat flame burner. Conditions are as noted in the figure.

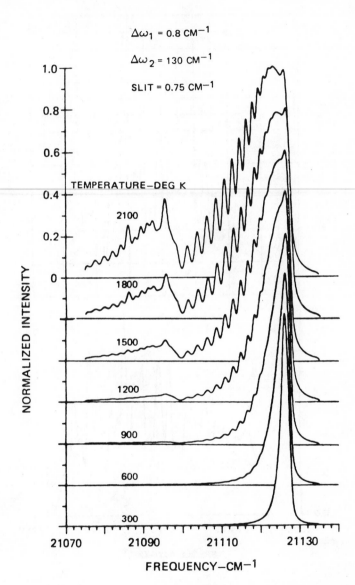

$\Delta\omega_1 = 0.8$ CM^{-1}

$\Delta\omega_2 = 130$ CM^{-1}

SLIT = 0.75 CM^{-1}

United Technologies Research Center

Figure 6.   *Plot of calculated CARS spectra of N_2 as a function of bulk gas temperature (7)*

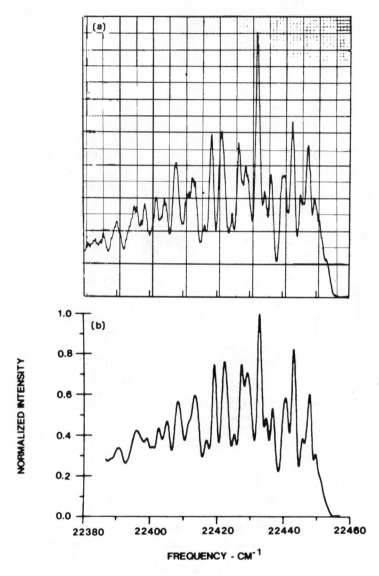

Optics Letters

Figure 7. (a) *Experimental CARS spectrum of* H_2O *in a premixed methane–air flame at one atmosphere.* (b) *Computed CARS spectrum of* H_2O *at 1675 K for a pump bandwidth of 0.8* cm^{-1} *and a triangular slit function of 1* cm^{-1} *FWHM. The measured flame temperature is 1675 K (8).*

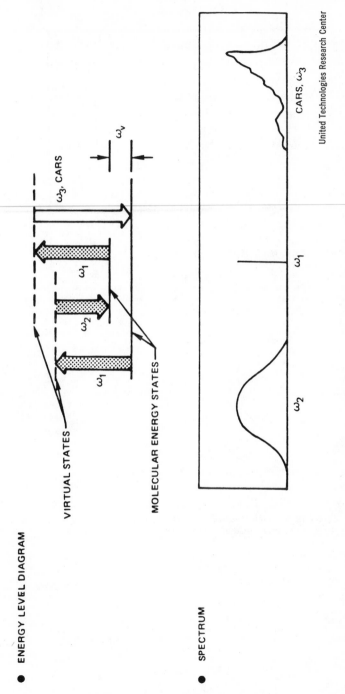

Figure 8.   Schematic to represent generation of broadband CARS signals. See the text for details (9).

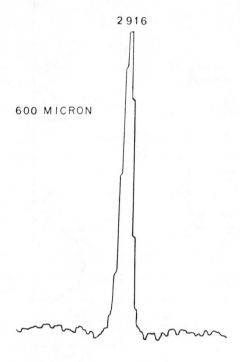

2916

600 MICRON

Elsevier

Figure 9.   Broadband CARS spectrum of CH$_4$ using OMA detection and a single-laser pulse: P$_1$ = 200 kw; P$_2$ = 2 kw.

measurements of diatomic species in flame systems. The computer
modelling sheme used to generate the spectra shown in Figure 6
uses a broad band CARS technique.

Noncolinear Phase Matching Techniques. Two serious problems
exist when one uses the natural colinear beam phase matching technique for measuring CARS spectra in gases. Because one is working
with beams which are overlapped in space at all points after recombination, CARS signals are generated at positions far removed
from the sample probe volume which one wishes to measure. This is
partially offset by tight waisting the beams using relatively
short focal length lenses. Because of the cubic dependence on
laser power most of the signal is generated near the focus of the
beams. In most practical applications the use of short focal length lenses is precluded. A second problem which results from colinear beams is signal generation within optical components in the
experimental system. At high laser powers this can contribute
considerable artifact distortion of experimental spectra. Obviously, a preferable technique would be the use of a crossed beam
technique.

One obvious solution is that shown in Figure 11. By inten-
tionally crossing and phase mismatching the beams the focal probe
volume is much better defined. The CARS signal beam then lies in-
termediately between the ω_1 & ω_2 beams and generation within
optical components due to ω_1 & ω_2 overlap is avoided. This
technique has been routinely used in our and other laboratories
for several years where the experimental conditions are not too
demanding. The disadvantage of this technique is the large loss
in signal resulting from the phase mismatch (11). Depending upon
the experimental conditions, phase mismatching by one degree can
result in an order of magnitude loss in CARS signal.

Workers at United Technologies have developed (9) a three
laser beam mixing technique referred to as BOXCARS which is shown
in Figure 12. Using this technique the beams can be crossed to
define the focal volume and avoid solid state generation and
simultaneously satisfy the phase matching condition required for
optimum CARS signal generation. This is done by splitting the ω_1
beam into two components. Then when the two quanta of ω_1 are
mixed with ω_2 under the crossing conditions specified in the figure the phase matching condition can be uniquely satisfied. As
explained in Reference (9) this can be conveniently accomplished
using α, ϕ and θ angles of a few degrees. This technique, of
course, adds one more level of complexity to the experiments because three (rather than two) beams must be spatically and tem-
porialy waisted at the focal volume. This extra level of complexity is more than offset in the quality of spectral information
which can be obtained by the use of the BOXCARS technique. Alan
Eckbreth in a later paper this morning will show convincing evidence for the use of this technique in hostile environments.

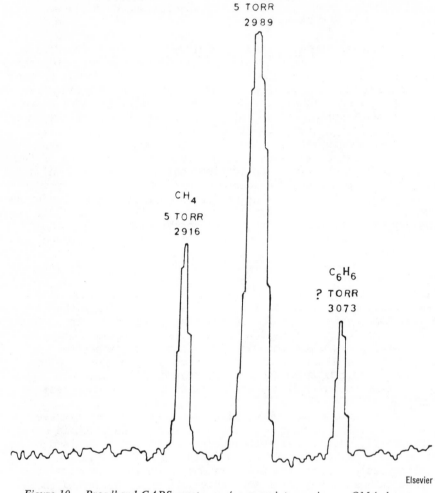

Elsevier

Figure 10.   Broadband CARS spectrum of a gas mixture using an OMA detector and a single-laser pulse (10)

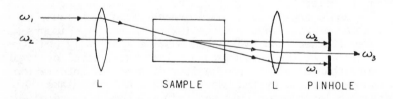

Figure 11.   Experimental diagram demonstrating CARS signal generation in gases using an intentional phase mismatch condition to minimize sample volume and spurious signal generation in optics.

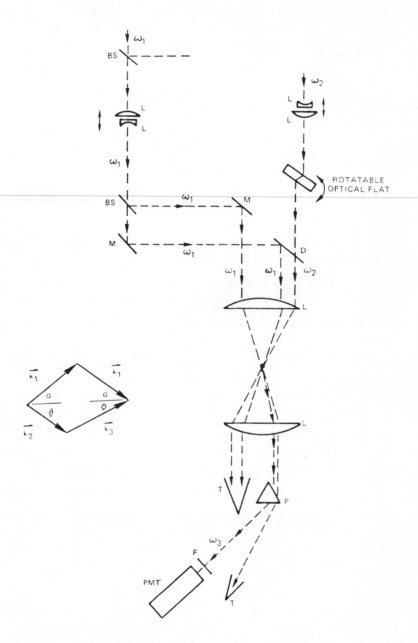

Figure 12. Experimental diagram from Ref. 9 demonstrating the experimental arrangement used to generate BOXCARS spectra. For legend symbols see Figure 2.

Dirty Flames. At this point one could well ask: so what happens in real combustors which are turbulent, soot and particle laden and are highly luminous? By the end of this morning's session you should be convinced that CARS can be applied to these systems. I don't want to steal all of Alan Eckbreth's slides so I will show only two more. Figure 13 shows the BOXCARS spectrum of N_2 with a computer fit to a temperature of $2000^{\circ}K$ in a laminar sooting propane diffusion flame (12). Figure 14 shows the vertical temperature profile for this same flame system. It should be pointed out that care must be taken under these conditions to account for the laser interaction with carbon in the flame which can generate laser induced Swan Band emission from C_2.

Future CARS Developments. There are many new developments in CARS which I have not discussed. Several of these techniques are designed to suppress background and nonresonant contributions to CARS signals through the use of time sequencing of the ω_1 and ω_2 beams or by use of polarization techniques. An example of the latter is shown in Figure 15. Rahn and coworkers (13) have shown that a signal to noise improvement of >25 is achievable for the CO CARS flame spectrum using the crossed polarization technique over the conventional CARS spectrum shown at the bottom of the figure. We should expect further developments along these lines in the near future for obtaining improved CARS measurements in flames.

Beattie and coworkers (14) have demonstrated that it is feasible to measure pure rotational CARS spectra using one atmosphere of air as a sample. We have been developing this technique in our laboratory with somewhat improved experimental conditions. Figures 16 and 17 show spectra of O_2 and air at one atmosphere down to ~ 20 cm^{-1} shifts measured in our laboratory. We are investigating this technique for combustion probing. The use of Rotational CARS has the potential for providing both temperature and concentration information for several species on a single shot basis because the required spectral information is contained in a narrow range of frequencies which could be probed by a single dye output.

Saturated Laser Induced Fluorescence Spectroscopy. The development of saturated laser induced fluorescence spectroscopy is more recent than CARS and is less published. Even though this is the case, this introductory review will not be comprehensive. I will likely miss some work and I apologize in advance to those authors. I will not attempt to discuss laser absorption experiments or laser induced fluorescence experiments in the low laser power, i.e., non-saturated, limit. There is much work in the latter area of merit and several important papers on LIF in this conference.

Historical Development. Saturation spectroscopy in flame systems dates back only to about 1972. The important early work

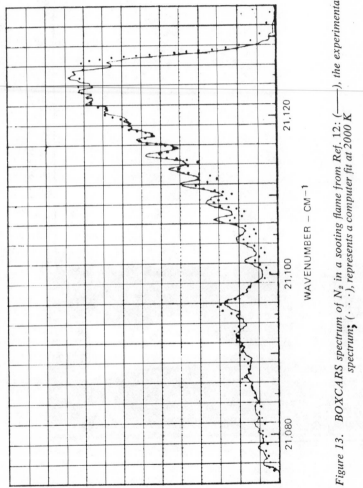

Figure 13. BOXCARS spectrum of N_2 in a sooting flame from Ref. 12: (———), the experimental spectrum; (· · ·), represents a computer fit at 2000 K

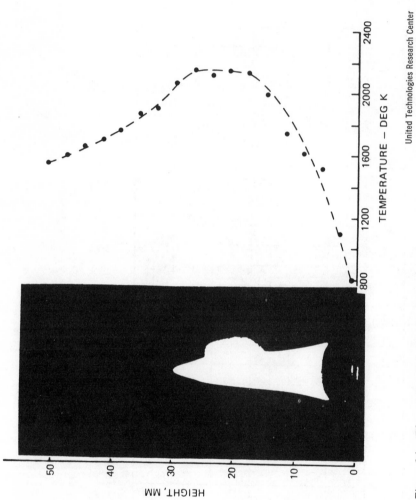

Figure 14.   The axial variation of temperature in a sooting flame with height above the tube burner using a propane–air diffusion flame (12)

United Technologies Research Center

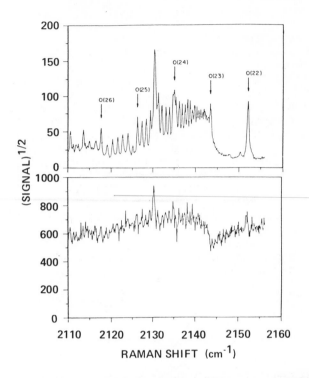

Spectrochimica Acta

Figure 15.   CARS spectra in the region of the CO Q-branch from a rich methane–air flat flame. Above is shown the background-free CARS spectrum, below the conventional CARS spectrum (15).

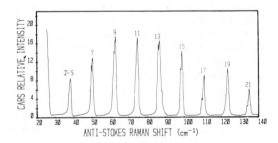

Figure 16.   A rotational CARS spectrum of O_2 at one atmosphere pressure. The spectrum is generated using a Nd:YAG laser and a scanning-amplified dye laser. Scattered-light discrimination is achieved using a scanning double nonochromator slaved to the dye laser scan.

was done on the simpler atomic systems. This work was primarily
from two laboratories (14, 15, 16) and culminated with the major
review paper by Omenetto and Winefordner (17). The extension of
saturation spectroscopy from atomic to molecular systems involves
several levels of added complexity. To extract concentration information in atomic systems one must be able to avoid having to
cope with the measurement of electronic quenching (i.e., determination of the fluorescence quantum yield). This is accomplishable
within the context of a two level model for atomic systems without
intermediate electronic metastable levels in the saturation limit.
This is aided by the fact that in atomic systems one can often
work with strong resonance transitions with natural lifetimes in
the region of one to a few nanoseconds. In a flame system such an
excited atom will suffer only 10 to a few score collisions during
its radiative lifetime.
   The situation in molecular systems is much more complex.
Among the factors which must be considered are the following:

(a)   Vibrational distributions and relaxation in the lower
      and upper states;

(b)   Rotational relaxation during the pumping pulse in the
      upper and lower electronic states;

(c)   Chemical reactivity differences in the lower and upper
      states;

(d)   Electronic quenching of the excited state level;

(e)   The existinence of intermediate metastable electronic
      levels between the two pumped levels; and

(f)   The radiative lifetime of the prepared upper state.

As we will see the use of saturated laser induced fluorescence
spectroscopy will allow us to ignore some of these effects. We
can infer the importance of others and for the time being the
remainder have to be evaluated on a case by case basis for each
molecular system and for the operating parameters of the
experiments.
   This technique was first considered theoretically for molecules by Daily (18, 19) in important papers in 1976 and 1977. In
these papers Daily extended the two level model used in atomic
systems to molecules and attempted to define the range of utility
of the model. Baronavski and McDonald reconsidered the model
and made the first experimental measurements in flames under saturated conditions to test the model (20, 21) for C_2 in acetylene
flames. During the same year workers at United Technologies began
experimental evaluation of the technique (22) and in the last two

years several other papers have appeared which further consider
the theory and application of the technique , (23-28).

Experimental Setup. The instrumentation (both optics and
electronics) for studying saturated laser induced fluorescence
spectroscopy is much less complicated than for CARS. The experimental setup shown in Figure 18, as used in our laboratory, is
typical for these studies. In some experiments it is advantageous to use a monochromator rather than band pass filters to isolate the laser induced fluorescence signal. The lasers used are
either flash lamp pumped systems or Nd:YAG pumped dye lasers.
The latter systems are characterized by a 10-20 nsec pulse while
most flash lamp pumped systems have 500-1000 nsec pulse widths.
Experimentors should carefully consider the use of 20 nsec pulsed
laser with respect to the steady state approximations made in the
theoretical treatments used to model the experiments. In some
cases upper and lower state relaxation processes may be too slow
to allow a steady state to be attained during the pumping pulse.
    Typically the laser beam is waisted through the portion of
the flame zone to be sampled and the probed volume is further
limited by the acceptance aperatures in the 90^{0} collection optics. Under tightly focussed conditions it is possible to probe
flame volumes on the order of 10^{-5} cm^{3}. The detection and
signal processing equipment consists of routinely used instruments in most spectroscopy laboratories. We have never found
it difficult to observe laser induced fluorescence signals for
most species while rejecting scattered light and background flame
luminosity.
    To begin experiments one must unambiguously identify and
characterize the species to be studied in the flame in order to
insure that the correct and only the correct molecule is contributing to the LIF signal. Moreover, for experimental measurements one must excite known rovibronic levels within a given
electronic system so that measured number densities can be scaled
to total number densities. This is most conveniently done by
scanning the dye laser frequency to plot out a fluorescence excitation spectrum. Figure 19 shows a partial excitation spectrum
of the Swan Band system in C_2 including the 0-0 and 1-1 rovibronic transitions. By using these spectra one can readily identify rovibronic levels to be used for excitation.
    Figure 20 shows an energy level diagram for another complex
molecular system, MgO, which we have been studying in an aspirating slot burner. In this system the excited $B^{1}\Sigma^{+}$ level is radi-
atively coupled to both the ground $X\Sigma^{+}$ state and a metastable $A^{1}\Pi$
level. In Figures 21 and 22 the excitation spectra are shown for
each of these two levels. We have analyzed the spectrum in Figure 22 and have shown that the flame temperature can be independently determined from the excitation spectrum. We do not recom-
mend this technique for measuring bulk flame temperatures.

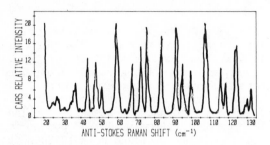

Figure 17.   A rotational CARS spectrum of air at one atmosphere pressure. Conditions are as described in Figure 16.

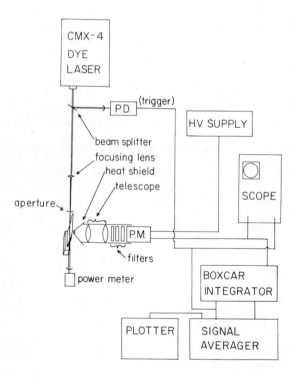

Figure 18.   Experimental arrangement used in the author's laboratory to measure laser-induced fluorescence signals from flame species

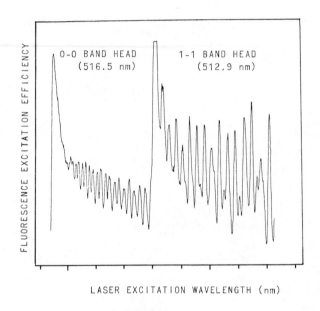

Applied Optics

Figure 19.   The laser-induced fluorescence excitation spectrum of the C_2 swan band system in an acetylene–air flame (21)

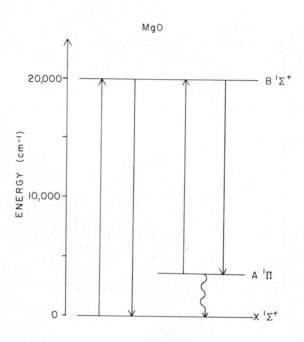

Journal of Chemical Physics

Figure 20.   An energy-level diagram for the low-lying electronic singlet states of MgO (24)

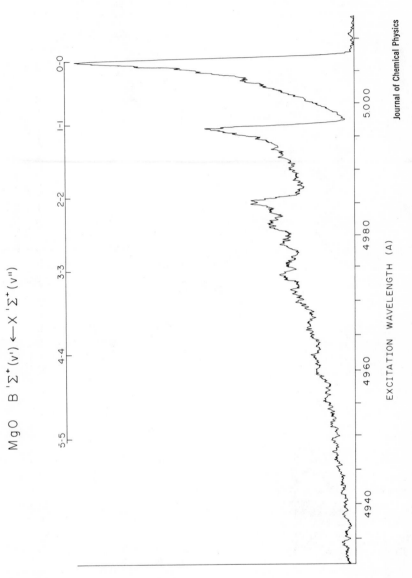

$$MgO \quad B\,^1\Sigma^+(v') \longleftarrow X\,^1\Sigma^+(v'')$$

FLUORESCENCE INTENSITY

EXCITATION WAVELENGTH (A)

Journal of Chemical Physics

Figure 21. The fluorescence excitation spectrum of the MgO B $^1\Sigma^+ \leftarrow X\,^1\Sigma^+$ transition in an C_2H_2- air aspirating slot burner. The magnesium is added as a salt solution via the aspirating port (24).

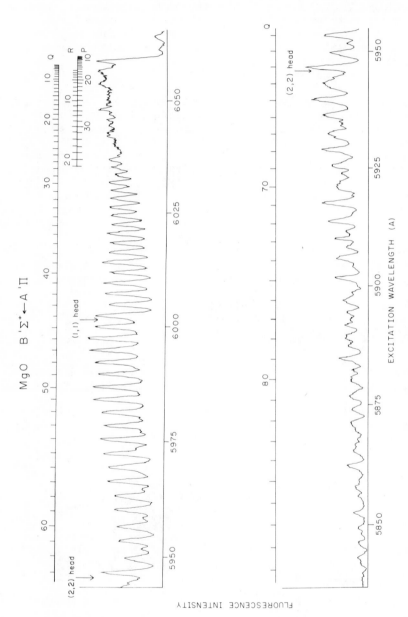

Figure 22.   The fluorescence excitation spectrum of the MgO B $^1\Sigma^+ \leftarrow$ A $^1\Pi$ transition under conditions described in Figure 21. Rotational analysis of the spectrum demonstrates that the A $^1\Pi$ metastable is thermalized relative to the bulk flame temperature.

The Theoretical Models.   To date most experiments have been analyzed using some form of the two level model. This model, shown in Figure 23, assumes laser excitation between levels 1 and 2. If a third intermediate level, 3, exists it is assumed that it relaxes back to the ground state on a time scale fast relative to the length of the laser pulse. Moreover, one assumes that a steady state equilibrium exists among the pumped levels (including the bath of rovibronic levels in state 1) during the laser pulse. The observed fluorescence signal intensity, S, is given by the radiative transfer equation shown in Figure 23. S is expressible in terms of the upper state population N_2, the Einstein A coefficient, the probed sample volume and the collection and detection efficiency. Daily ([18], [19]) has shown that in the steady state approximation the observed fluorescence intensity can be recast in terms of the number density of the pumped lower state level N_1^0. Here, I_v is the laser power, the A and B terms are spectroscopic constants which must be independently known. The L, ω, and A_c terms are instrumental and experimental constants defined in the Radiative Transfer equation. The remaining term, Q, is the sum of all nonradiative collision induced quenching rates. This includes, electronic, vibrational, rotational and chemical quenching. In complex chemical systems such as flames it is not possible to independently evaluate this term Q, which is inversely proportional to the fluorescence quantum efficiency.

In the limit of high laser power, $(B_{12}+B_{21})\, I_v \gg Q+A_{21}$ the expression reduces to the form:

$$S = \frac{h\nu\, A_{21}}{4\pi}\, L\Omega\, A_c \left(\frac{B_{12}}{B_{12}+B_{21}} \right) N_1^0 . \tag{6}$$

Under these conditions of complete saturation the fluorescence signal becomes independent of laser power and the species number density N_1^0 can be theoretically evaluated with only the knowledge of the spectroscopic and instrumental constants.

Figure 24 shows a plot of fluorescence intensity vs. laser power for the Swan Band system of C_2 which we showed in Figure 19. It is apparent that fluorescence response becomes non-linear at laser powers on the order of 1 joule/m^2. However, it is equally apparent that the signal never reaches complete saturation (independent of laser power) even at 15 joules /m^2. We therefore reconsidered the model for this power region where $(B_{12}+B_{21})\, I_v > Q+A_{21}$ but where the pumping rate is not adequate to completely saturate the transition. Under these conditions we have shown ([20], [21]) that:

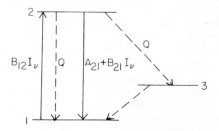

RADIATIVE TRANSFER FOR 90° COLLECTION OPTICS:

$$S = h\nu \frac{A_{21}}{4\pi} L \Omega A_c N_2$$

STEADY STATE APPROXIMATION:

Figure 23. Schematic of the two-level model used to treat saturation spectroscopy data. See the text for details.

$$S = h\nu \frac{A_{21}}{4\pi} L \Omega A_c \frac{N_1^0 B_{12} I_\nu}{Q + A_{21} + (B_{12} + B_{21})I_\nu}$$

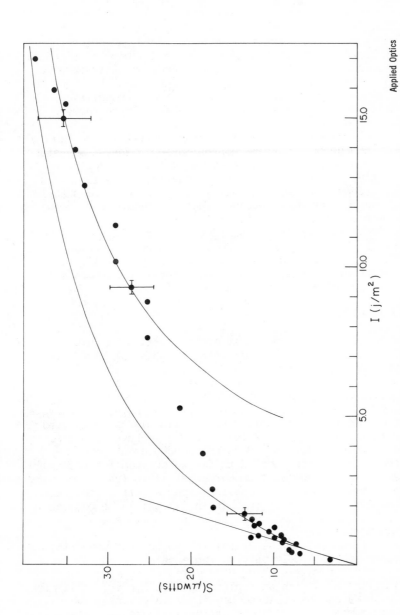

Figure 24. A plot of the fully corrected C_2 fluorescence emission signal vs. I_ν. The straight line passing through the origin is the predicted signal based on a linear response. The lower curve is the calculated fit to the high I_ν data shown in Figure 24. The curved line passing through the origin is that theoretically predicted by the steady-state equation shown in Figure 23 (21).

$$S = \frac{h\nu \, A^{21}}{4\pi} L\Omega \, A_c \left(\frac{B_{12}}{B_{12}+B_{21}} \right) \, N_1^0 - \frac{N_1^0 \, (Q + A_{21})}{(B_{21}-B_{12})I_\nu} + \cdots . \quad (7)$$

Through two terms in the expansion, one predicts an inverse dependence on laser power. The data in Figure 24, between 5 and 15 joules/m^2 is replotted in Figure 25 as a function of inverse laser power. The data fits the predicted functional form and the intercept (I → ∞) can be used to calculate the C_2 population. Our evaluation in this system yields a C_2 lower state number density of 4.5×10^{15} cm^{-3}.

These evaluations are made within the context of the two level model and the steady state approximation. The steady-state approximation is probably valid for this experiment. C_2 is not (electronically, vibronically, or rotationally) a two level system. Other groups, particularly Daily (23) and Berg and Shackleford (18) have developed expressions which allow for the inclusion of more levels and provide for incomplete relaxation. Lucht and Laurendeau (28) have carefully considered the effect of rotational equilibration. There is not time here to discuss these models in detail. The theoretical models which include specifically more than two electronic levels require experimental measurements independently of the radiation coupling the various levels. We have not found a system experimentally tractible for testing the three electronic level model.

Experiments With Sodium. Most groups working in this field have studied Na for various experimental reasons. I would like to refer you to the recent work by Muller, et al. (25) for a study of the flame chemistry of this system. This system is very complex in flames and dominated by unexpected effects for what one would expect to be a simple atomic system. Figure 26 shows a 1/I$_\nu$ plot for sodium. The solid straight line is the predicted fit within the context of the 2 level theory we have been discussing. This assumes a flat-topped sharp edged beam shape. The dashed curve is the theoretical data fit we derived for the truncated gaussian beam shape used in our experiments (24). The dot-dash curve is the theoretical behavior expected for a true Gaussian beam. This demonstrates that it is impossible to completely saturate with a gaussian beam because of the diminishing intensity in the wings of the beam.

Figure 27 shows the use of this technique for measuring Na concentrations over a widely varying range of number densities. The concentration of Na was changed by varying the Na concentration of solutions used with an aspirating slot burner. This plot indicates that Na can be measured down to ∿ 10 ppt relative to the flame gas number density and that the response is linear over the concentration range measured.

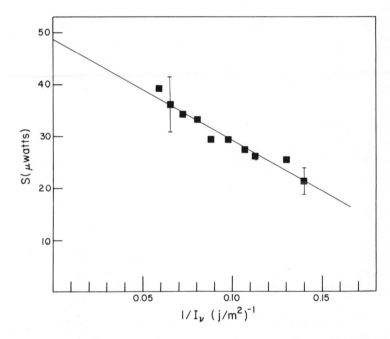

Applied Optics

Figure 25.   A plot of the C_2 fluorescence signal vs. $1/I_v$. The slope of this plot gives the total quenching rate for the $^3\Pi_g$ state while the intercept is used to evaluate the $^3\Pi_u$ state number density. $C_2 = 4.5 \times 10^{15}$ cm^{-3}; $Q = 1.2 \times 10^{12}$ sec^{-1} (21).

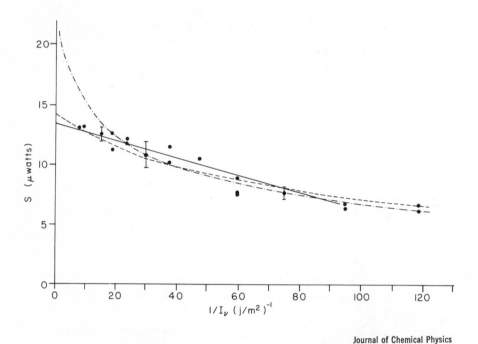

Figure 26.   Plot of the Na emission signal vs. $1/I_v$: (————), best fit to the data for a rectangular beam profile; (— — —), the best fit assuming a truncated Gaussian profile; (· — ·), the best fit for a Gaussian profile (24).

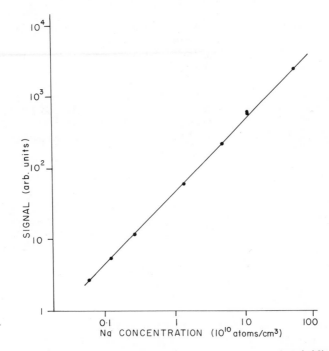

Figure 27.   Plot of fluorescence signal vs. Na concentration in a C_2H_2–air aspirating slot burner (24)

Comparison With Independent Measurement. We, at NRL, and
Alan Eckbreth and coworkers at United Technologies have made experimental measurements on several systems by saturated laser induced fluorescence spectroscopy which we can simultaneous measure independently by other techniques. The results are shown
in Figure 28. Most of the independent measurements were made by
long path length absorption measurements in homogeneous burners.
There are considerable uncertainties in the absorption measurements made at high concentrations because of experimental and
spectroscopic difficulties. In all cases the saturated fluorescence measurements give concentration measurements which are
lower than the absorption measurements. For Na and MgO the differences are slightly larger than the combined uncertainties of
the two measurements. In all cases these have been treated as
two level systems for fluorescence measurements. This is certainly not the case for CN, C_2 and MgO. The MgO was chosen
as a worst possible intermediate metastable non-two-level system.
    While the independent techniques do not give the exact same
concentration measurements they are similar enough to be highly
encouraging. The laser fluorescence technique is many orders of
magnitude more sensitive than the best absorption measurements
and it is a point sampling technique.

Conclusions

    Laser induced saturation fluorescence spectroscopy has been
demonstrated to be a sensitive technique for measuring low level
atomic and molecular concentrations in flames. Most atoms have
electronic transitions[17] amenable to detection by this technique.
Depending upon the luminosity of the flame system and the line
strength of the atomic transition, the detection limits for most
atoms in flames should be at or below 1 ppt. Careful experiments
should give absolute concentrations for these species to within
a factor two and relative measurements can be made with much higher precision. The application of this technique to the selected
group of molecules noted in Figure 29 can also give very sensitive measurements of concentrations. The sensitivity varies
considerably with the molecular system being studied because of
the molecular electronic structure and, more importantly, because
of the applicability of the theoretical two level model used to
evaluate concentrations. I feel that the present state of the art
should give absolute number densities which can be trusted to within a factor of 3 or 4. The sensitivity limits for most of these
molecular species using the techniques I have described is in the
region of 100 ppt. Except in rather specialized applications such
as two level atomic systems laser induced fluorescence will probably not be significantly exploited for temperature measurements.
    Coherent Antistokes Roman Spectroscopy has been used to make
both concentration and temperature measurements in flame systems.
The accuracy and detection limits vary depending upon the type of

SPECIES	CONCENTRATION Sat. Fluor. (cm^{-3})	CONCENTRATION Absorption (cm^{-3})	FLAME	LASER
C_2	5×10^{15}	$\simeq 10^{16}$	C_2H_2/O_2	1μs
CH	7.1×10^{13}	1.6×10^{14}	C_2H_2/O_2	250ns
CN	8.1×10^{13}	3.8×10^{14}	C_2H_2/NO	250ns
Na	8.5×10^{9}	4.8×10^{10}	C_2H_2/Air	1μs
MgO	2.9×10^{11}	$\simeq 2.1 \times 10^{12}$	C_2H_2/Air	1μs

Figure 28. *Comparison of concentration measurements made by saturation fluorescence and absorption measurements. See the text for details.*

Selected Atomic Species	Aromatic Hydrocarbons
CH	SO
CN	SH
C_2	CS
OH	S_2
NH	S_2O
NO	C_3
	HCN

Figure 29.   Partial list of species that are likely amenable to laser flame measurements

flame system under study and the type of experimental CARS
arrangement which must be used to study the system. In homogeneous flames which can be studied using scanning techniques and extensive time averaging, species concentration detection limits
are better than 0.5% with detection limits for selected species
near the 100 ppm level. Under these conditions absolute temperature measurements based upon such species as N_2 and H_2O have an
accuracy of 1-2% although most workers quote their measurements
to $\pm$ 50°K. In turbulent, luminous, sooty flames using single
shot Boxcars techniques most workers claim temperature measurements to $\pm$ 100°K. Under these conditions the current detection
limit for species such as CO is ∿10% where polarization techniques combined with Boxcars and OMA detection is used. We can expect improvements in CARS both experimentally and theoretically in
the future. The use of new techniques which we described earlier
involving the use of background supression techniques and the use
of rotational CARS are under intense development.

In summary the recently developed fields of CARS and laser
induced saturation fluorescence spectroscopy offer considerable
potential as diagnostic techniques for combustion systems. The
techniques are complimentary. CARS has its best application for
relatively high concentration flame gases and for temperature
measurement. The fluorescence technique is well suited for low
concentration measurements of atoms and radicals and flame
transients.

Literature Cited

1. Tolles, W.M.; Nibler, J.W.; McDonald, J.R.; and Harvey, A.B.
    Appl. Spect., 31, 253 (1977)
2. Nibler, J.W.; and Knighten, G.V. Chapter 7 in "Topics in
    Current Physics", ed. Weber, A.; Springer Verlag, Stuttgart,
    Germany (1977)
3. Druet, S.; and Taran, J.P. in "Chemical and Biological Appli-
    cations of Lasers", Ed. Moore, C.B.; Academic Press, New York
    (1979)
4. Regnier, P.; and Taran, J.P. Appl. Phys. Lett., 23, 240
    (1973).
5. Shaub, W.M.; Lemont, S.; and Harvey, A.B. Computer Physics
    Comm. 16, 73 (1978)
6. Shaub, W.M.; Lemont, S.; and Harvey, A.B. Appl. Spectroscopy,
    33, 268 (1979)
7. Eckbreth, A.C. "Cars Investigations In Sooting and Turblent
    Flames", Technical Report R79-954196-3 United Technologies
    Research Center, East Hartford, Connecticut, p. 48. Project
    SQUID Report, UTRC-5-PU, March 1979.
8. Hall, R.J.; Shirley, J.A.; and Eckbreth, A.C. Opt. Lett., 4,
    87 (1979)

9. Eckbreth, A.C.; Bonczyk, P.A.; and Shirley, J.A. "Experimental Investigations of Saturated Laser Fluorescence and CARS Spectroscopic Techniques for Practical Combustion Diagnostics", Technical Report R78-952665-18 United Technologies Research Center, East Hartford. Conn., p. 58. EPA-600/7-78-104, June 1978.

10. Nibler, J.W.; Shaub, W.M.; McDonald, J.R.; and Harvey, A.B. in "Vibrational Spectra and Structure", Ed. Durig, J.R. Chapter 3, p. 174-225 Elsevier Scientific Publishing Co. New York (1977).

11. Tolles, W.M., Nibler, J.W.; McDonald, J.R.; and Harvey, A.B. Appl. Spec. 31, 253 (1977).

12. Shirley, J.A.; Eckbreth, A.C.; and Hall, R.J. "Investigation of the Feasibility of CARS Measurements in Scram jet Combustion", Technical Report R79-954390-8 United Technologies Research Center, East Hartford Conn. (1979)

13. Ralm, L.; Zych, L.; and Mattern, P.; "Polarization CARS Demonstrated in Methane-Air Flame", Combustion Research Facility News, Vol. 1, No. 2, Sandia Laboratories, Livermore, May 1979.

14. Piepmeier, E.H. Spectrochim Acta, 27B, 431, 445 (1972)

15. Omenetto, N.; Benetti, P.; Hart, L.P.; Winefordner, J.D.; and Alkemade, C.T.J. Spectrochim Acta, 28B, 289 (1973)

16. Smith, B.; Winefordner, J.D.; and Omenetto, N. J. Ap. Phys., 48, 2676 (1977)

17. Omenetto, N.; and Winefordner, J.D. "Analytical Laser Spectroscopy", Chapt. 4., p. 167, Wiley (1979)

18. Daily, J. Appl. Opt., 15, 955 (1976)

19. Daily, J. Appl. Opt., 16, 568 (1977)

20. Baronavski, A.P.; and McDonald, J.R. J. Chem Phys., 66, 3300 (1977)

21. Baronavski, A.P.; and McDonald, J. R. Appl. Opt., 16, 1897 (1977).

22. Eckbreth, A.C.; Bonzyk, P.A.; and Verdieck, J.F. Appl. Spec. Rev., 13, 15 (1977)

23. Daily, J.W.; Appl. Opt., 17, 225 (1978)

24. Pasternack, L.R.; Baronavski, A.P.; and McDonald, J.R. J.Chem. Phys., 69, 4830 (1978)

25. Muller, C.H.; Schofield, K.; and Steinberg, M.S. Chem. Phys. Lett., 57, 364 (1978)

26. Bonczyk, P.A.; and Shirley, J.A. Comp. and Flame, 34, 253 (1979)

27. Berg, J.O.; and Shackleford, W.L. Appl. Opt., 18, 2093 (1979)

28. Lucht, R.P.; and Laurendeau, N.M. Appl. Opt., 18, 856 (1979)

RECEIVED March 25, 1980.

LASER-INDUCED FLUORESCENCE: MOLECULES

Laser-Induced Fluorescence Spectroscopy in Flames

JOHN W. DAILY

Department of Mechanical Engineering, University of California, Berkeley, CA 94720

The purpose of this paper is to review the use of laser induced fluorescence spectroscopy (LIFS) for studying combustion processes. The study of such processes imposes severe constraints on diagnostic instrumentation. High velocities and temperatures are common, as well as turbulent inhomogeneities, and there is a need to make space and time resolved species concentration and temperature measurements. The development of LIFS has reached the point where it is capable of making significant contributions to experimental combustion studies.

Fluorescence is spontaneous radiation that arises because of the stimulation of an atomic or molecular system to energies higher than equilibrium. This is illustrated in Figure 1 for a simple two-level atom. The atom is excited by absorption of a photon of energy $h\nu$. If the fluorescence is observed at 90° to a collimated excitation source, then a very small focal volume may be defined resulting in fine spatial resolution. The fluorescence power an optical system will collect is

$$P_F = h\nu \, \frac{A_{21}}{4\pi} \, \Omega_c V_c N_2, \tag{1}$$

where V_c is the effective focal volume, Ω_c the solid angle of the collection optics, and A_{21} is the Einstein coefficient for spontaneous emission, the probability of decay in any direction.

The fluorescence signal can be used in a number of ways. Most simply it provides a measure of the population of the excited state or states through Equation 1. In addition, if a relationship can be found between the number density of all the quantum states under excitation conditions, then the total number density of the species can be deduced. Unfortunately, collisional decay process can cause redistribution of population from the excited level, complicating interpretation.

0-8412-0570-1/80/47-134-061$10.00/0
© 1980 American Chemical Society

By examining the excitation spectrum of a molecular species one can deduce a ground state Boltzmann temperature. Also, as will be discussed below, if one can predict the population distribution in the atom or molecule under excitation conditions, then one can use the observed fluorescence spectrum to recover the gas temperature.

Finally, energy transfer and chemical processes can be studied by observing the transient and steady state response of a molecular system to laser excitation.

Lasers are used as an excitation source for three reasons. Because the laser output is coherent it offers special advantages in directionality and focusing. Tunable lasers allow the possibility of examining several species. Finally, lasers provide significantly higher power levels than conventional light sources.

In the following we consider the nature of LIFS in more detail. The theoretical foundations of laser excitation and fluorescence are outlined and such issues as detectability and dynamic range are discussed. Finally the status of LIFS is summarized and a prognosis for future development given.

Theoretical Considerations

As discussed above, a LIFS signal is proportional to the excited state number density of the species being excited. This information is not itself normally useful. What is desired is a measure of the total population, or the temperature. Often one seeks the population of individual ground states. To be able to relate the observed signal to variables of interest one must be able to describe the dynamics of the excitation process.

The Rate Equations. As illustrated in Figure 1, molecules are excited by photon absorption in the process

$$N_k + h\nu \xrightarrow{B_{k\ell}\rho_\nu} N_\ell \, , \, \ell > k \qquad (2)$$

where $B_{k\ell}$ is the Einstein B coefficient for absorption and ρ_ν is the spectral energy density due to laser excitation. Likewise, molecules can be de-excited by the induced emission process

$$N_\ell + h\nu \xrightarrow{B_{k}\rho_\nu} N_k \, , \, \ell > k \qquad (3)$$

and by spontaneous emission to lower levels

$$N_\ell \xrightarrow{A_{\ell i}} N_i + h\nu, \, \ell > i = 1,2,3.... \qquad (4)$$

In addition, collisions can cause both excitation and de-excitation in the process

$$N_j + M \xrightarrow{Q_{ji}} N_i + M , \quad i \neq j. \tag{5}$$

Collisional de-excitation is called quenching because it competes with spontaneous emission, and if significant the fluorescent signal will be reduced, or quenched.   Chemical decay can also be important in some circumstances.

Taking into account these processes one may write rate equations (Daily, 1) for the individual energy levels.   For a simple two-level system one has

$$\frac{dN_2}{dt} = (Q_{12} + B_{12}\rho_\nu) N_1$$

$$- (Q_{21} + A_{21} + B_{21}\rho_\nu) N_2, \tag{6a}$$

and

$$N_{TOT} = N_1 + N_2 . \tag{6b}$$

The steady state solution to this system is (using the detailed balance relation $g_1 B_{12} = g_2 B_{21}$):

$$N_2 = \frac{[Q_{12} + B_{12}\rho_\nu] N_T}{[Q_{12} + Q_{21} + A_{21} + (B_{12} + B_{21})\rho_\nu]} . \tag{7}$$

The first term represents the equilibrium excited state number density N_2^*, when $\rho_\nu = 0$.   Generally $N_2 \gg N_2^*$ and $Q_{12} \ll Q_{21}$ so that

$$N_2 \simeq \frac{B_{12}\rho_\nu}{[Q_{21} + A_{21} + (B_{12} + B_{21})\rho_\nu]} N_T . \tag{8}$$

It is conventional to define a "saturation" energy density as

$$\rho_\nu^s \equiv (Q_{12} + Q_{21} + A_{21})/(B_{12} + B_{21}) \tag{9}$$

so that (noting $g_1 B_{12} = g_2 B_{21}$)

$$N_2 \simeq \frac{N_T}{(1 + g_1/g_2)} \left\{ \frac{\rho_\nu}{\rho_\nu + \rho_\nu^s} \right\} . \tag{10}$$

There are two limits of interest.

    Low intensity limit. In this case Equation 10 becomes

$$N_2 \simeq \frac{N_T}{(1 + g_1/g_2)} \left\{ \frac{\rho_\nu}{\rho_\nu^s} \right\} . \tag{11}$$

This may be rewritten as

$$N_2 = (B_{12}/A_{21})(A_{21})/(Q_{21} + A_{21})\rho_\nu N_T$$

$$= (B_{12}/A_{21}) Y \rho_\nu N_T \tag{12}$$

so that the collected fluorescence power becomes

$$I_F = \frac{h\nu}{4\pi} B_{12} \Omega_c V_c Y \rho_\nu N_{TOT} . \tag{13}$$

Y is called the fluorescence yield, and for combustion conditions is typically 10^{-2}-10^{-3}; that is, quenching is significant.

    Given knowledge of the atomic parameters B_{12} and A_{21} and the collisional rates, one can directly relate the observed N_2 to the total number density of the species. If calibration is possible, only the temperature dependence of the yield, Y, need be known.

    The difficulty with Equation 13 is that under conditions of turbulent combustion the temperature and composition, and thus Y, may vary in an unknown manner.

    For certain special cases, going to the high intensity limit provides a remarkably simple solution to the quenching problem.

    High intensity limit (saturation). Consider the rate equation for level 2 of our simple atom, Equation 6a. In the limit of large energy density, this equation reduces to

$$N_2 = (B_{12}/B_{21}) N_1 \tag{14}$$

which because of detailed balancing $(g_1 B_{12} = g_2 B_{21})$ becomes

$$N_2 \cong (g_2/g_1) N_1 . \tag{15}$$

This remarkable result states that if the laser intensity is high enough, then N_1 and N_2 will occur in a fixed, known ratio. Equation 10 also reduces in this limit to

$$N_2 = \left\{ \frac{1}{1 + g_1/g_2} \right\} N_{TOT} \qquad (16)$$

Atomic Systems. Many atomic species may be modeled as three-level systems. Figure 2 illustrates the energy level diagram for sodium. Other alkali and alkaline metals behave in a similar manner.

For excitation from level 1 to 2 the steady state rate equations become

$$Q_{13}N_1 + Q_{23}N_2 = (Q_{32} + Q_{31} + A_{32} + A_{31}) N_3, \qquad (17a)$$

$$(Q_{12} + B_{12}\rho_\nu)N_1 + (Q_{32} + A_{32}) N_3 = (Q_{21} + Q_{23} + A_{21} + B_{21}\rho_\nu) N_2,$$

and $\qquad\qquad\qquad\qquad\qquad\qquad\qquad\qquad\qquad\qquad (17b)$

$$N_{TOT} = N_1 + N_2 + N_3 . \qquad (17c)$$

If one defines

$$(N_3/N_2)^* \equiv (Q_{13} + Q_{23})/(Q_{32} + Q_{31} + A_{32} + A_{31}) \qquad (18)$$

as a quasi-equilibrium population ratio, then one may show that

$$N_2 = \frac{N_T}{[1 + g_1/g_2 + N_3/N_2)^*]} \left\{ \frac{\rho_\nu}{\rho_\nu + \rho_\nu^s} \right\}, \qquad (19)$$

where

$$\rho_\nu^s = \frac{Q_{21} + Q_{23} - (Q_{32} + A_{32}) \; (N_3/N_2)^*}{B_{12} \; [1 + g_1/g_2 + (N_3/N_2)^*]} \qquad (20)$$

Because of the difficulty in knowing $(N_3/N_2)^*$, experimentally one would measure N_2 and N_3. Under saturation conditions $g_1 N_1 = g_2 N_2$ so that

$$N_{TOT} = (1 + g_2/g_1) N_2 + N_3 , \qquad (21)$$

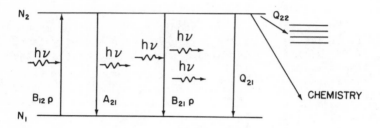

Figure 1.   Energy-level diagram

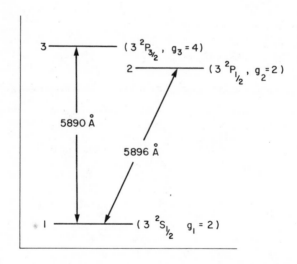

Figure 2.   Sodium energy-level diagram

These expressions can be easily generalized for multi-level systems.

Molecular Systems. Molecules present a considerably more complex picture. Illustrated in Figure 3 is the energy level diagram for OH, the hydroxyl radical. The structure consists of several electronic states, each of which supports a number of vibrational states. Rotational motion is superimposed on each electronic-vibrational state as illustrated in Figure 3b. OH is an attractive molecule for analysis because of its dominant importance in combustion kinetic schemes and because its structure, while more complicated than any atom's, is fairly simple compared to many other molecules.

The complexity of analysis depends on the relative values of collisional relaxation rates and the degree to which excited vibrational levels become populated thermally or under excitation conditions. For OH it is advantageous to pump into the $^2\Sigma^+$ electronic level to either the ground or first excited vibrational level.

The strategy we have been following is to pump the $\nu'' = 0 \rightarrow \nu' = 0$ vibrational band in the UV and observe the resulting fluorescence. This method reduces complexity caused by eliminating the need to consider vibrational relaxation and results in most of the fluorescence signal appearing in the 0-0 vibrational band. Moreover, the energy levels, transition frequencies and transition probabilities for this band have been studied extensively and can be found in the literature.

We have assumed that all vibrational levels other than the two ground levels for the $^2\pi$ and $^2\Sigma^+$ states are not present. (This is not strictly true. At a flame temperature of 2000°K about 10% of the molecules will be in the first excited vibrational state.) Since the transition probability for the $\nu' = 0$ to $\nu'' = 1$ is relatively small, we also assume that all transitions are between the two ground vibration states only.

The rate equations. The steady state rate equations for the number density of any rotational state in the $^2\Sigma^+$ state other than the one involved in the laser excitation can be written as

$$\frac{dN_2(i)}{dt} = 0 = -N_2(i)[\sum_{j \neq i} Q_{22}(i,j) + Q_{21}(i) + A_{21}(i)]$$

$$+ \sum_{j \neq i} N_2(j)\, Q_{22}(j,i) + \sum_k N_1(k)Q_{12}(k), \qquad (22)$$

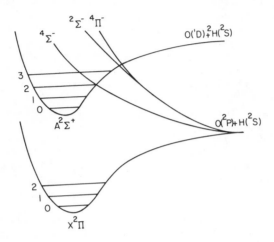

Figure 3a.   OH energy-level diagram–electronic and vibrational structure

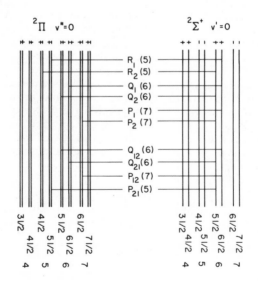

Figure 3b.   OH energy-level diagram—rotational structure

where
$N_2(i)$ is the number density in the ith level of the $^2\Sigma^+$ state,

$Q_{22}(i,j)$ is the rotational relaxation rate for transition $i \rightarrow j$,

$Q_{21}(i)$ is the total quenching rate of the ith level, and $A_{21}(i)$ is the Einstein A coefficient of the ith level.

The steady state rate equation for the level excited by the laser can be written as

$$\frac{dN_2(e)}{dt} = 0 = -N_2(e) \left[\sum_{j\neq e} Q_{22}(e,j) + Q_{21}(e) + A_{21}(e) + B_{21}\rho_\nu \right]$$

$$+ N_1(e) B_{12}\rho_\nu + \sum_{j\neq e} N_2(j) Q_{22}(j,e) + \sum_k N_1(k) Q_{12}(k), \qquad (23)$$

where B_{21} is the Einstein B coefficient and ρ_ν is the spectral irradiance. A similar set of equations describes the ground electronic vibrational-rotational states.

These equations have been analyzed in some detail by Lucht and Laurendeau (2) and Chan and Daily (3). When the laser is turned on, molecules in the $^2\pi$ state are pumped to the $^2\Sigma$ state. Since the laser selectively pumps from one rotational sublevel to another, the other rotational sublevels in the $^2\Sigma$ state can be populated only by rotational relaxation. If quenching, or electronic de-excitation, is fast compared to rotational relaxation, then only the laser coupled state will be populated. On the other hand, if rotational relaxation is fast then all the rotational states will be populated and in Boltzmann equilibrium.

For OH, experimental evidence indicates the intermediate case as illustrated in Figure 4, which shows the deviation of each rotational sublevel from its normalized Boltzmann value. Not surprisingly, the laser coupled state is overpopulated compared to other states.

For molecules, the energy density required to saturate the excited transition can be as much as three orders of magnitude higher than for atoms. This may be seen from Equation 23 in terms of a saturation spectral energy density with a result similar to but more complicated than that of Equation 20. Both expressions may then be more conveniently written as

$$\rho_\nu^s = Q^{eff}/B ,$$

where Q^{eff} is an effective quenching rate.

Saturation. For the simple two-level model the saturation energy density was shown to be

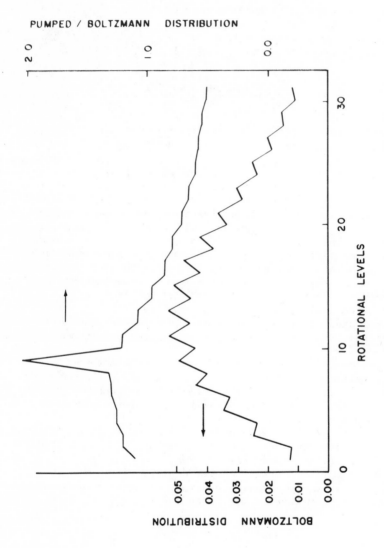

Figure 4.   Calculated excited-state population distribution for OH: T = 2000 K; $Q_{22}/Q_{21} = 15$.

$$\rho_\nu^s = \frac{Q_{21}/B_{12}}{1 + g_1/g_2} = \frac{(8\pi h/\lambda^3)(Q_{21}/A_{21})}{1 + g_1/g_2} \tag{24}$$

where the Einstein relation $B_{12} = c^3 A_{21}/8\pi h\nu^3$ has been used to replace B_{12}. For a laser beam, the relationship between power density and energy flux is $P/A = 4\pi c\rho_\nu \delta\nu_s$, where $\delta\nu_s$ is the spectral width of the laser source. Thus the saturation energy flux is

$$(P/A)^s = \frac{(32\pi^2 hc/\lambda^3)}{(1 + g_1/g_2)} \delta\nu_s \, (Q_{21}/A_{21}) \; . \tag{25}$$

Typical tunable laser line widths are about 0.1 cm^{-1} and for atoms $Q_{21}/A_{21} \sim 10^2 - 10^3$, so $(P/A)^s$ atoms $\sim 3 \times 10^4 - 10^5$ w/cm^2. Likewise for molecules $(Q_{21}/A_{21})^{eff} \sim 10^3 - 10^6$ so that $(P/A)^s_{molec} \sim 3 \times 10^5 - 10^6$w/cm^2.

There is no difficulty in focusing a laser beam to a diameter of 0.1 mm. A one-watt laser could then provide an irradiance of about 10^4w/cm^2 and a one-kilowatt laser an irradiance of about 10 MW/cm^2. Since one watt and one kilowatt are powers typical of CW and pulsed dye lasers respectively, it may be seen that if saturation is a goal, then CW laser sources are not appropriate for studying molecular species.

Detectability and Dynamic Range

The useful range over which LIFS can be used to measure the number density of some excited state is determined at the low end by noise and uncertainty considerations and at the high end by radiative trapping effects.

Detectability Limits. Recall that the actual measurement is that of a fluorescence energy or photon flux

$$P_F = h\nu \, \frac{A_{21}}{4\pi} \, \Omega_c V_c N_2 \; . \tag{26}$$

What we seek is

$$N_2 = \frac{P_F}{h\nu \left(\dfrac{A_{21}}{4\pi}\right) \Omega_c V_c} \tag{27}$$

The fractional statistical uncertainty with which N_2 can be measured is thus

$$\frac{W_{N_2}}{N_2} = \left\{ \left(\frac{W_{A_{21}}}{A_{21}} \right)^2 + \left(\frac{W_{\Omega_c}}{\Omega_c} \right)^2 + \left(\frac{W_{V_c}}{V_c} \right)^2 + \left(\frac{W_{P_F}}{P_F} \right)^2 \right\}^{\frac{1}{2}} . \qquad (28)$$

The detectability limit is that value of N_2 for which the fractional uncertainty becomes too large, say 10%.

The critical source of uncertainty will be in the measurement of P_F. For low values of P_F Poisson statistics apply and the fractional uncertainty in P_F becomes

$$\frac{W_{P_F}}{P_F} = (\text{photon count})^{-\frac{1}{2}} . \qquad (29)$$

The photon count may be written as

$$\varepsilon \eta \left(\frac{A_{21}}{4\pi} \right) \Omega_c V_c \Delta t N_2 \qquad (30)$$

where ε is the detector efficiency, η an optical efficiency and Δt a measurement of time.

Thus

$$\frac{W_{P_F}}{P_F} \simeq \left\{ \varepsilon \eta (A_{21}/4\pi) \Omega_c V_c \Delta t N \right\}^{-\frac{1}{2}} \qquad (31)$$

Equation 31 is plotted in Figure 5 for typical combustion conditions and with the combined uncertainty in A_{21}, Ω_c and V_c as a parameter. The sample time is 1 μsec, typical of flash-lamp pumped dye laser pulses, and thus remarkably low detectability limits are achievable with single pulse sampling.

Equation 31 is used to determine the smallest concentration of a species that can be detected. The dynamic range of the instrument will then be determined by the effects of radiative trapping at large number densities.

Interferences. There are a number of interferences that must be taken into account and which may limit detectability or even destroy the possibility of using the diagnostic.

Rayleigh scattering is an ever present interference for any resonant fluorescence component and will set the ultimate lower limit on detectability. It is generally very small though and and has the property of being highly polarized. This property can be used to reject the Rayleigh scattered light with a ratio of about 10^{-7}. As a result, Rayleigh scattering is rarely a problem.

Mei scattering can be a severe interference, although also only for resonance fluorescence. The presence of solid particles can completely eliminate the possibility of using an optical method. For smaller particle densities, however, one may overcome particle problems with optical and processing tricks. Each case must be examined individually.

Eckbreth (4) has pointed out that, when high power laser sources are being used, the laser may heat small particles in the flow to the point where they contribute a significant amount of blackbody radiation to the signal. Again, the severity of the effect depends on local circumstances, but in sooting flames can be quite bad. Interferences of these sorts are discussed in detail by Daily (5).

Radiative Trapping. The radiative trapping problem is illustrated in Figure 6.

When the emitted radiation leaves the laser focal volume the possibility exists that it will be absorbed, or trapped, by gas molecules along the path to the collection optics. This effect will manifest itself as self-reversal of individual fluorescence lines or band absorption of the fluorescence spectrum.

If thermal emission outside the focal volume is ignored, the specific intensity observed outside the flame is

$$I_\nu = e^{-\tau_\nu} I_\nu(0) , \qquad (32)$$

where $I_\nu(0)$ is the specific intensity leaving the focal volume, and

$$\tau_\nu = \int_0^L \alpha_\nu(x)\,dx \qquad (33)$$

is the optical depth, where $\alpha_\nu(x)$ is the local absorption coefficient.

The integrated intensity for an individual line is then

$$I_L = \int_{line} \varepsilon^{-\tau_\nu} I_\nu(0)\,d\nu \qquad (34)$$

For bands, the expression for the integrated band intensity is identical but with the integration over the whole band.

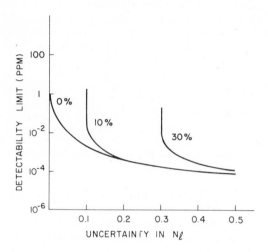

Figure 5.   Detectability limits for typical flame conditions: $\Delta t = 1\,\mu sec.$

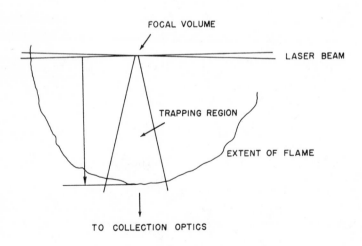

Figure 6.   The radiative trapping effect

In the absence of interferences, α_ν arises from the absorption spectrum of the test species. For atoms and certain molecules, the absorption spectrum consists of individual lines or small groupings of lines in which each line is described by a Voigt-type profile. For most molecules, pressure broadening causes a merging together of the lines into absorption bands. The absorption coefficient for an individual line may be written as

$$\alpha_\nu = h\nu N_k B_{k1} \phi(\nu),$$ (35)

or

$$\alpha_\nu = \frac{N_k e^2 f_{k1}}{4\varepsilon_0 M_e c} \phi(\nu), \quad 1 > k,$$ (36)

where h is the Planck's constant, B_{k1} the Einstein coefficient for absorption, $\phi(\nu)$ the normalized absorption line shape parameter, e the electron charge, ε_0 the permittivity of free space, M_e the electron mass, c the speed of light, and f_{k1} the oscillator strength for the transition.

The importance of trapping must, of course, be assessed for each experiment and test species considered. If one adopts a criteria for the maximum absorption, an upper limit is placed on the number density of the absorbing energy level which can be allowed. We can roughly estimate this by assuming an oscillator strength of unity for atomic transitions, and of 10^{-3} for molecular transitions. This leads to line center ground state absorption coefficients of the order of $\alpha_{atom} \sim 10^{-16} N(m^{-3})m^{-1}$ and $\alpha_{mole} \sim 10^{-19} N(m^{-3})m^{-1}$ for a 2000°K atmospheric pressure flame. For a 1-m pathlength and an optical depth of unity, this corresponds to an upper limit in mole fraction of about 0.01 PPM and 10 PPM for atoms and molecules, respectively. Of course, for absorption that originates in higher energy levels, both the oscillator strength and the number densities drop rapidly.

Figure 7 illustrates the trapping effect for sodium (6). The measurements were made across the top of a flat flame burner, and as can be seen, trapping is significant for mole fractions larger than about 0.15 PPM.

Near Resonant Rayleigh Scattering. One potential method for overcoming the problem of radiative trapping that appears to work well for atoms is near resonant Rayleigh scattering (7). If an atom is excited near a resonant line, part of the light is scattered as enhanced Rayleigh scattering. If the atom being excited also undergoes collisions then the possibility exists that a second component of light will be emitted at the resonant frequency. This process is called

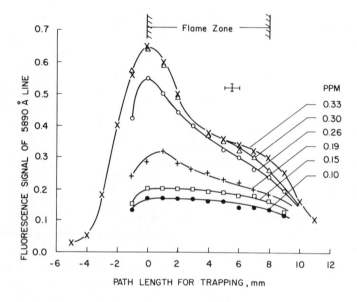

Figure 7.   Radiative trapping of sodium in a methane–air flame

collisional redistribution. The importance of these processes is that the Rayleigh component is outside the absorption line and cannot be trapped.

Figure 8 shows a typical fluorescence spectrum for near resonant excitation of sodium. The large peak arises from collisional redistribution, the other is the enhanced Rayleigh component. The relative intensities agree with the theory of Mollow (8). Mollow's theory can also be used to predict the ratio of the Rayleigh signal to the resonant fluorescence signal one would obtain if there were no trapping. This ratio is shown in Figure 9 which illustrates that the Rayleigh component can be quite large even at detunings several Angstroms from line center.

Figure 10 illustrates the effect in sodium, showing that the Rayleigh component was not trapped.

Applications

Use of Saturation. Because of the potential for simplification of the population balance equations, much recent work has concentrated on studying saturation phenomena. First proposed by Piepmeier (9), and elaborated on by Daily (10), saturation in atomic species can lead to complete elimination of the need to know any collisional rates, and in molecular species may provide substantial simplification of the balance equation analysis.

The approach to saturation in sodium has been studied in detail, with several early workers reporting anomalous results. Such results seem to be explained by taking account of the laser beam intensity distribution (11, 6, 12). In controlled measurements, van Calcar, et al. (13) and Blackburn (14) have demonstrated saturation of sodium in flames under pulsed and CW laser operations respectively.

Saturation in molecular species is more difficult due to syphoning of population to other levels. Thus higher laser powers are required. Baronavski and McDonald (15) have studied the approach to saturation of C_2 and suggested means to use the saturation curve to extract collisional rate information. Eckbreth, et al. (16) have studied saturation in CH and CN and verified that under saturation conditions reasonable estimates of molecular number density can be obtained.

Currently it appears that there are no difficulties in saturating atomic species, while molecular species may be saturated with sufficient laser power. There are some difficulties associated with saturation. Because of chemistry, the quasi-equilibrium population of a species may change substantially when excited. See, for example, Daily and Chan (7), and Muller, et al. (17).

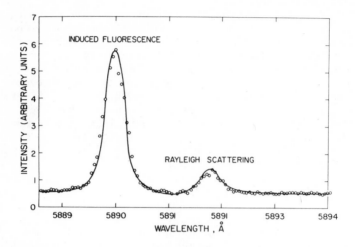

Figure 8. Fluorescence spectrum resulting from near-resonant excitation

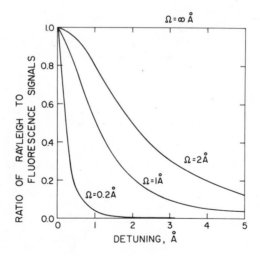

*Figure 9.   Ratio of Rayleigh to resonance fluorescence signal (Ω is the Rabi fre-
quency; at $\Omega = 1$ Å the radiation density is $\rho_v = 8 \times 10^{-17} J/m^3$).*

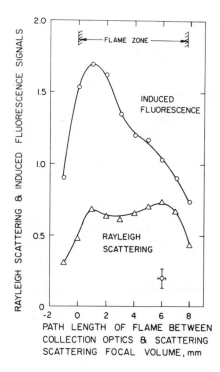

Figure 10.   Comparison of Rayleigh and fluorescence trapping in sodium

Excitation Dynamics. The response of atomic and molecular systems to exciting radiation has long been of interest and work has been going on to understand such phenomena for over one hundred years (18). Recent work has involved the use of lasers and modern detection systems to observe and measure individual radiative and collisional rates.

The choice of suitable species is dictated to a large extent by the availability of rate data, and although a great deal of work has been done, little has been directed at the problems of applying LIFS to the study of the turbulent combustion environment. Chan and Daily (3) and Chan (19) have studied OH dynamics in atmospheric flames and found useful the low pressure data of Lengel and Crosley (20). Lucht and Laurendeau (2) have naalyzed OH numerically. Stepowski and Cottereau (21) have used pulsed fluorescence (22) to measure decay rates in lower pressure flames and their cross section data is of direct interest to higher pressure combustion applications. Little other work has appeared although physical chemists are increasingly becoming interested in providing appropriate data.

Concentration Measurements. The use of LIFS to measure atomic species' concentrations in flames has been demonstrated repeatedly in analytical applications and the field is well reviewed by Winefordner and Elser (23) and Winefordner (24).

For atomic species the saturation approach appears to be most fruitful, although care must be taken to avoid chemical effects. Daily and Chan (7) have measured sodium concentrations in flames using saturated LIFS with a pulsed laser source and compared the results with absorption measurements. Smith, et al. (25) and Blackburn, et al. (14) have done the same under CW laser excitation.

Molecular measurements in flames have been made of C_2 by Baronavski and McDonald (15) and of CH and CN by Eckbreth, et al. (16). Chan and Daily (3) have worked with OH and Chan (19) has done more extensive measurements in OH.

Temperature Measurements. There are a number of techniques for measuring temperature using LIFS which show promise.

The first, called two-line fluorescence by Omenetto, et al. (26), involves seeding the flow with an appropriate atomic species, such as indium or thallium, which has two excited electronic states, one of which is close to the ground state. The seed is selectively and sequentially pumped with a light source at two wavelengths and the non-resonant fluorescence is observed in each case. The ratio of the two fluorescence signals is related to the temperature. The method was first demonstrated by Haraguchi, et al. (27), who measured temperatures in a variety of flames and whose work has been extended by

Bradshaw, et al. (28). In their experiments, a continuum light
source was used although they have since used pulsed laser
sources. We have also performed some preliminary experiments
(29). The result of these experiments show the necessity of
using laser excitation sources if there is to be adequate signal
noise to perform measurements in turbulent flows. We are
currently assembling a CW laser system for two-line fluorescence
in our laboratory.

The second promising method is the use of the spectrum of
a diatomic or larger molecule. As discussed in Section II-C,
if one can describe accurately the population distribution for
the molecule under excitation conditions, then the temperature
can be extracted from the measured spectrum. The difficulty
lies in capturing the spectrum in a sufficiently short time
period. This can be accomplished through the use of a multiple
detector array, or Optical Multichannel Analyzer such as is
manufactured by Princeton Applied Research Co. (30).

There is another approach which can be used in suitable
circumstances. Developed by Kowalik and Kruger (31), it involves
measuring the population of an excited atomic state by LIFS. If
the ground state population is known to be uniform in the flow
field, then information about temperature can be inferred. They
have used the method to measure electron number density in MHD
plasma flows.

Summary and Conclusions

We have examined the nature of LIFS in some detail. The
response of an atomic or molecular system is described in terms
of appropriate rate (or balance) equations whose individual
terms represent the rate at which individual quantum states are
populated and depopulated by radiative and collisional processes.
Given the response of a system to laser excitation, one may use
the rate equations to recover information about total number
density, temperature and collision parameters.

The detectability limit for any given measurement is defined
in terms of measurement uncertainty and for LIFS can be quite
small. This limit, however, can be affected by interferences of
various kinds and care must be taken in instrument design to
avoid difficulties. The dynamic range for LIFS is generally
controlled by radiative trapping effects.

The phenomena of saturation has also been examined and
satisfactorily described. It has also been shown that LIFS is
suitable for studying excitation and collision dynamics, and
for measuring species concentrations and temperatures.

LIFS is now ready to begin being seriously applied to
turbulent flows. For some species, sufficient information
already exists to obtain quantitative results of direct
applicability, although a major effort to collect and collate
collision data must continue. Reliable equipment is available

which can be used to build measuring systems that interface to conventional analog or digital data processing systems.

There are several areas in which future development will concentrate. Laser systems which provide more power and flexibility than current systems are needed. Higher power frequency doubled CW lasers and high rep rate pulsed systems would both be useful. Optical multichannel analyzers that are faster reading and easy to interface would be especially useful for rapid spectra recording and interpretation. Methods for increasing instrument dynamic range without sacrificing detectability limits will be useful in studying radical species.

It seems inevitable that LIFS will start to be used by more and more researchers. Combined with a technique such as coherent Anti-Stokes Raman Scattering (Eckbreth, et al., 16), which is best suited for measuring major species concentrations, a common laser and detection system provide a wide range of measurement possibilities.

Acknowledgment

The work reported that was performed by the Authors was supported by Air Force Office of Scientific Research Grant No. 77-3357.

Literature Cited

1. Daily, J. W. Applied Optics, 1977a, 16, 2322.
2. Lucht, R. P.; Laurendeau, N. M. Applied Optics, 1979, 18, 856.
3. Chan, C.; Daily, J. W. "Measurement of OH Quenching Cross-Sections in Flames using Laser Induced Fluorescence Spectroscopy," Western States Section/Combustion Institute Paper No. 79-20.
4. Eckbreth, A. C. "Laser Raman Thermometry Experiments in Simulated Combustor Environments"; AIAA Paper No. 76-27, 1976.
5. Daily, J. W. Applied Optics, 1978a, 17, 1610.
6. Daily, J. W.; Chan, C. Combustion and Flame, 1978, 33, 47.
7. Chan, C.; Daily, J. W. J. Quant. Spectrosc. Radiat. Transfer, 1979, 21, 527.
8. Mollow, B. R. Phys. Rev., 1977, A15, 1023.
9. Piepmeir, E. H. Spectroschem. Acta., 1972, 27B, 431.
10. Daily, J. W. Applied Optics, 1977b, 16, 568.
11. Rodrigo, A. B.; Measures, R. M. J. Quant. Elec., 1973, QE-9, 972.
12. Daily, J. W. Applied Optics, 1978b, 17, 225.
13. van Calcar, R. A.; van de Ven, M. J. M.; van Vitert, B. K.; Biewenga, K. J.; Hollander, Tj.; Alkemade, C. Th. J. J. Quant. Spectrosc. Radiat. Transfer, 19 , 21, 11.

14. Blackburn, M. B.; Mermet, J. M.; Boutitier, G. D.;
    Winefordner, J. D. Applied Optics, 1979, 18, 1804.
15. Baronavski, A. P.; McDonald, J. R. Applied Optics, 1977,
    16, 1897.
16. Eckbreth, A. C.; Bonczyk, P. A.; Shirley, J. A. "Investi-
    gation of Saturated Laser Fluorescence and CARS Spectro-
    scopic Techniques for Combustion Diagnostics"; EPA Report
    600/7-78-104, 1978.
17. Muller, C. H1; Schofield, K.; Steinberg, M.; Broida, H. P.
    "Sulphur Chemistry in Flames"; 17th Symposium on Combustion,
    Leeds, England, 20-25 August, 1978.
18. Kirchoff, G.; Bunson, R. Pogg. Ann., 1860, 110, 161.
19. Chan, C. "Measurement of OH in Flames using Laser Induced
    Fluorescence Spectroscopy"; Ph.D. thesis, Department of
    Mechanical Engineering, University of California, Berkeley,
    1979.
20. Lengel, R. K.; Crosley, D. R. J. Chem. Phys., 1977, 67, 2085.
21. Stepowski, D.; Cottereau, M. J. Applied Optics, 1979, 18,
    354.
22. Daily, J. W. Applied Optics, 1976, 15, 955.
23. Winefordner, J. D.; Elser, R. C. Anal. Chem., 1971, 43, 24A.
24. Winefordner, J. D. Chemtech, 1975, 128, February.
25. Smith, B.; Winefordner, J. D.; Omenetto, N. J. Appl. Phys.,
    1977, 48, 2676.
26. Omenetto, N.; Benetti, P.; Rossi, G. Spectrochem. Acta,
    1972, 27B, 453.
27. Haraguchi, H.; Smith, B.; Weeks, S.; Johnson, D. J.;
    Winefordner, J. D. Applied Spectroscopy, 1977, 31, 156.
28. Bradshaw, J.; Bower, J.; Weeks, J.; Johnson, D. J.;
    Winefordner, J. D. "Application of the Two Line Fluor-
    escence Technique to the Temporal Measurement of Small
    Volume Flame Temperature"; 10th Material Research Symposium
    on Characterization of High Temperature Vapors and Gases,
    NBS, Gaithersburg, Maryland, 18-22 September, 1978.
29. Pitz, R. W.; Daily, J. D. "Measurement of Temperature in
    a Premixed Methane-Air Flame by Two-Line Atomic Fluor-
    escence"; Western States Section/The Combustion Institute
    Spring Meeting, 1977.
30. Princeton Applied Research. "OMA Vidicon Detectors"; PAR:
    Princeton, N.J., 1978.
31. Kowalik, R. M.; Kruger, C. H. "Experiments Concerning
    Inhomogeneity in Combustion MHD Generators"; 18th
    Symposium on the Engineering Aspects of MHD, Butte,
    Montana, 1979.

RECEIVED March 14, 1980.

Laser Probes of Premixed Laminar Methane-Air Flames and Comparsion with Theory

JAMES H. BECHTEL

Physics Department, General Motors Research Laboratories, Warren, MI 48090

The measurement of temperature and species composition profiles in premixed, laminar flames plays a key role in the development of detailed kinetic models of hydrocarbon combustion. One of the few hydrocarbon fuels for which a detailed reaction mechanism with air has been postulated is methane. The flame models for methane combustion include both species diffusion and thermal conduction, and they are restricted to laminar propagation only (1-4). Most previous measurements of the details of flame structure have been done on low-pressure (a few kPa) flames (5) or flames with very slow burning velocities. These flames are usually much thicker than near-stoichiometric atmospheric pressure flames. Species concentration profiles have been derived from sampling sonic microprobes (6), absorption spectroscopy, or supersonic molecular beam sampling with mass spectrometer detection (7, 8). Although molecular beam sampling with mass spectrometric analysis has very good sensitivity, recent results clearly demonstrate that these probes can significantly perturb the flame (9).

By contrast, laser scattering methods now permit temperature, composition and flow measurements that are both nonintrusive and give very high spatial resolution. These light scattering methods include laser Raman spectroscopy, laser-induced fluorescence, coherent Raman spectroscopy as well as laser velocimetry.

One of the important aspects of these laser scattering methods is the very high spatial resolution that can be achieved by the focusing of the laser beam. The primary reaction zone of atmospheric-pressure hydrocarbon-air flames may now be probed to give accurate temperature and composition profiles. These primary reaction zones are typically only a fraction of a mm thick. This extension of combustion diagnostics to higher pressures will give new insights into flames that have significantly different radical mole fractions and burning velocities at different pressures.

0-8412-0570-1/80/47-134-085$05.75/0
© 1980 American Chemical Society

The experimental results that are presented here are temperature and species profiles for premixed, laminar CH_4-air flames. The kinetic mechanism of CH_4-air combustion can be schematically represented as a sequence of pathways for carbon evolution e.g., $CH_4 \rightarrow CH_3 \rightarrow CH_2O \rightarrow CHO \rightarrow CO \rightarrow CO_2$. Each step in this series evolves by various parallel reactions, and a specific mechanism (1) is given in Table I. This is only one possible scheme and alternative methods may be found elsewhere (2-5). In addition to the reactions involving carbon containing species there are also chain branching, chain propagating, and termolecular recombination reactions that involve only species that contain hydrogen and/or oxygen. The complexity of methane combustion is demonstrated when it is realized that there are still many unresolved problems associated with methane-air flames. These include the fates of both the CH_3 radical and the CHO radical, the importance of HO_2, accurate rate constants for many of the reactions, and accurate high temperature species diffusivities. In spite of these uncertainties, models for methane-air flames have been developed, and the central objective of this contribution is to compare the results of the model of Ref. 1 to experiments that use modern laser probes.

Experimental Apparatus and Methods

One of the more novel aspects of these experiments is the slot burner that supports the flame. A schematic diagram of this burner is exhibited in Figure 1. The geometry of the flame is such that a focused laser beam can probe the center of the flame where the flame geometry is approximately one-dimensional. Steady, laminar flows were maintained by metering both fuel and air through critical flow orifices. These flows were thoroughly premixed before combustion, and the burner assembly was mounted on a two-dimensionally translatable stage. This assembly allowed positioning the burner with a precision of 10 μm.

The laser used for the Raman scattering experiments was a frequency doubled Nd:YAG (neodymium doped yttrium aluminum garnet) laser Q-switched at 2 kHz. Temperature and species concentration profile measurements were facilitated by a two-channel photon counter. Channel A was gated immediately before the laser pulse and detected both the Raman scattering signal and the luminous, flame-produced background signal. Channel B was gated for an identical duration several microseconds after the laser pulse and detected only the background signal. The gatewidths were 300 ns and counts accumulated in both channels for a 5s integration time. The difference signal, A-B, provides a measure of the true laser Raman scattering signal.

The beam spot size was measured to contain 90% of the laser power within a diameter of 40 μm even in the region of maximum flame temperature gradients. Typical average laser powers were 0.25 W.

TABLE I. Postulated mechanism for a methane-air flame.[1]

No.	Reaction	Forward Rate Constant (cgs units)*		
		A	n	E cal/g mol
A1	$CH_4 + OH \rightleftarrows CH_3 + H_2O$	3×10^{13}	0	5,000
A2	$CH_4 + H \rightleftarrows CH_3 + H_2$	2×10^{14}	0	11,900
A3	$CH_4 + O \rightleftarrows CH_3 + OH$	2×10^{13}	0	6,900
B1	$CH_3 + O \rightleftarrows CH_2O + H$	7×10^{13}	0	1,000
B2	$CH_3 + O_2 \rightleftarrows CH_2O + OH$	3×10^{13}	0	17,500
C1	$CH_2O + M \rightleftarrows CO + H_2 + M$	2×10^{16}	0	35,000
C2	$CH_2O + OH \rightleftarrows CHO + H_2O$	2.5×10^{13}	0	1,000
C3	$CH_2O + O \rightleftarrows CHO + OH$	3×10^{13}	0	0
C4	$CH_2O + H \rightleftarrows CHO + H_2$	1.7×10^{13}	0	3,000
D1	$CHO + O_2 \rightleftarrows CO + HO_2$	3×10^{13}	0	0
D2	$CHO + OH \rightleftarrows CO + H_2O$	1×10^{14}	0	0
D3	$CHO + O \rightleftarrows CO + OH$	5.4×10^{11}	1/2	0
D4	$CHO + M \rightleftarrows CO + H + M$	2×10^{12}	1/2	28,800
E1	$CO + OH \rightleftarrows CO_2 + H$	5.5×10^{11}	0	1,080
E2	$CO + O + M \rightleftarrows CO_2 + M$	3.6×10^{18}	-1	2,500
F1	$HO_2 + O \rightleftarrows O_2 + OH$	2.5×10^{13}	0	0
F2	$HO_2 + OH \rightleftarrows O_2 + H_2O$	2.5×10^{13}	0	0
F3	$HO_2 + H \rightleftarrows OH + OH$	2×10^{14}	0	2,000
F4	$HO_2 + H \rightleftarrows O_2 + H_2$	6×10^{13}	0	2,000
F5	$H + O_2 + M \rightleftarrows HO_2 + M$	1.4×10^{16}	0	-1,000
G1	$H + O_2 \rightleftarrows OH + O$	2.2×10^{14}	0	16,800
G2	$O + H_2 \rightleftarrows OH + H$	1.7×10^{13}	0	9,460
G3	$OH + H_2 \rightleftarrows H_2O + H$	2.2×10^{13}	0	5,200
G4	$OH + OH \rightleftarrows H_2O + O$	6×10^{12}	0	780
H1	$H + OH + M \rightleftarrows OH + M$	7×10^{19}	-1	0
H2	$O + H + M \rightleftarrows OH + M$	4×10^{18}	-1	0
H3	$H + H + M \rightleftarrows H_2 + M$	2×10^{19}	-1	0
H4	$O + O + M \rightleftarrows O_2 + M$	4×10^{18}	-1	0

*$k = AT^n \exp(-E/RT)$ g-mol, sec, K units

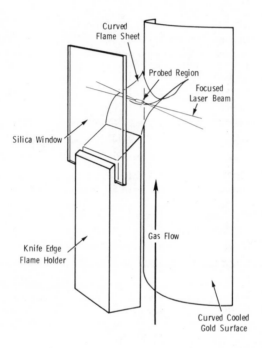

Curved
Flame Sheet

Probed Region

Focused
Laser Beam

Silica Window

Knife Edge
Flame Holder

Gas Flow

Curved Cooled
Gold Surface

Applied Optics

Figure 1.   Schematic of the slot burner used in this experiment (13)

All data were processed by a central computer to derive both temperature and species concentrations. The details of these computer fits are identical to those described elsewhere (10, 11) with the exception that a vibrational partition function correction was included in the analysis of the H_2O data. The absolute mole fractions of fuel, O_2, CO, H_2, CO_2, and H_2O were determined by flowing known concentrations of these gases mixed with known concentrations of N_2 through the burner. A comparison of the intensity of the N_2 Raman spectrum intensity to the Raman spectrum intensity of any of the other gases provided an absolute calibration for all laser Raman scattering flame studies.

For the species profiles obtained by laser Raman spectroscopy several profiles were measured in a specific flame and the average mole fraction is reported at a given flame position. For CO as many as five profiles were averaged to obtain a composite profile. The precision of the results depends on both the temperature and the number of composition profiles that are averaged. The precision is better in the cooler, leading edge of the flame because the Raman scattering signal depends on the number of molecules/cm^3. For all cases the precision was better than ±0.01 mole fraction and much better than this both in the leading edge of the flame and when several profiles were averaged to obtain a composite final profile for a specific species.

Hydroxyl temperatures and concentrations were measured in these flames by laser-induced fluorescence. The fluorescence was excited by a frequency-doubled, tunable dye laser. The measured fluorescence induced from an individual electronic absorption line was scaled to absolute concentrations by laser absorption measurements of the hydroxyl concentration along a homogeneous hydroxyl concentration path length in the post combustion zone. The collisional quenching of the laser excited state was determined throughout the flame by mesuring the concentration of the major quenching species, by using literature values of quenching cross sections (12), and by determining the relative collision velocities of hydroxyl with the other species. The $^2\Pi(v''=o)$ electronic ground state rotational temperature was also measured by laser fluorescence. The laser frequency was tuned to various P and Q branch transitions of the $^2\Pi(v''=o) \rightarrow {}^2\Sigma^+(v'=o)$ series (see Figure 2). The spectrometer, however, detected only the fluorescence of a large number of emission lines in the R_1 and R_2 bands. The temperature is determined by plotting the fluorescence intensities of a given laser-excited transition divided by the transition strength for the absorption versus the energy of the absorbing rotational state in the $^2\Pi(v''=o)$ electronic state. A requirement for the validity of this method is that the laser beam is not significantly attenuated before it reaches the scattering or probed region. An example of such a plot is given in Figure 3. The temperature is determined by the slope of the line through the

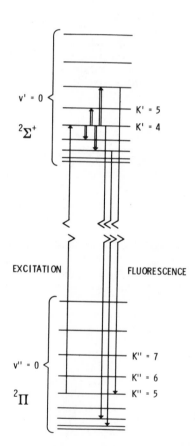

Figure 2. Schematic of the energy levels for the OH molecule. The collision-induced energy-transfer transitions are denoted by double-line arrows. The rotational quantum number is denoted by K' or K''. Both spin doubling and lambda doubling have been suppressed for clarity.

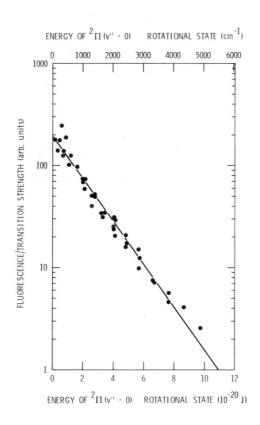

Figure 3. Plot of the laser-induced fluorescence intensity per transition strength vs. energy of initial rotational state in $^2\Pi(v'' = 0)$ electronic state. The slope of the line gives the OH rotational temperature (13).

data points. Additional experimental details are found elsewhere (13).

Theory

The general procedure for theoretically predicting flame temperature and concentration profiles has been described in detail in Ref. 1. In this method the unsteady conservation equations of total mass, momentum, species mass and energy are simultaneously solved by finite difference methods. To solve these coupled differential equations several assumptions were made, and these assumptions are listed in Table II. The coupled conservation equations were solved in a transformed coordinate system by a method similar to that used by Spalding et al. (14). The program required the following input: enthalpies, heat capacities, thermal conductivities, species diffusivities and reaction rate constants. The program typically contained 50 grid points in the direction of the flame, and approximately 600 time iterations were required for a convergence to a steady state.

Since there are many input parameters associated with these types of computer programs, one needs an estimate of the total uncertainty of the theoretical predictions. This uncertainty was obtained by combining the theoretical profiles of Tsatsaronis (2), Kelly and Kendall (3), and those computed here for the 40 torr flame of Peeters and Mahnen (5). Figure 4 demonstrates that the theoretical uncertainty for CH_4 CO, CO_2, H, H_2O, O, O_2, OH and temperature are approximately $\pm 10\%$ and that the experimental results fall nearly within these bounds. This is not true for the species HO_2 and CH_3.

One should note that some of the kinetic rate constants in all of these models are derived from Peeters and Mahnen mass spectrometric results; therefore, it is not surprising that the theoretical fits to this data are rather good. It is reassuring that the model of Ref. 1 also exhibits overall good agreement with the following laser probe results that are free of mass spectrometer calibration estimates and flame perturbation.

Experimental results

The theoretical and experimental results for a fuel-lean methane-air flame are given in Figures 5-7. These results include temperature and major species compositions. The experimental and theoretical results are compared by matching the abcissas of the temperature profiles. The model very accurately predicts the slope of the temperature profile but predicts a larger final flame temperature than is measured. This is a consequence of heat lost to the cooled, gold-coated burner wall that is 1.5 mm away from the positions where data were taken. One should note that it has been calculated that there is very

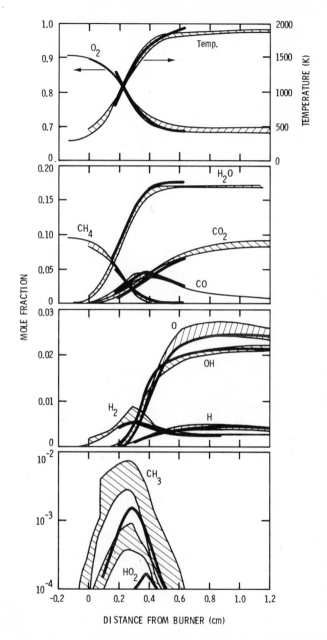

Figure 4. Comparison of the predicted (shaded areas) and measured (heavy line) CH_4–O_2 flame (P = 40 torr, 9.5% CH_4) species concentrations and temperature. The shaded areas are the composite predictions bounded by the results of Refs. 1, 2, and 3. The experimental data are from Ref. 5(a) O_2 and temperature; (b) CH_4, H_2O, CO_2, and CO; (c) OH, O, H, and H_2; (d) CH_3 and HO_2.

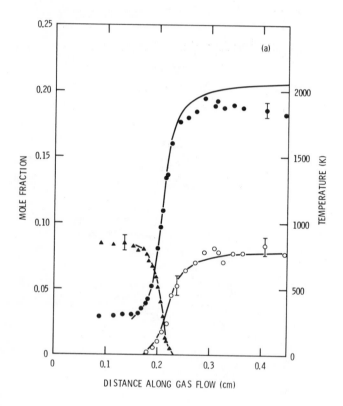

Figure 5.   Temperature, CH$_4$, and CO$_2$ profiles for a fuel-lean ($\phi = 0.86$) atmos-pheric-pressure, premixed, laminar CH$_4$– air flame.   The experimental data are from laser Raman scattering and the theoretical predictions are from the computer code of Ref. 1 (———), theory; (●), temperature (N$_2$); (○), CO$_2$; (▲), CH$_4$.

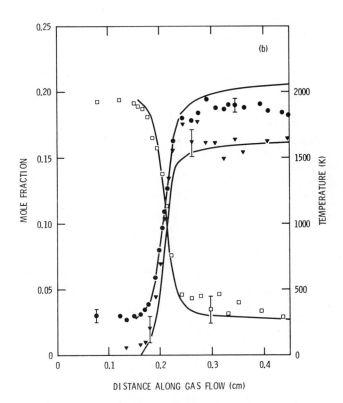

Figure 6.   Temperature, H₂O and O₂ profiles for a fuel-lean (φ = 0.86) atmospheric-pressure, premixed, laminar CH₄–air flame. The experimental data are from laser Raman scattering and the theoretical predictions are from the computer code of Ref. 1: (———), theory; (●), temperature; (▼), H₂O; (□), O₂.

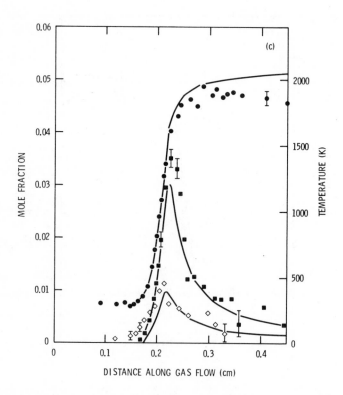

Figure 7.   *Temperature, CO, and H₂ profiles for a fuel-lean ($\phi = 0.86$) atmos-pheric-pressure, premixed, laminar CH₄–air flame. The experimental data are from laser Raman scattering and the theoretical predictions are from the computer code of Ref. 1: (———), theory; (●) temperature; (■), CO; (◊), H₂.*

TABLE II. Assumptions made in flame model.

1. Laminar flow

2. One dimensional

3. Constant static pressure

4. Negligible viscous dissipation

5. Negligible external forces

6. Negligible radiative heat transfer

7. Negligible heat lost to the surroundings

8. Negligible Soret and Dufour effects

little difference in species profiles for adiabatic and non-adiabatic cases, even though the final temperatures differ by approximately 200 K (15).

If one compares the composition profiles in Figures 5 and 6, one finds a good agreement between experiment and theory. Any differences between the experimental data and the theoretical predictions can be attributed to experimental scatter in the data.

For the CO and H_2 profiles some noticeable differences between experiment and theory occur. The experimental peak CO concentration is systematically greater than the model prediction. This same qualitative feature is also observed when one compares the Smoot model (Ref. 1) with the low-pressure experimental results (5, 16). This discrepancy between the predicted and measured CO concentration seems to result from insufficient formation rates for CO in this model. The CO disappearance rate is determined almost totally by reaction E1. The rate constant, although non–Arrhenius in temperature dependence, has been studied extensively, and more recent values suggest, if anything, a slightly larger rate constant should be used (17). This would decrease the theoretical profile and also shift its position. Moreover, the rate constant for reaction C1 is very large and shock tube results (18) indicate a smaller value is more accurate. If one were to use the shock tube results for the rate constant for reaction C1, the disagreement between theory and experiment would also be enhanced.

The profile for H_2 also shows a discrepancy between theory and experiment in the leading edge of the flame. The H_2 concentration is much greater than the model prediction. There are several possible explanations for this difference; however, a more accurate and complete treatment of diffusion is a distinct

possible explanation. Tsatsaronis (2) has calculated profiles
for a stoichiometric CH_4-air flame, and he shows larger preflame
H_2 concentrations than the corresponding profiles using the
predictions of Smoot et al. (1). A major difference between
these two models is that Tsatsaronis uses a more complete
description of the multicomponent diffusion coefficients.

   The hydroxyl concentration profile for a stoichiometric
CH_4-air flame is presented in Figure 8. Here the maximum mole
fraction observed and the predicted mole fraction are equal to
better than 10% accuracy. The abscissas of the theoretical and
the experimental results were matched by setting the theoretically predicted temperature equal to the measured hydroxyl
rotational temperature. At all positions in the flame the
hydroxyl $^2\Pi$(v''=o) state exhibited a Boltzmann distribution of
rotational states. This rotational temperature is equal to the
N_2 vibrational temperature to within the ± 100 K precision of the
laser induced fluorescence and laser Raman scattering experiments. An example of this comparison is given in Figure 9.

   One should parenthetically note that the measurement of OH
concentration and temperature in the flame recombination zone
provides a method of determining the O atom concentration as
well. The reactions G1 - G3 are usually fast; consequently, if
one assumes partial equilibrium,

$$[O] = \frac{K_{G3}}{K_{G2}} \frac{[OH]^2}{[H_2O]} .$$

Here K_{G3} and K_{G2} are the temperature-dependent equilibrium constants for reactions G3 and G2 respectively.

   In conclusion, laser probes have been demonstrated to give
an excellent way of measuring species compositions and temperatures in laminar flames. The comparisons between a model and
the concentrations of fuel, O_2, H_2O, CO_2, and OH are in good
agreement with the model predictions. The maximum in the CO
profile is, however, systematically larger than the model
predictions for the fuel-lean flame. Moreoever, the H_2 concentration is also systematically larger in the preflame region
than the model prediction. Finally, the application of more
powerful lasers for Raman scattering, and the extension of
laser fluorescence to other wavelengths should make laser
probes extremely useful for the detection of the composition
profiles of other dilute species as well.

Acknowledgments

   The author would like to acknowledge his close collabora-
tion with his colleagues R. J. Blint, C. J. Dasch and R. E. Teets
in much of the research reviewed here. He would also like to
acknowledge the technical assistance of Louis Green and Doreen

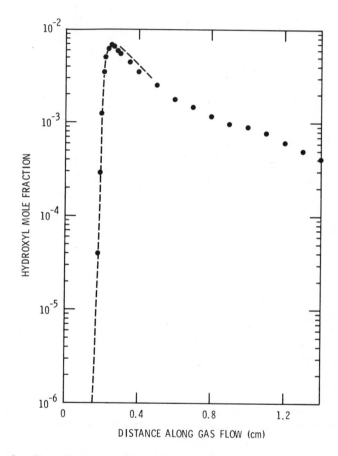

Figure 8. Comparison of hydroxyl concentration as measured by laser-induced fluorescence and the theoretical predictions of Ref. 1. The hydroxyl concentration uncertainty is ±30%. Stoichiometric CH₁–air flame. (●), Experiment; (– – –), theory.

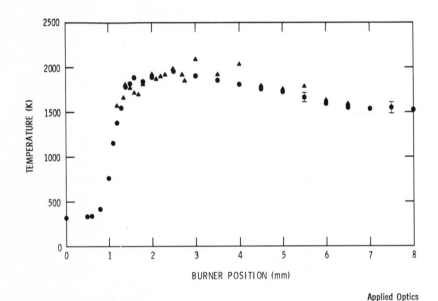

BURNER POSITION (mm)

Applied Optics

Figure 9.   Comparison of OH and N₂ temperatures in both primary reaction zone and recombination regions. The fuel–air equivalence ratio was ϕ = 0.93. The probed region was 1.5 mm from the curved wall. The uncertainty in the hydroxyl temperature is ±100 K. CH₄–air flame; (●), N₂; (▲), OH; (13).

Weinberger. Finally, the support and encouragement of D. L. Fry and J. C. Tracy is greatly appreciated.

Abstract

The measurements of temperature and species concentrations profiles in premixed, laminar flames play a key role in the development of detailed models of hydrocarbon combustion. Systematic comparisons are given here between a recent laminar methane-air flame model and laser measurements of temperature and species concentrations. These results are obtained by both laser Raman spectroscopy and laser fluorescence. These laser probes provide nonintrusive measurements of combustion species for combustion processes that require high spatial resolution. The measurements reported here demonstrate that the comparison between a model and the measured concentrations of CH_4, O_2, H_2O, CO_2, and OH are in good agreement with the model predictions. The maximum CO concentration is, however, systematically larger than the model prediction for a fuel-lean flame. Moreover, the H_2 concentrations is also systematically larger in the preflame region than the model prediction. Finally, the rotational temperature of the electronic ground state of OH exhibits a Boltzmann distribution throughout the flame and is thermally equilibrated with the vibrational temperature of N_2 throughout the flame.

Literature Cited

1. Smoot, L. D., Hecker, W. C. and Williams, G. A., Combust. Flame (1976) 26, 323.
2. Tsatsaronis, G., Combust. Flame (1978) 33, 217.
3. Kelly, J. T. and Kendall, R. M., Proceedings of the Second Stationary Source Combustion Symposium, Vol. IV. EPA-600/7-77-073d (1977) p. 311.
4. Creighton, J. R. and Lund, C. M., Lawrence Livermore Laboratory preprint UCRL-80832 (unpublished).
5. Peeters, J. and Mahnen, G., Fourteenth Symposium (International) on Combustion, (The Combustion Institute, Pittsburgh, 1973) p. 133.
6. Fristrom, R. M. and Westenberg, A. A., Flame Structure (McGraw-Hill, New York, 1965).
7. Lazzara, C. P., Biordi, J. C. and Papp, J. F., Combust. Flame (1973) 21, 371.
8. Foner, S. N. and Hudson, R. L., J. Chem. Phys. (1953) 21, 1374.
9. Revet, J. M., Puechberty, D. and Cottereau, M. J., Combust. Flame (1978) 33, 5.
10. Stephenson, D. A. and Blint, R. J., Appl. Spectrosc. (1979) 33, 41.
11. Blint, R. J., Stephenson, D. A. and Bechtel, J. H., J. Quant. Spectrosc. Radiat. Transfer (to be published).

12. Bechtel, J. H. and Teets, R. E., Appl. Opt. (to be
     published).
13. Bechtel, J. H., Appl. Opt. (1979) 18, 2100.
14. Spalding, D. B., Stephenson, P. L. and Taylor, R. G.,
     Combust. Flame (1971) 17, 55.
15. Smoot, L. D., Combust. Flame (1978) 31, 325.
16. Fristrom, R. M., Grunfelder, C. and Favin, S., J. Phys.
     Chem. (1960) 64, 1386.
17. Vandooren, J., Peeters, J. and van Tiggelen, P. J.,
     Fifteenth Symposium (International) on Combustion, (The
     Combustion Institute, Pittsburgh, 1975), p. 745.
18. Dean, A. M., Craig, B. L., Johnson, R. L., Schultz, M. C.
     and Wang, E. E., Seventeenth Symposium (International)
     on Combustion, (The Combustion Institute, Pittsburgh) (to
     be published).

RECEIVED February 1, 1980.

Laser-Induced Fluorescence: A Powerful Tool for the Study of Flame Chemistry

C. H. MULLER, III,[1] KEITH SCHOFIELD, and MARTIN STEINBERG

Quantum Institute, University of California, Santa Barbara, CA, 93106

The recent availability of tunable dye lasers has markedly enhanced our ability to inquire into the chemistry and physics of combustion systems. The high sensitivity, spectral and spatial resolution, and non-perturbing nature of laser induced fluorescence makes this technique well suited to the study of trace chemistry in complex combustion media. A barrier to the quantitative application of fluorescence to species analysis in flames has been the need to take into account or bypass the effects of quenching. The use of saturated fluorescence eliminates quenching as a problem and has the further advantage that fluorescence intensity is insensitive to variations in laser power (1,2). However, the generation of high concentrations of excited states under saturated excitation in an active flame environment opens up the possibilities for laser induced chemistry effects that also must be taken into account or avoided (3,4,5).

In the following we present an application of laser induced fluorescence to a study of the chemistry of sulfur in rich hydrogen/oxygen/nitrogen ($H_2/O_2/N_2$) flames and demonstrate a simple rationale for taking quench effects into account. Fluorescence measurements for S_2, SH, SO_2, SO, and OH along with measurements of flame temperature and H-atom (in sulfur free flames) have been employed to develop a kinetic model for the highly coupled flame chemistry of sulfur. The kinetic aspects of the study already have been presented in considerable detail (6). This presentation will accent the experimental techniques and results in an effort to complement the earlier report.

Experimental

A series of fuel rich $H_2/O_2/N_2$ premixed flames were burned at atmospheric pressure on a 2 cm dia. Padley-Sugden (7) burner constructed of bundled sections of stainless steel hypodermic

---

[1] Current Address: General Atomic Co., P.O. Box 81608, San Diego, CA 92138.

0-8412-0570-1/80/47-134-103$07.00/0
© 1980 American Chemical Society

tubing to produce one-dimensional flows in the post flame gases
above the burner. The hypodermic tube bundle was manifolded so
as to allow separate premixed supplies to an inner tube bundle
1 cm dia. and to the surrounding annular tube bundle. A sectional
sketch of the burner is shown in Figure 1. Identical $H_2/O_2/N_2$
mixtures were burned on the two burner sections and H_2S or a NaCl
aerosol was added to the gases flowing to the central burner core.
The outer flame ring serves as a shield to maintain the one-
dimensional character of the central flow for several centimeters
above the burner. The H_2/O_2 and N_2/O_2 ratios were varied to
generate a series of flames with varying composition and temperature. Flame temperatures varied from 1700 to 2350 K with burnt
gas flow velocities ranging from 4 to 24 m s^{-1}.

A schematic of the optical system is seen in Figure 2. A
flash-lamp pumped tunable dye laser (Chromatix CMX-4) is used to
excite the fluorescence of the species of interest. The laser is
equipped with intracavity doubling crystals to carry its operation
into the ultra-violet (UV) down to 265 nm. The absorption bands
of S_2, SH, SO, SO_2 and OH all lie in the UV. The laser linewidth in this region is about 5.4 cm^{-1}. With an etalon, built
into the unit, the UV line-width can be narrowed to about
0.27 cm^{-1}, slightly greater than the doppler widths of the individual lines for the systems under study. The high selectivity
of the narrowed line configuration was required for use with the
SH and OH fluorescence measurements to minimize excitation of
interfering species. The laser beam width was narrowed to 0.1 cm
for the study. The laser pulse duration is slightly greater than
1 μs at the half-intensity point and extends out to about 2 μs.
In order to avoid any influence of laser induced chemistry, laser
power was limited to about 1% of the saturation parameter for
each of the species monitored. The laser beam passed horizontally
through the flame and fluorescence was monitored at 90° to the
beam. The fluorescence was collected by a spherical mirror,
passed through a 90° image rotator and imaged with unit magnification onto the entrance slit of the monochromator. The
collection optics were matched to the monochromator aperture.
With a slit height of 1 cm and slit width of 50 μm the detection
system monitored a 1 cm length of laser beam centered on the burner
core with a vertical spatial resolution, including depth of field
effects, no greater than .01 cm. The 50 μm slit width corresponds
to a spectral resolution of 0.13 nm (FWHM). The monochromator
output was detected with a photomultiplier (EMI 9558 QBM), the
output of which was coupled to current/voltage and voltage/voltage
amplifiers and passed into a boxcar averager. The laser was
operated at 15 pulses/s, and the boxcar was triggered to center a
1 μs gate about the peak of the fluorescence profile which matched
the laser profile. Good signal to noise ratios were obtained
with a 3 s time constant for most test conditions. Fluorescence
intensities were normalized to a constant laser power to correct
for any drifts in laser power for which laser dye decay was the

WATER

GAS
MIXTURE

GAS
MIXTURE

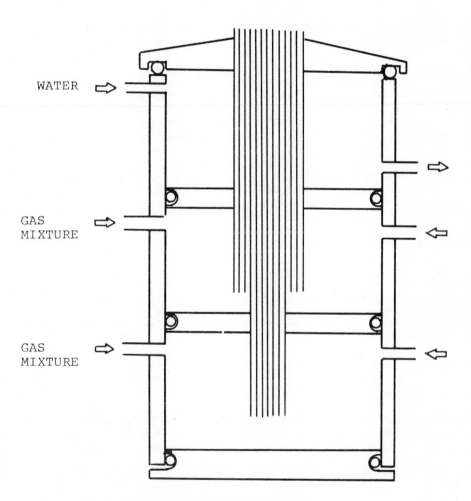

Figure 1.   *Sectional sketch of the burner*

dominant cause. Fluorescence variations as a function of height in the one-dimensional flows above the burner were monitored by raising or lowering the burner relative to the optical axes.

Flame temperatures were monitored using the sodium line reversal method (8). Sodium was added to the central core flow as an aerosol of a sodium chloride solution. The aerosol, genera-ted in an ultrasonic nebulizer (9), was swept along with a small flow of air and passed through a pyrex tube heated to 475 K to dry the aerosol before it was passed into the gas stream leading to the burner core. The NaCl aerosol is rapidly dissociated on passage through the flame front to yield elemental sodium. Line reversal measurements were made using a tungsten ribbon lamp that had been calibrated for brightness temperature at 589.3 nm as a function of lamp current.

Absolute H-atom measurements also were made using the Na/Li method (10) in sulfur free flames. An aerosol of an equimolar solution of NaCl and LiCl was added to the central core flow through the nebulizer. Relative intensity measurements were made of the Na 589.0 nm and Li 670.8 nm emission from which the H-atom concentrations were calculated. The H-atom measurements could only be made in the sulfur free flames. Reaction of Na or Li with sulfur species would render the technique inoperative.

Fluorescence Measurements

The fluorescence intensities, under low power excitation conditions, have a complex dependence on several factors which can be represented in the form

$$I_f = AFI_L n_i \sigma_k l \ \tau^{-1} / \left(\tau^{-1} + k_d + \sum_{i,j} k_{ij} [M_j] \right) ,$$ (1)

where A includes geometric and transmission factors through the flame and detection system, F is the fraction of the fluorescence falling within the detection bandpass, I_L is the laser intensity, n_i is the number density of the species in the particular quantum state involved in the absorption process, σ_k is the effective absorption cross section including effects of laser and absorbing line overlap, and l is the length of beam monitored by the detect-ion system. The fluorescence efficiency expression, $\tau^{-1}/(\tau^{-1}+k_d+\Sigma_{i,j}k_{ij}[M_j])$, is a measure of the fraction of the molecules excited by the laser which fluoresce. The terms are defined as follows with A* representing the excited state.

$$
\begin{array}{lll}
A^* \rightarrow A \ + \ h\nu & k=\tau^{-1} & \\
A_i^* + M \rightarrow A_m^* + M & k_M & \text{(vib. or rot. relaxation)} \\
A^* + M \rightarrow A \ + M & k_q & \text{(electronic quenching)} \\
A^* \rightarrow B \ + C & k_d & \text{(predissociation)}
\end{array}
$$

Excitation conditions were selected to avoid predissociation effects. The quenching term in the fluorescence efficiency sums over all of the collisional partners, M_j, of which N_2, H_2O, and H_2 are dominant in the $H_2/O_2/N_2$ flames studied.

Under laser excitation conditions it is often possible to excite a single transition and selectively populate particular vibrational and rotational quantum levels in the excited electronic state. Collisional quenching by vibrational and rotational relaxation decreases the population of this state, redistributing the molecule among adjacent vibrational and rotational states which may radiate or suffer further quenching collisions. Under such conditions the fluorescence efficiency is dependent on the spectral bandwidth of the detection system. With a broad band detection system of 10 to 20 nm that might be represented by a filter-photomultiplier combination, collision induced vibrational or rotational relaxation in the excited state may still lead to fluorescence that falls into the detection bandwidth. As the bandwidth of the detection system is decreased, vibrational and rotational relaxation leads to radiative transitions that fall outside of the detection bandwidth. Rotational and vibrational relaxation increasingly contributes to the quenching process. The quench term can be expanded to include such collision induced vibrational and rotational transitions

$$\sum_{i,j} k_{ij} [M_j] = \sum_j (k_q^j + k_{vib}^j + k_{rot}^j) [M_j] \; . \tag{2}$$

By decreasing the detection bandwidth as much as possible, consistent with maintaining a good signal to noise ratio, a limiting condition can be approximated for which the quenching summation varies in a simple manner from flame to flame. In the limit in which only one transition is monitored from the v'J' state populated by the laser, almost every vibrational or rotational relaxation from that state is an effective quenching collision. Under these conditions the quench summation term approximates to a gas kinetic quench rate,

$$\sum_j (k_q^j + k_{vib}^j + k_{rot}^j) [M_j] \rightarrow \sum_j k_{gas\ kinetic}^j [M_j] \; . \tag{3}$$

Since $k_{gas\ kinetic} \propto T^{1/2}$ and at a given pressure $[M_j] \propto T^{-1}$, the quenching rate varies from flame to flame as $T^{-1/2}$.

The experimental conditions for the excitation and detection of all the species are listed in Table I along with the radiative lifetimes of the excited states. Under the narrow detection bandwidth conditions for these measurements the quench term is much greater than τ^{-1} for the species studied and the fluorescence efficiency varies as $\tau^{-1}T^{1/2}$. Thus with fixed geometry, laser excitation wavelength, and detection parameters, the fluorescence intensity in Equation (1) simplifies to

$$I_f = \alpha I_L n_i \sigma_k T^{1/2} \tag{4}$$

Table I. LASER FLUORESCENCE EXCITING AND DETECTING WAVELENGTHS

Species	Exciting Wavelength, nm	Absorbing Level	Excited Level	Fluorescence Monitored, nm	τ
OH	281.14	$X^2\Pi, v''=0, N''=6$ $R_1(6)$ $J''=13/2$	$A^2\Sigma^+, v'=1, N'=7$ $J'=15/2$	314.69 $(1,1)Q_1(7)$	0.76 μs
S_2	296.0	$X^3\Sigma_g^-, v''=0, N''\simeq44$	$B^3\Sigma_u^-, v'=5, N'\simeq44$	302.5 $(5,1)$	20/40 ns
SH	323.7	$X^2\Pi, v''=0, J''=6.5$ $^RQ_{21}(6.5)$	$A^2\Sigma^+, v'=0, J'=6.5$	328.0 $(0,0)$ $Q_2(0.5-5.5)$ $^QR_{12}(0.5-5.5)$	0.55 μs
SO	266.5	$X^3\Sigma^-, v''=4$	$B^3\Sigma^-, v'=1$	283.4 $(1,6)$	16.6 ns
SO_2	266.5	$^1A_1$ $E\geq4000$ cm^{-1}	$^1B_2$ $(E\geq0)$	279.3	12/35 ns

where α includes all the invariant factors. The absorption
cross section, σ_k, varies slightly from flame to flame due to
changes in the absorption line shape. This effect is almost
negligible in the present study. The factor n_i, the population
of the absorbing level in the electronic ground state, is assumed
to be thermally equilibrated. For a given molecule or radical,
relative values of n_i can be represented by an appropriate
function of temperature from flame to flame and point to point
within each flame. When normalized for laser power variation,
absorbing state population, and temperature, the fluorescence
intensities are proportional to the number density of the species
under study. This simplified fluorescence relation appears to
describe the conditions existent in the $H_2/O_2/N_2$ flames.

If the quenching was dominated by some single or few collision partners as might be the case if He replaced N_2 as the
diluent, the fluorescence intensity relation would become

$$I_f = \beta I_L n_i \sigma_k T^{-1/2} \left[\sum_j [M_j]\right]^{-1} . \tag{5}$$

The measurements are placed on an absolute scale by including
a high temperature flame ($H_2/O_2/N_2 = 4/1/2$ with 1% H_2S, 2350 K)
which reaches thermal equilibrium rapidly. Measurement of the
fluorescent intensity in the equilibrium plateau a few centimeters
above the burner along with a calculation of the equilibrium concentration of each of the species at the temperature of this
flame permits evaluation of the proportionality constant α (or β).
In this manner absolute concentrations can then be calculated
using the relative fluorescence intensity inputs for each of the
species.

Results

Selected results illustrating the development of the
diagnostic conditions for making the routine measurements on
the flame series will be presented in the following discussion
together with examples of the data acquisition for selected flame
conditions.

Temperature Measurements. Sodium line reversal temperature
profile measurements were made on the flame series with varying
additions of H_2S. Results for $H_2/O_2/N_2$ (3/1/4,5,6) are shown in
Figure 3. The increase in temperature with distance above the
burner is due to the slow recombination of the radicals H and OH.
In the stoichiometric flames the temperature reaches a plateau in
a few centimeters above the burner. In the richer flames the
temperature gradient is steeper indicating a larger departure of
the radical concentration from equilibrium values. The equilibrium temperatures decrease with H_2S addition. However, the
presence of sulfur compounds enhances radical recombination (6,11)
producing almost equivalent temperature profiles, independent of
H_2S addition.

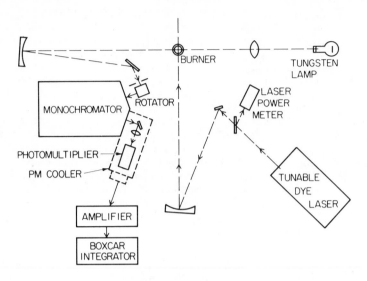

Figure 2.   Optical system for laser-fluorescence measurements in flames

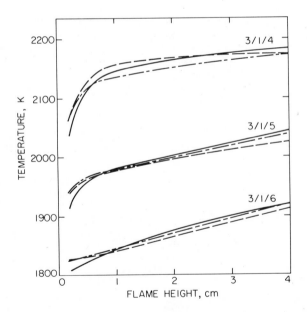

Figure 3.   Sodium line reversal temperature profiles above the burner surface in rich H_2–O_2–N_2 flames with added H_2S: (———), 0% H_2S; (— — —), 0.5% H_2S; (– · –), 1.0% H_2S.

OH Measurements. ([12](12)) Addition of H_2S to some of the $H_2/O_2/N_2$ flames exhibited strong S_2 $B^3\Sigma_u^-$ - $X^3\Sigma_g^-$ fluorescence from 300 to 400 nm. To avoid possible S_2 interference with the OH $A^2\Sigma^+$ - $X^2\Pi$ (0,0) band at 306 nm it was decided to excite in the OH A-X (1,0) band at 281 nm and monitor the fluorescence in the (1,1) band. Little difficulty was experienced in identifying favorable conditions for monitoring OH. Two excitation scans of the OH R_1 band head in a slightly rich $C_2H_2/O_2/N_2$ flame are shown in Figure 4 with and without added H_2S. The scans exhibit identical features, indicating that there is no interference by S_2 under these conditions. Excitation at 281.14 nm with detection at 314.69 nm was found to be favorable. Figure 5 shows some OH fluorescence decay profiles for a rich $H_2/O_2/N_2$ flame with varying amounts of added H_2S. It clearly exhibits the catalytic effect of sulfur on the OH decay rate.

H-atom measurements were made in the sulfur free rich $H_2/O_2/N_2$ flames using the Na/Li method. By this means it becomes possible to check on the method for taking account of quenching with the OH data. The radical balance reaction

$$H_2 + OH = H_2O + H \qquad K = [H_2O][H]/[H_2][OH] \qquad (6)$$

is known to be equilibrated in microseconds at flame conditions ([13](13)). Since H_2 and H_2O are major products in the rich $H_2/O_2/N_2$ flames these concentrations are essentially constant in each flame and equal to the thermodynamic equilibrium values. The equilibrium concentration ratios for Reaction 6 were evaluated and are plotted against T^{-1} in Figure 6 for six rich flames. The agreement of the experimental equilibrium ratio K_{exp} with the values K_{eq}, calculated from the JANAF thermochemical tabulation ([14](14)), is most gratifying and constitutes a validation of the OH measurements and the data reduction method.

Recently Stepowski and Cottereau ([15,16](15,16)) measured the quenching rates of OH $A^2\Sigma^+$ (v=0) in several low pressure propane/oxygen flames of different stoichiometry. They found that the quenching rate remained nearly constant through the flame front and well into the post flame gases for each flame and did not vary much from flame to flame. Extrapolation of their quenching rates to atmospheric pressure gives a value $\simeq 10^9$ s^{-1} which approximates to the gas kinetic value. This relative invariance of the OH quenching is an indication that the effect is caused by the major products, the sum of whose concentrations do not vary much. The minor constituents, whose concentrations may vary markedly, appear to make negligible contributions to quenching.

S_2 Measurements. ([17](17)) The rich spectrum of the S_2 $B^3\Sigma_u^-$-$X^3\Sigma_g^-$ system offers many options for excitation and detection free of interference by other species. Excitation in the B-X (5,0) band

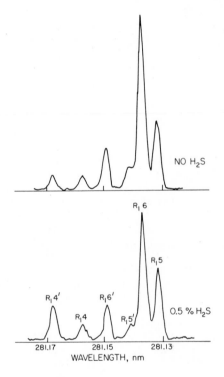

Figure 4. Laser-excitation spectra for OH $A^2\Sigma^+ - X^2\Pi$ in a C_2H_2–O_2–N_2 (1.2:2.5:10) flame with 0% and 0.5% H_2S. Fluorescence detected at 314.69 nm.

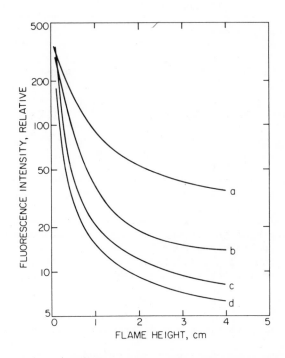

Figure 5.   OH $A^2\Sigma^+ - X^2\Pi$ fluorescence profiles above the burner surface in a H_2–O_2–N_2 (4:1:5) flame with added H_2S: (a), 0%; (b), 0.25%; (c), 0.5%; and (d), 1.0%.

at 296 nm with detection in the (5,1) band at 302.5 nm appears to
be quite suitable. A portion of the S_2 B-X (v'=5) progression is
seen in Figure 7. The background on which the S_2 progression
rides is due to vibrationally relaxed S_2 $B^3\Sigma_u^-$. The S_2 fluorescence intensities were measured from the baseline. S_2 fluorescence profiles for a rich $H_2/O_2/N_2$ flame with varying additions
of H_2S are shown in Figure 8.

SH Measurements (17). Predissociation of SH $A^2\Sigma^+$ above v=0
(18) limits fluorescence monitoring of SH $A^2\Sigma^+$-$X^2\Pi$ to the (0,0)
band with excitation in one branch and detection in another.
Strong S_2 B-X bands overlay the SH system and severely complicate
the SH fluorescence measurements. However, with excitation and
fluorescence scans using the narrow line laser output and with
the use of deuterium substitution for hydrogen it was possible
to identify a highly selective set of conditions for monitoring
SH. Figure 9a shows a laser excitation scan, with the detector
set at 328.0 nm, of a rich $H_2/O_2/N_2$ flame containing 1% H_2S. The
spectrum consists of both SH A-X and S_2 B-X contributions. Substitution of D_2 for H_2 as the fuel produces the changes indicated
in Figure 9b. The SD lines are shifted out of the 0.1 nm range
of the scan leaving the S_2 B-X bands. Four lines are labelled
in Figure 9a that can be assigned to SH. Of these, the $^RQ_{21}$ (6.5)
transition at 323.755 nm appears best suited for excitation of
the SH fluorescence. The small peak at the Q_{21} (6.5) transition
wavelength in Figure 9b is probably caused by the use of H_2S
rather than D_2S. In addition, for economy, D_2 was used only in
the central core of the burner. The possibility exists that
some H_2 diffused into the core from the annular guard flame and
formed SH. Scans of the fluorescence spectra for SH are shown in
Figures 10a and 10b for the same flame conditions as in Figure 9.
Again H_2 was used as the fuel in Figure 10a and D_2 in Figure 10b.
The substitution of D_2 for H_2 markedly decreases the SH fluorescence intensity at 328.0 and 326.9 nm. The residual SH signal with
D_2 substitution probably derives from the H_2 sources as discussed
above.

A fluorescence spectrum of SH excited at 323.76 nm is compared in Figure 11 with a synthetic emission spectrum of SH calculated using Ramsay's (18) line assignments and Earls' (19)
assessments of line strengths. Two points are notable. First,
there are no features in the fluorescence spectrum which do not
appear in the synthetic emission spectrum, strong evidence that
S_2 interference has been eliminated. Secondly, the labelled
transitions in the fluorescence spectrum originate from the
directly pumped level SH $A^2\Sigma^+$ (v'=0, J'=6.5). All other lines
result from rather extensive rotational relaxation which, however,
is incomplete over the duration of the laser pulse. It should
be noted that the strong fluorescence labeled Q_2(6.5) and $^QR_{12}$
(5.5) at 328 nm encompasses the Q_2 and $^QR_{12}$ branch heads and thus
includes contributions from Q_2 (0.5-6.5) and $^QR_{12}$ (0.5-6.5).

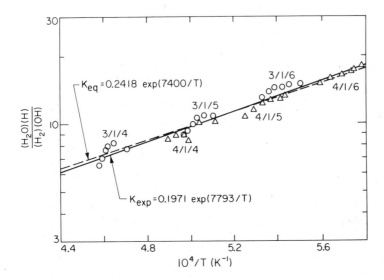

Figure 6.   A test of the equilibrium of $H_2 + OH = H_2O + H$ in six fuel-rich $H_2-O_2-N_2$ flames. Experimental points are based on OH fluorescence data and Na–Li data for H-atom at 0.5, 1.0, 1.5, 2.0, 2.5, and 3.0 msec, with temperatures increasing slightly with time in each flame.

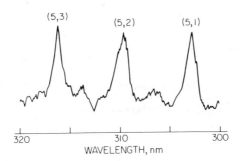

Figure 7.   Fluorescence spectrum for S_2 $B^3\Sigma_u^- - X^3\Sigma_g^-$ in a $H_2-O_2-N_2$ (4:1:6) flame containing 1% H_2S. Laser excitation at 296.0 nm.

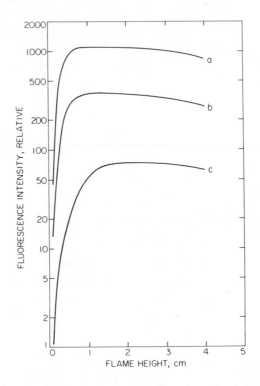

Figure 8. $S_2 B^3\Sigma_u^- - X^3\Sigma_g^-$ *fluorescence profiles above the burner surface in a* H_2–O_2–N_2 *(4:1:6) flame with added* H_2S: *(a), 1%, (b), 0.5%; (c), 0.25%. Laser excitation at 296.0 nm with detection at 302.5 nm.*

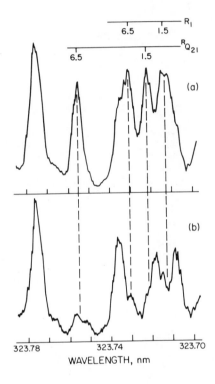

Figure 9. SH $A^2\Sigma^+ - X^2\Pi$ laser-excitation spectra in flames with 1% H_2S added to the unburnt gas: (a), $H_2-O_2-N_2$ (4:1:6); (b), D_2 substituted for H_2 in the burner core. Fluorescence detected at 328.0 nm.

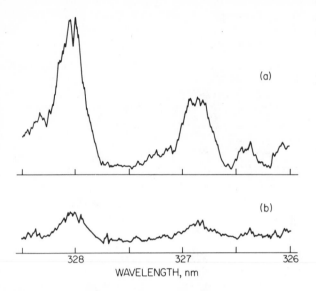

Figure 10.   SH $A^2\Sigma^+ - X^2\Pi$ fluorescence spectra in flames with 1% H_2S added to the unburnt gas: (a), $H_2-O_2-N_2$ (4:1:6); (b) D_2 substituted for H_2 in the burner core.   Laser excitation at 323.76 nm.

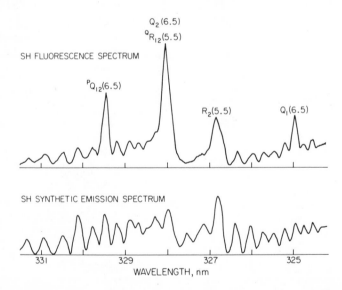

Figure 11.   Comparison of SH $A^2\Sigma^+ - X^2\Pi$ fluorescence and synthetic emission spectra for a $C_2H_2-O_2-N_2$ (2:2.5:10) flame with 0.5% H_2S added to the unburnt gas.   Fluorescence excited at 323.76 nm.

Fluorescence profiles for SH in a rich $H_2/O_2/N_2$ flame containing varying amounts of H_2S are shown in Figure 12. The similarity with the profiles for S_2 in Figure 8 suggests that S_2 and SH are chemically coupled.

SO and SO_2 Measurements (20). The SO $B^3\Sigma^- - X^3\Sigma^-$ absorption system extends from 240 to 400 nm with the strongest transitions lying in the ultraviolet where they overlay an SO_2 absorption continuum in sulfur bearing flames. Our Chromatix CMX-4 laser has a short wavelength limit of 265 nm. Thus it was necessary to excite the SO B-X (1,4) transition at 266.5 nm. Under these conditions the SO_2 continuum also was excited. At flame temperatures, about 4% of the SO population is in the v"=4 level. A low resolution fluorescence spectrum for SO and SO_2 in a rich $H_2/O_2/N_2$ flame with added H_2S is shown in Figure 13. The SO_2 continuum extends from 250 to beyond 370 nm. Between 260 and 300 nm the banded structure is an overlay of the SO B-X system on the continuum. A fluorescence spectrum at higher resolution (detection bandwidth=0.13 nm) is reproduced in Figure 14. A portion of the SO B-X (v'=1) progression is now clearly seen lying on the continuum. Detection of SO, for flame analyses, was made at 283.4 nm in the SO B-X (1,6) band by measuring the amplitude above the continuum background. SO_2 was monitored using the intensity of the continuum at 279.3 nm.

The SO_2 fluorescence pulse shape matched that of the laser indicating that the excited state lifetime was short compared to 1 μs. This implied that the short lived SO_2 $^1B_2$ ($\tau < 35$ ns) state was being excited by pumping hot bands of the SO_2 $^1A_1$ ground state at 266.5 nm.

Fluorescence profiles for SO in a rich flame with added H_2S are presented in Figure 15. The corresponding profiles for SO_2 are shown in Figure 16. Similarity of the SO and SO_2 profiles suggest that they are chemically coupled.

Concentration Profiles. The relative fluorescence intensity profiles for OH, S_2, SH, SO, and SO_2 were converted to absolute number densities according to the method already outlined. Resulting concentration profiles for a rich, sulfur bearing flame are exhibited in Figure 17. H-atom densities were calculated from the measured OH concentrations and H_2 and H_2O equilibrium values for each flame according to Equation 6. Similar balanced radical reactions were used to calculate H_2S and S concentrations (6). Although sulfur was added as H_2S to this hydrogen rich flame, the dominant sulfur product at early times in the post flame gas is SO_2.

Discussion

These data have already been employed to develop a chemical model of the sulfur chemistry in fuel rich $H_2/O_2/N_2$ flames (6).

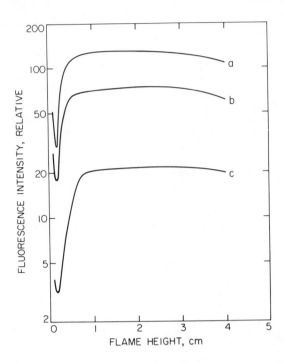

Figure 12. SH $A^2\Sigma^+ - X^2\Pi$ fluorescence profiles above the burner surface for H_2–O_2–N_2 (4:1:6) flames with added H_2S:(a),1.0%;(b),0.5%;(c),0.25%. Laser excitation at 323.76 nm with detection at 328.0 nm.

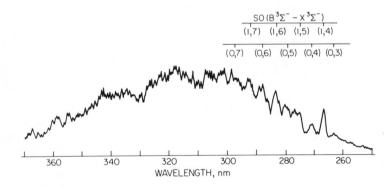

Figure 13. Fluorescence spectrum for SO_2 and SO in a H_2–O_2–N_2 (3:1:5) flame with 1% H_2S added to the unburnt gas. Laser excitation at 266.5 nm.

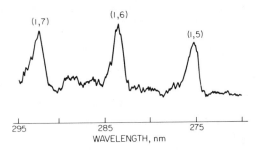

Figure 14.   SO $B^3\Sigma^-$ — $X^3\Sigma^-$ fluorescence spectrum for H_2–O_2–N_2 (4:1:6) flame with 1% H_2S added to the unburnt gas. Laser excitation at 266.5 nm.

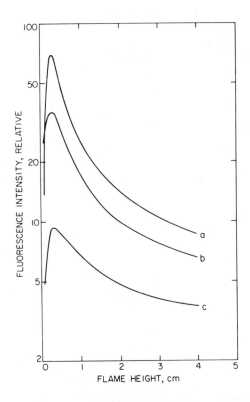

Figure 15.   SO $B^3\Sigma^-$ — $X^3\Sigma^-$ fluorescence profiles above the burner surface in H_2–O_2–N_2 (4:1:6) flames with added H_2S: (a), 1%; (b), 0.5%; (c), 0.25%. Laser excitation at 266.5 nm with detection at 283.4 nm.

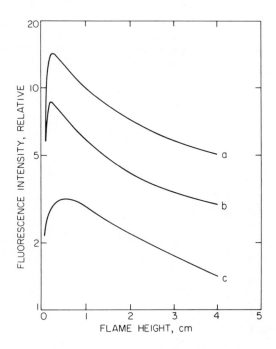

Figure 16.   SO$_2$ ^{1}B$_2$ — ^{1}A$_1$ fluorescence profiles above the burner surface in H$_2$–O$_2$–N$_2$ (4:1:6) flames with added H$_2$S: (a), 1%; (b), 0.5%; (c), 0.25%. Laser excitation at 266.5 nm with detection at 279.3 nm.

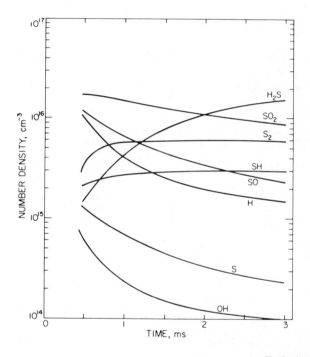

The Combustion Institute

Figure 17.   Concentration profiles in a N_2–O_2–N_2 (4:1:6) flame with 1% H_2S (6)

In that earlier study an examination of the available sulfur chemical kinetics lead to the identification of the following 8 fast coupled radical reactions that could account for the measured concentration profiles.

$$H + S_2 = SH + S \qquad\qquad (7)$$

$$S + H_2 = SH + H \qquad\qquad (8)$$

$$SH + H_2 = H_2S + H \qquad\qquad (9)$$

$$S + H_2S = SH + SH \qquad\qquad (10)$$

$$OH + H_2S = H_2O + SH \qquad\qquad (11)$$

$$H + SO_2 = SO + OH \qquad\qquad (12)$$

$$S + OH = SO + H \qquad\qquad (13)$$

$$SH + O = SO + H \qquad\qquad (14)$$

Reaction (7) couples S_2 and SH, as was noted from their fluorescence profiles. Similarly, reaction (12) links SO to SO_2. Reactions (13) and (14) connect oxidized and reduced species, SO with S_2 and SH. The model relates all sulfur bearing species in the flames. The non-equilibrium concentrations of H and OH radicals generated in the flame front by the fast radical chain branching reactions

$$H + O_2 = OH + O \qquad\qquad (15)$$

$$O + H_2 = OH + H \qquad\qquad (16)$$

controls the specific distribution of sulfur among the sulfur bearing species at a particular flame position. As H and OH recombine by slow 3-body reactions the distribution among the sulfur bearing species correspondingly shifts toward their thermodynamic equilibrium values.

We would expect the radical balance reactions (6) through (16) to be equilibrated at all points in these flames. Tests of the equilibration have been made, as for example in Figure 6, by evaluating the equilibrium concentration ratios using experimentally measured concentration values. Since O, H, S, and H_2S concentrations were not measured directly we can indirectly evaluate the equilibration of the radical balance process by using reactions that are sums of the above listed processes. Four such reactions are listed below with an indication of a combination of reactions (6) through (16) that is chemically equivalent.

$$S_2 + H_2 = SH + SH \qquad\qquad (7) + (8)$$

$$H_2 + SO_2 = SO + H_2O \qquad\qquad (6) + (12)$$

$$SH + OH = SO + H_2 \qquad\qquad (13) - (8)$$

$$SO_2 + 2H_2 = SH + OH + H_2O \qquad\qquad (6)+(8)+(12)-(13)$$

As was noted earlier (6), the combination of reactions on the right is not unique. Other reaction paths could connect the left and right sides of the four equations listed above. Nonetheless, these reactions can serve our purpose. The equilibrium ratios are evaluated in Figures 18 to 21 using experimentally measured values for T, [OH], [S_2], [SH], [SO], and [SO_2]. Equilibrium flame concentrations were used for the major products H_2 and H_2O. The equilibrium constants evaluated using JANAF thermodynamic data are shown in the figures for comparison.

From Figures 6, 18, and 20 we see that relative fluorescence measurements for OH, SH, S_2, and SO along with the method for data reduction leads to reasonable agreement with the equilibrium expectations. In Figures 19 and 21 there is a somewhat wider spread of the data about the equilibrium expectation. This is probably caused by the use of non-optimal measuring conditions and data reduction for SO_2 which has a very complex spectrum at flame temperatures. We are expecting a Nd-Yag laser shortly which will operate deeper into the UV than our present flash lamp pumped dye laser and will permit a more extensive characterization of SO_2 fluorescence in the flame environment.

Summary

Low power laser fluorescence measurements of OH, S_2, SH, SO and SO_2 have been made in a series of sulfur bearing $H_2/O_2/N_2$ flames. A simple generally applicable method for taking account of quench effects has been employed to convert relative fluorescence data into absolute concentrations. The technique has been employed to develop a kinetic model for the coupled chemistry of sulfur in rich $H_2/O_2/N_2$ flames (6).
This work is part of an on-going program. Analysis of the effects of sulfur on radical decay, further examination of the stoichiometric $H_2/O_2/N_2$ data, and analysis of sulfur chemistry in rich $C_2H_2/O_2/N_2$ flames are underway. The laboratory program is continuing with fluorescence measurements of NO, NO_2, NH, NH_2 and CN in an effort to develop a unified kinetic model for fuel nitrogen chemistry in flames.

Acknowledgments

This research has been supported by the Department of Energy under Contract Number DOE E-34 PA 372.

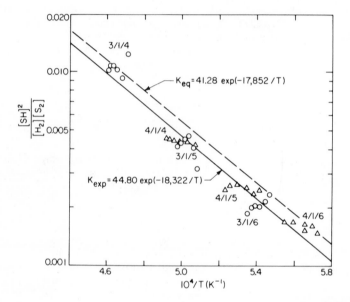

The Combustion Institute

Figure 18. A test of the equilibration of $S_2 + H_2 = SH + SH$ in fuel-rich H_2–O_2– N_2 flames with 1% H_2S. Experimental points are based on S_2 and SH fluorescence data at 0.5, 1.0, 1.5, 2.0, 2.5, and 3.0 msec, temperatures increasing slightly with time in each flame (6).

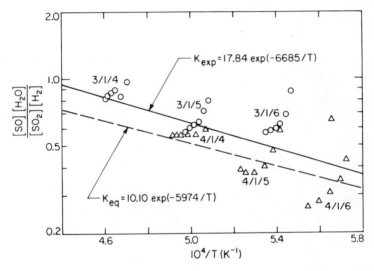

The Combustion Institute

Figure 19. A test of the equilibration of $H_2 + SO_2 = SO + H_2O$ in fuel-rich $H_2-O_2-N_2$ flames with 1% H_2S. Experimental points are based on SO and SO_2 fluorescence data at 0.5, 1.0, 1.5, 2.0, 2.5, and 3.0 msec, temperatures increasing slightly with time in each flame (6).

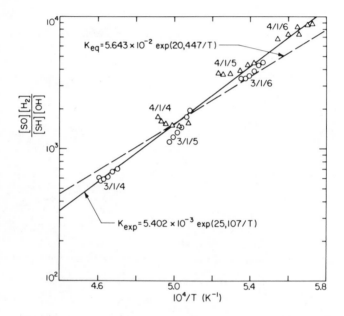

The Combustion Institute

*Figure 20.   A test of the equilibration of SH + OH = SO + H₂ in fuel-rich
H₂–O₂–N₂ flames with 1% H₂S. Experimental points are based on SH, OH, and
SO fluorescence data at 0.5, 1.0, 1.5, 2.0, 2.5, and 3.0 msec, temperatures increasing
slightly with time in each flame (6).*

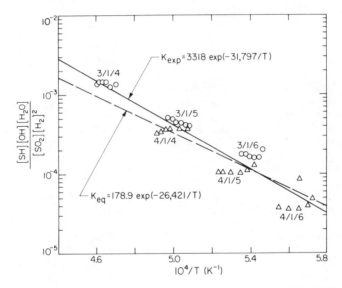

The Combustion Institute

Figure 21. A test of the equilibration of $SO_2 + 2H_2 = SH + OH + H_2O$ in fuel-rich H_2–O_2–N_2 flames with 1% H_2S. Experimental points are based on SO_2, SH, and OH fluorescence data at 0.5, 1.0, 1.5, 2.0, 2.5, and 3.0 msec, temperatures increasing slightly with time in each flame (6).

Literature Cited

1. Daily, J.W., _Appl. Opt._, 1977, 16, 568.
2. Baronavski, A.P., McDonald, J.R., _Appl. Opt._, 1977, 16, 1897.
3. Muller, C.H., III, Schofield, K., Steinberg, M., _Chem. Phys. Lett._, 1978, 57, 364; 1979, 61, 212.
4. Muller, C.H., III, Schofield, K., Steinberg, M., _10th Materials Research Symposium on Characterization of High Temperature Vapors and Gases_, Natl. Bur. Stand. (U.S.) Spec. Publ. 561/2, 1979, 855.
5. Muller, C.H., III, Steinberg, M., Schofield, K., Present Symposium.
6. Muller, C.H., III, Schofield, K., Steinberg, M., Broida, H.P., _Symp. (Int.) Combust._, (Proc.), 1979, 17, 867.
7. Padley, P.J., Sugden, T.M., _Proc. Roy. Soc. London_, 1958, A248, 248.
8. Gaydon, A.G., Wolfhard, H.G., "Flames,Their Structure, Radiation, and Temperature" Chapman and Hall, London, 1979; p. 268.
9. Denton, M.B., Swartz, D.B., _Rev. Sci. Instrum._, 1974, 45, 81.
10. Bulewicz, E.M., James, C.G., Sugden, T.M., _Proc. Roy. Soc. London,_ 1956, A235, 89.
11. Fenimore, C.P., Jones, G.W., _J. Phys. Chem._, 1965, 69, 3593.
12. Muller, C.H., III, Schofield, K., Steinberg, M., "The Fluorescence Measurement of Trace Constituents in Flames. I. OH," _J. Quant. Spectrosc. Radiat. Transfer_, (In preparation).
13. Baulch, D.L., Drysdale, D.D., Horne, D.G., Lloyd, A.C., "Evaluated Kinetic Data for High Temperature Reactions," Vol. I, Butterworths, London, 1972.
14. "JANAF Thermochemical Data" The Dow Chemical Company, Thermal Research Laboratory, Midland, Michigan.
15. Stepowski, D., Cottereau, M.J., _Appl. Opt._, 1979, 18, 354.
16. Stepowski, D., Cottereau, M.J., Present Symposium.
17. Muller, C.H., III, Schofield, K., Steinberg, M., "The Fluorescence Measurement of Trace Constituents in Flames. II. S_2 and SH," _J. Quant. Spectrosc. Radiat. Transfer_, (In preparation).
18. Ramsay, D.A., _J. Chem. Phys._, 1952, 20, 1920.
19. Earls, L.T., _Phys. Rev._, 1935, 48, 423.
20. Muller, C.H., III, Schofield, K., Steinberg, M., "The Fluorescence Measurement of Trace Constituents in Flames. III. SO and SO_2," _J. Quant. Spectrosc. Radiat. Transfer_, (In preparation).

RECEIVED February 1, 1980.

Laser-Induced Fluorescence Spectroscopy Applied to the Hydroxyl Radical in Flames

M. J. COTTEREAU and D. STEPOWSKI

Laboratoire de Thermodynamique, L.A. C.N.R.S. N° 230, Faculté des Sciences et des Techniques de Rouen, B.P. 67 76130 Mont-Saint-Aignan, France

One of the main problems met in Laser Induced Fluorescence measurements is the excited population dependence on the quenching due to collisional deexcitation. The saturation mode proposed to avoid this dependence is very difficult to achieve (1)(2) particularly with molecular species and moreover the very strong laser pulses required may cancel the non-perturbing characteristic of the method. Therefore precise knowledge of the quenching is necessary in some experimental circumstances.

This paper is devoted to the work we are pursuing at the University of Rouen to study this problem. It consists essentially in using low pressure flames and short duration laser pulses to obtain :

1) direct measurements of the quenching rate in various experimental conditions,

2) direct local concentration measurements in single pulse mode.

Until now the work has been performed on OH in flames but it can easily be used for any other species.

Theory

For molecular species measurement the exciting laser pulse (with a spectral energy density U_ν) is tuned on a rotational line of an electronic transition. If N_1 is the population of the lower vibrational level, the population N_2 of the excited vibrational level increases according to

$$\frac{dN_2}{dt} = \frac{N_1}{\alpha_1} B_{12} U_\nu - \frac{N_2}{\alpha_2} B_{21} U_\nu - N_2 (A+Q) \qquad (1)$$

where B_{12} and B_{21} are the probability coefficients for stimulated absorption and emission respectively ; A and Q are the probability coefficients of the radiative and collisional relaxation for the excited vibrational level. The coefficients α_1 and α_2 take into

0-8412-0570-1/80/47-134-131$05.00/0
© 1980 American Chemical Society

account the redistribution among the neighbouring rotational levels not directly connected by the pumping ; they may be rather complicated under strong laser pulses (3) but for weak pulses a Boltzmann equilibrium can be assumed and with Z_R the rotational partition function :

$$\alpha_i = \frac{Z_R}{2J_i+1} \exp \frac{E_{Ri}}{kT}$$

Writing $N_1+N_2 = N_0$, we obtain :

$$N_2(t) = \frac{N_0}{\alpha_1} B_{12}U_\nu \left[\left(\frac{B_{12}}{\alpha_1} + \frac{B_{21}}{\alpha_2}\right) U_\nu + A + Q\right]^{-1} \left[1 - \exp\left[-\left[\left(\frac{B_{12}}{\alpha_1} + \frac{B_{21}}{\alpha_2}\right) U_\nu + A + Q\right] t\right]\right]$$

(2)

The conditions to reach a saturated steady state with a pulse of duration Δt,

$$\left(\frac{2BU_\nu}{\alpha} + A + Q\right) \Delta t \gg 1 \quad \text{and} \quad \frac{2BU_\nu}{\alpha} \gg A + Q$$

are more difficult to achieve, because of the α factor, than for a simple two level model.

We have proposed an alternative to this saturation approach (4) : if

$$\Delta t \left(\frac{2BU_\nu}{\alpha} + A + Q\right) \ll 1$$

the expansion of the exponential in equ. 2 can be approximated by the first term and we have

$$N_2(\Delta t) = \frac{N_0 B_{12} U_\nu \Delta t}{\alpha_1}$$

After the excitation the population N_2 decreases and the whole fluorescence decay, if exponential, gives a mean quenching rate Q according to :

$$\phi(t) = \frac{N_0 B_{12} U_\nu \Delta t}{\alpha_1} \frac{V\Omega A}{4\pi} \exp\left[-(A+Q)t\right]$$

The local concentration N_0 can be determined either by measuring ϕ_{max} (at $t = 0$) or by integrating the total number of fluorescence photons N_F. In this latter case, with N_L the total number of photons in the laser pulse, we obtain :

$$N_0 = \frac{N_F \alpha_1 \, 4\pi(A+Q) \, C \, \Delta\lambda}{N_L \, \Delta x \, \Omega \, A \, B_{12} \, h\nu}$$

(3)

where Δx is the length of the excited detected zone.

Experimental

We have studied first the OH radical in a low pressure flame (15 torr < p < 80 torr) to obtain quenching times longer than our pulse duration of 4 ns. This exciting radiation is derived from the second harmonic of a dye laser pumped by a nitrogen laser. An intracavity Fabry-Perot etalon assures a laser spectral width ($\Delta\lambda = 2\ 10^{-12}$ m) closely matching the absorption line. We have excited the $Q_1 7$ line ($\lambda = 308.9734$ nm) of the $^2\Sigma^+(v'=0) \leftarrow {}^2\Pi(v''=0)$ transition of OH.

The flat flame is set up by a porous burner ($\phi = 2$ cm) supplied with a premixed O_2-C_3H_8 flow.

Quenching studies in low pressure flames (5)

The fluorescence signal emanating from all the rotational levels of the excited state is detected by a photomultiplier connected to an oscilloscope or to a boxcar analyzer.

Fig. 1 shows how the quenching rate in the burnt gases zone of a stochiometric flame increases linearly with the pressure. Beyond 50 torr the observed deviation from linear dependence is due to the exciting pulse duration itself. Extrapolation to 1 Atm. leads to a quenching rate $Q = 10^9$ s^{-1} which is close to the value found by Bonczyk and Shirley (6) for CH and CN in flames.

On Fig. 2 we have plotted the quenching rate versus the height of the incident beam above the burner plane for several oxygen-propane flames of different stochiometries. The quenching remains nearly constant through each flame in spite of temperature variation from 1300 to 2000 K. In the reaction zone where drastic concentration changes occur the quenching is surprisingly nearly constant. This result confirms first that the quenching is not very sensitive to the temperature ; it suggests secondly that water, the concentration of which rises quickly in the reaction zone, remains the most efficient quencher ; its mole fraction determines for a large part the global quenching rate which changes with the stochiometry. Quenching by radicals which could have been thought very strong appears to be equal or slightly higher than the quenching by H_2O so as to compensate for the decrease of the temperature.

These experiments have been made using the $Q_1 7$ line ; when exciting several other lines (P or Q) no noticeable variation of the quenching was found. This result validates temperature measurements from Boltzmann plots of relative fluorescence intensities.

[OH] measurement

In a low pressure flame a direct local and instantaneous measurement of [OH] can be obtained by use of equ. 3. Fig. 3 shows a profile of [OH] through a stochiometric C_3H_8-O_2 flame at 20 torr.

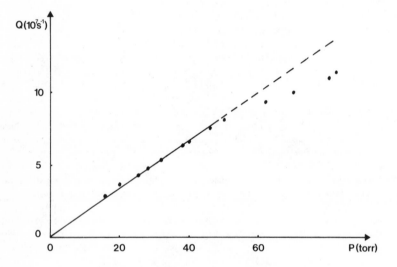

Figure 1. Quenching rate vs. flame pressure

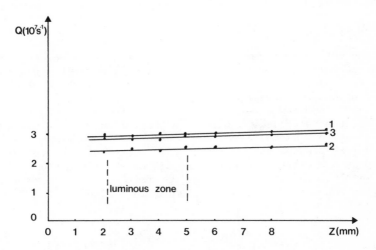

Figure 2. Quenching rate along flames of different stoichiometries: Molar ratio $|C_3H_8|/|O_2| = \phi$; Flame 1, $\phi = 0.20$; Flame 2, $\phi = 0.11$; Flame 3, $\phi = 0.25$.

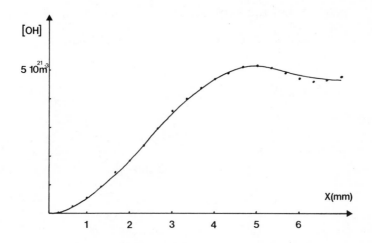

Figure 3. $|OH|$ profile through the low-pressure flat flame

Assuming that the quenching behaviour in an atmospheric flame is the same as in a low pressure flame, the fluorescence efficiency may be deduced from this low pressure study and then $[OH]$
may be measured.

Conclusions

The method proposed allows direct absolute measurement of
local concentration at the instant of the laser pulse in a low
pressure flame. We believe that this method could be applied to
higher pressure flames by the use of ultrashort duration laser
pulses with the new mode locked dye laser technique. But until
the detector technology allows such short time resolutions we
think that collisional lifetimes studies must be pursued to
obtain more precise evaluation of the fluorescence efficiency,
and to have a better understanding of the redistribution phenomena involved in optical pumping. For this purpose we are now
studying the decay of resolved fluorescence lines.

Literature cited

(1) J.W. DAILY, Applied Optics, vol. 16, n° 3, 568-571, March
    1977.
(2) BARONAVSKI and McDONALD, Applied Optics, vol. 16, n° 7.
(3) R.K. LENGEL, D.R. CROSLEY : J. Chem. Phys., vol. 67, n° 5,
    sept. 1977.
(4) D. STEPOWSKI and M.J. COTTEREAU, Applied Optics, vol. 18, n°3,
    Feb. 1979.
(5) D. STEPOWSKI and M.J. COTTEREAU (To be published in Combus-
    tion and Flame)
(6) P. BONCZYK and J. SHIRLEY, Combustion and Flame, vol. 34, n°3,
    (1979).

RECEIVED March 19, 1980.

A Multilevel Model of Response to Laser-Fluorescence Excitation in the Hydroxyl Radical

ANTHONY J. KOTLAR—Ballistic Research Laboratory,
Aberdeen Proving Ground, MD 21005

ALAN GELB—Physical Sciences Incorporated, Woburn, MA 01801

DAVID R. CROSLEY—SRI International, Menlo Park, CA 94025

Optical Saturation in Laser-Induced Fluorescence

Experiments using the technique of laser-induced fluorescence (LIF) in flames have provided ample demonstration of its selectivity and sensitivity, and hence of its applicability as a probe for the reactive intermediates present in combustion systems. The relationship between the measured fluorescence intensity and the concentration of the molecule probed, however, must take into account the collisional quenching of the electronically excited state pumped by the laser. Because the flame contains a mixture of species, each with different quenching cross sections, it may be difficult to estimate the total quenching rate even if many of these cross sections are known.

One solution ($\underline{1}$) to this problem is to increase the laser intensity I to a value where stimulated emission becomes faster than collisional quenching. For a two-level system, a steady-state equation can then be written for N_e:

$$dN_e/dt = 0 = B_u IN_g - (B_d I + Q + A)N_e,$$

where the symbols are defined in Table 1; the fluorescence intensity S is proportional to N_e. The ratio N_e/N_g, extrapolated to infinite I, is equal to the ratio of their degeneracies. The ratio Q/A may be obtained from the slope-to-intercept ratio of a plot of S^{-1} vs. I^{-1}, which is linear ($\underline{2},\underline{3}$). Atomic sodium and several molecular systems have been investigated under saturation conditions in flames.

Any real molecule can only be approximated as a two-level system, since each electronic state involved in the transition contains a number of vibrational and rotational levels. In the (probably unrealistic)limit where quenching directly back to the single lower level pumped by the laser is much faster than rotational or vibrational relaxation, a two-level system with the

0-8412-0570-1/80/47-134-137$05.00/0
© 1980 American Chemical Society

TABLE 1. NOTATION USED IN TEXT

N_e, N_g...population of excited, ground states in two-level systems
B_u, B_d...Einstein coefficient for absorption, stimulated emission
A........Einstein coefficient for spontaneous emission
Q, σ_Q....Quenching rate, cross section; excited to ground state
R, σ_R....Rotational transfer rate, cross section
V, σ_V....Vibrational transfer rate, cross section; in ground state

correct value of Q/A obtained from the S^{-1} vs. I^{-1} plot would apply. In the opposite limit where $R,V \gg Q$, the plot would yield an apparent quench rate equal to Q/A times an upper state partition function.

In actuality, the molecular populations are collisionally transferred among these internal levels at rates of the same general magnitude as the quenching, so that an appreciable fraction of the original population of the two levels connected by the laser can reside in the remainder of the levels. The fraction in the upper laser-pumped level, and thus S, is in general dependent on the detailed state-dependent rates of transfer within and between the two manifolds of vibrational and rotational levels.

The Multilevel Model

The present study is a computer model of the time evolution of individual level populations of the OH molecule under the influence of laser excitation. The environment simulates that of the burnt gases of an atmospheric pressure methane-air flame at 2000°K. OH is studied because of its importance in combustion chemistry and suitability for LIF, which have made it the most popular molecule for LIF investigations in flames; in addition, it has a small enough number of significantly populated levels to be computationally tractable.

The model includes 30 rotational levels (N=0-15) in the v=0 vibrational level in each of the excited $A^2\Sigma^+$ and ground $X^2\Pi$ electronic states. A single "dummy" vibrationally excited level is used to represent all levels in the X state not belonging to v=0. The pumping transition ($Q_1 3$ line) connects the N=3, J=7/2 levels of the v=0 levels. Relative state-to-state rotational transfer rates in the A-state are constructed from an information theoretic analysis of experimental values, and this same energy dependence of the X-state rotational relaxation rates is assumed. The final-state dependence of σ_Q is calculated assuming an a priori statistical distribution of the released energy among all available degrees of freedom, assuming a quenching partner having only translational modes. This results in 44% of the molecules quenching

back to v=0. From available information on X-state vibrational
transfer rates, we estimate a nominal effective cross section for
vibrational transfer from the dummy level to v=0, σ_V, at 0.4 Å.
Nominal values of σ_Q and σ_R are 14 $Å^2$ and 50 $Å^2$, respectively.

Results

Using this single laser excitation, the equations have been
integrated for a variety of values of I and of σ_Q, σ_V, σ_R near the
nominal ones. The goal of the study is not merely to attain a description in terms of these parameters, but to explore the limits
of applicability of more simplified approaches for OH and other
molecules. In particular, a two-level approximation is found to
be clearly inadequate for OH. More detailed aspects are described
in what follows.

1. <u>Time Dependence</u>. The pumped level population does not
settle to within 90% of its steady state value until ≥30 nsec
after the laser is turned on. Thus, while a steady state approximation is valid for a flashlamp pumped laser with a pulse length
of the order of 1 μsec, the signals using a Nd:YAG pumped dye
(~10 nsec) may exhibit an observable time dependence.

2. <u>Rotational Population Distributions</u>. As expected, there
is an overpopulation in the A state level and an underpopulation
in the X state level connected by the laser, compared to a thermal
distribution (see Fig. 1). Higher rotational levels (N ≥ 6) in the
A-state are described by a Boltzmann distribution with T~940°K,
well less than the gas temperature and reflecting the energy dependence of the σ_R. The high-N levels of the X-state are described by a Boltzmann distribution with very high T (3200°K) but this
may be an artifact of the model, due principally to the assumed
N-dependence of the σ_Q and σ_V.

3. <u>Populations as I Becomes Large</u>. The ratio of the populations in the levels connected by the laser, plotted vs. I^{-1}, is
a straight line with unit intercept, as shown in Fig. 2. Plots
are made for three assumed σ_Q, shown in $Å^2$, with all other transfer rates held constant. Analyzed as a two-level system, the apparent σ_Q is considerably larger than the σ_Q inserted into the
model; see Table 2. The difference lessens as σ_Q increases. This
is in accord with the simple limiting picture noted earlier.

However, as the upper pumped level population increases,
that of the lower pumped level decreases (see Fig. 1), and so does
that of the pair as a whole. That is, under the influence of the
laser pumping, population of the pair is depleted. The reason for
this is that the rate of energy transfer out of the upper pumped
level (depletion of the pair population) is significantly faster
than energy transfer into the lower pumped level (repletion).

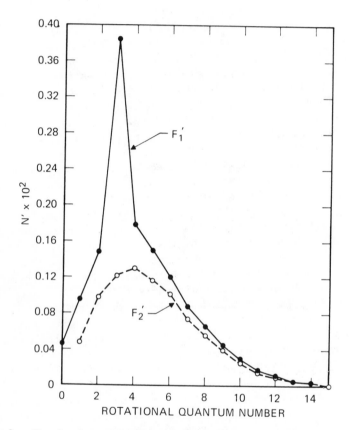

Figure 1a. *Fractional populations as a function of rotational quantum number;*
$F_2(3)$ *is pumped.* $A^2\Sigma^+$ *state:* ($\bigcirc$), F_2 *levels,* ($\bullet$), F_1 *levels.*

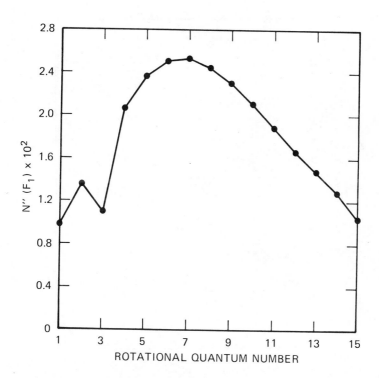

Figure 1b. *Fractional populations as a function of rotational quantum number; $F_2(3)$ is pumped. $X^2\Pi$ state, F_1 levels.*

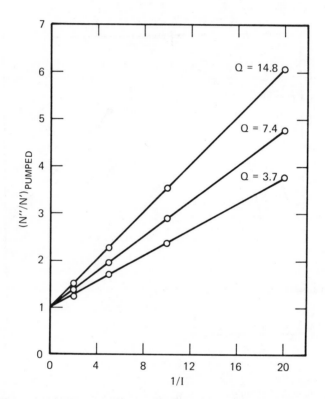

Figure 2.   Ratio of populations (ground:excited) in the two levels connected by the laser as a function of inverse laser intensity for three assumed values of σ_Q in square angstroms

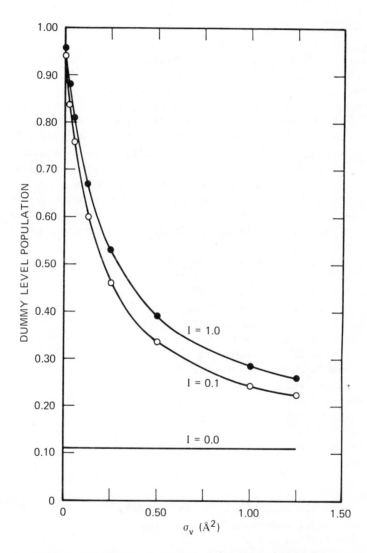

Figure 3. Dummy (vibrationally excited)-level population, as a fraction of the total, as a function of assumed transfer cross section σ_V (in square angstroms) to $v = 0$ of $X^2\Pi$ for three values of laser intensity

TABLE 2. VALUES OF σ_Q, FROM FIG. 2, $\overset{\circ}{A}{}^2$

σ_Q, input to model	σ_Q, from slope	ratio
3.7	97	.038
7.4	140	.053
14.8	183	.081

Thus, at high I, the pair population is a considerably smaller fraction of the total OH population than the initial fraction given by a Boltzmann distribution at the flame temperature. For example, for the nominal values of 14 and 0.4 $\overset{\circ}{A}$ for σ_Q and σ_V, the infinite-intensity fraction is < 1% of the total while the zero-intensity value is ~4%. This result is generally valid for the entire range of parameters inserted into the model, which represent physically realistic energy transfer rates. However, the precise numerical values depend sensitively on the actual parameters inserted. These facts form the central conclusions of this study (4). A steady state model with no dummy level and a different set of rate constants and level structure (5) shows some similar features.

4. <u>Dummy level population</u>. With no laser, the population of the dummy level is set at 11% of the total, the thermal equilibrium fraction in v=1 at 2000°K. Because vibrational energy transfer rates are generally slow, the laser excitation causes a sizeable fraction of the total to be pumped into the dummy level. Fig. 3 shows the dummy level population for three laser intensities as a function of assumed σ_V. (In the dimensionless notation used in the computer, I=1 corresponds to 10^4 erg sec^{-1} cm^2 Hz^{-1}, or that of the unfocussed output of the fundamental from an efficient dye pumped by a powerful doubled Nd:YAG laser). At the nominal 0.4 $\overset{\circ}{A}$, nearly 40% of the population is driven into the dummy level at high I. Clearly the value of σ_V, a poorly known parameter, is important for a quantitative description of fluorescence saturation.

Literature Cited

1. Daily, J.W., Appl. Opt. <u>16</u>, 568 (1977).
2. Allen, J.E., Anderson, W.R., and Crosley, D.R., Optics Lett. <u>1</u>, 118 (1977).
3. Allen, J.E., Anderson, W.R., Crosley, D.R., and Fansler, T.D., Seventeenth Symposium (Int.) on Combustion, 1979, p. 797.
4. Kotlar, A.J., Crosley, D.R., and Gelb, A., to be published.
5. Lucht, R.P., and Laurendeau, N.M., Appl. Opt. <u>18</u>, 856 (1979).

RECEIVED April 15, 1980.

8

Saturated-Fluorescence Measurements of the Hydroxyl Radical

ROBERT P. LUCHT, D. W. SWEENEY, and N. M. LAURENDEAU

School of Mechanical Engineering, Purdue University, W. Lafayette, IN 47907

Laser-induced fluorescence is a sensitive, spatially resolved technique for the detection and measurement of a variety
of flame radicals. In order to obtain accurate number densities
from such measurements, the observed excited state population
must be related to total species population; therefore the population distribution produced by the exciting laser radiation must
be accurately predicted. At high laser intensities, the fluorescence signal saturates (1, 2, 3) and the population distribution
in molecules becomes independent of laser intensity and much less
dependent on the quenching atmosphere (4). Even at saturation,
however, the steady state distribution is dependent on the ratio
of the electronic quenching to rotational relaxation rates (4, 5,
6, 7). When steady state is not established, the distribution is
a complicated function of state-to-state transfer rates.

   The OH radical has been selected for preliminary saturated
molecular fluorescence studies. A Nd:YAG pumped dye laser is
used to excite an isolated rotational transition, and the resulting fluorescence signal is analyzed both spectrally and temporally in order to study the development of the excited state rotational distribution. It is found that steady state is not
established throughout the upper rotational levels, although the
directly excited upper rotational level remains approximately in
steady state during the laser pulse. The fluorescence signal
from the directly excited upper level exhibits considerable saturation.

Frozen Excitation Model
   A molecular fluorescence model is presented which is particularly appropriate for short pulse excitation. The frozen excitation model treats the two rotational levels which are directly excited by the laser as an isolated system with constant total
number density. Consider the four level molecular model illustrated in Fig. 1. The four level model was solved by Berg and
Shackleford (5) for the case where steady state is established
throughout all molecular levels. Levels 1e and 2e are the single

0-8412-0570-1/80/47-134-145$05.00/0
© 1980 American Chemical Society

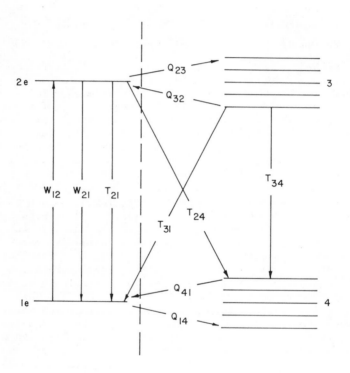

Figure 1.   Four-level molecular model. Q_{ij} *is the collisional-transfer rate constant from level i to level j,* T_{ij} *is the sum of the electronic quenching and spontaneous emission rate constants,* W_{12} *is the absorption rate constant, and* W_{21} *is the stimulated emission rate constant.* W_{12} *and* W_{21} *are proportional to the laser power* P_L. *The dashed vertical line separates levels 1e and 2e, which are treated as an isolated system, from those levels not affected directly by the laser radiation.*

upper and lower rotational levels which are directly connected by the laser radiation. Levels 3 and 4 include the rest of the upper and lower rotational levels, respectively. The rate equations for the level populations N are

$$\dot{N}_1(e) = -N_1(e)[W_{12}+Q_{14}] + N_2(e)[W_{21}+T_{21}] + N_3 T_{31} + N_4 Q_{41}, \qquad (1)$$

$$\dot{N}_2(e) = N_1(e)W_{12} - N_2(e)[W_{21}+T_{21}+T_{24}+Q_{23}] + N_3 Q_{32}, \qquad (2)$$

$$\dot{N}_3 = N_2(e)Q_{23} - N_3[T_{31}+T_{34}+Q_{32}], \qquad (3)$$

$$N_T = N_1(e)^0 + N_4^0 = N_1(e) + N_2(e) + N_3 + N_4. \qquad (4)$$

The superscript 0 refers to level populations prior to laser irradiation. The rate constant notation is described in Fig. 1.

At laser intensities sufficient to saturate the 1e-2e transition, the stimulated emission and absorption processes which couple the levels are fast relative to collisional transfer processes, and a quasi-equilibrium balance $[N_2(e)/N_1(e)]_{ss}$ is quickly established. If the total population of levels 1e and 2e is approximately constant during the laser pulse, the upper level population $N_2(e)$ can be reliably related to $N_1(e)^0$ using an analysis similar to a two level atomic model (1, 2, 3). $\Delta[N_1(e) + N_2(e)]$, the net population transfer into or out of levels 1e and 2e during the laser pulse, will be nearly zero if the laser pulse length τ_L is less than or comparable to the characteristic collisional transfer time $\tau_c \sim (Q_{23}+T_{24})^{-1} \sim Q_{14}^{-1}$, simply because few collisions will occur during the laser pulse. If $\tau_L > \tau_c$, $\Delta[N_1(e)+N_2(e)]$ will still be much less than $N_1(e)^0$ if the population transfer rate into levels 1e and 2e, $N_3(Q_{32}+T_{31})$ + $N_4 Q_{41}$, is comparable to the transfer rate out of levels 1e and 2e, $N_1(e)Q_{14} + N_2(e)[T_{21}+T_{24}+Q_{23}]$. This assumption can be justified on the basis of detailed balancing considerations, provided that the rotational relaxation rates in the upper and lower sets of rotational levels are not greatly different. For $\Delta[N_1(e) + N_2(e)] \simeq 0$, $N_1(e)^0 = N_1(e) + N_2(e)$ throughout the laser pulse, and Eq. (2) becomes

$$N_2(e) = N_1(e)^0 W_{12} - N_2(e)\{W_{12}+W_{21}+T_{21}+T_{24}+Q_{23}-[N_3/N_2(e)]Q_{32}\}. \qquad (5)$$

For a Boltzmann distribution in the upper levels, $N_3 Q_{32} = N_2(e)Q_{23}$. However, for a short laser pulse, level 3 will be significantly underpopulated. Therefore, the factor $[N_3/N_2(e)]$ Q_{32} is negligible compared to Q_{23}.

To show that a quasi-equilibrium ratio $[N_2(e)/N_1(e)]_{ss}$ is established very quickly, Eq. (5) is solved for a step function laser pulse (laser power $P_L = 0$, $t < 0$; P_L = constant, $t > 0$),

$$N_2(e) = N_1(e)^0 W_{12} \tau_{ss}[1-\exp(-t/\tau_{ss})], \qquad (6)$$

$$\tau_{ss} = [W_{12}+W_{21}+T_{21}+T_{24}+Q_{23}]^{-1} . \tag{7}$$

At near saturation conditions for OH, $\tau_{ss} < 10^{-10}$ s. Thus, after approximately 100 ps, levels 1e and 2e reach steady state, and $N_2(e)$ is given by

$$N_2(e) = N_1(e)^0[W_{12}/(W_{12}+W_{21})][1+(T_{21}+T_{24}+Q_{23})/(W_{12}+W_{21})]^{-1}. \tag{8}$$

At full saturation Eq. (8) becomes

$$N_2(e) = N_1(e)^0[W_{12}/(W_{21}+W_{12})] = N_1(e)^0[1+g_1(e)/g_2(e)]^{-1}, \tag{9}$$

where $g_1(e)$ and $g_2(e)$ are the rotational degeneracies for levels 1e and 2e. Steady state is established much more slowly throughout the rest of the levels via a succession of collisional transfers. For a short laser pulse, $\dot{N}_3 \neq 0$; hence, N_3 cannot be reliably related to a lower level population.

$N_1(e)^0$ is related to N_T by

$$N_T = N_1(e)^0 + N_4^0 = N_1(e)^0/F''_{1B}(e), \tag{10}$$

where $F''_{1B}(e)$ is the Boltzmann fraction for level 1e. Level 1e can be chosen so that $F''_{1B}(e)$ is a weak function of temperature (8). Consequently, if fluorescence is observed only from level 2e and the flame temperature can be estimated, N_T can be calculated using a simple two level analysis (1, 2, 3).

Experimental System and Results

Fluorescence is induced by a Molectron Nd:YAG pumped dye laser. The laser repetition rate is 10 Hz, the bandwidth is ~0.01 nm, and the maximum pulse energy and peak power at 309 nm are 3 mJ and .5 MW, respectively. The laser is focused into the flame by a 15 cm focal length lens; the focused spot size is about 100 μm, as determined from burn patterns on thermal paper. A mirror is placed past the focusing lens to reflect the beam vertically into the flame and parallel to the spectrometer slits. The fluorescence is collected by a 10 cm focal length lens and focused onto the entrance slit of a Spex 1800-II spectrometer operated in second order. The spectral resolution is ~0.1 nm. A 1P28 photomultiplier wired for fast response (9) (rise time ≃ 2 ns) is placed at the exit slit. The photomultiplier signal is processed by a Tektronix 5S14N sampler in a 5440 mainframe. The fluorescence spectrum is analyzed by fixing the sampling window of the oscilloscope at a given point in the pulse waveform and scanning the spectrometer; the temporal behavior of fluorescence from individual lines is investigated by setting the spectrometer at the appropriate wavelength and scanning the sampler across the pulse waveform.

The isolated $Q_1(4)$ line of the (0,0) band of the $A^2\Sigma^+-X^2\Pi$ electronic transition of OH is directly excited by the laser. To

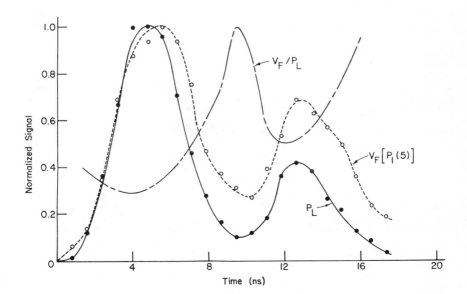

Figure 2. Normalized laser pulse and fluorescence signal from the directly excited upper level vs. time. The laser is tuned to the $Q_1(4)$ line (308.42 nm) and fluorescence is observed from the $P_1(5)$ line (310.21 nm). The laser-pulse energy and peak power are 0.63 mJ and .13 MW, respectively: (— — —), the ratio of the fluorescence signal to laser power.

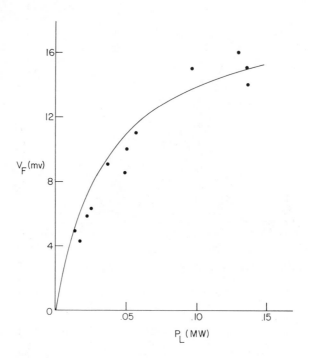

Figure 3.   Fluorescence signal vs. laser power.   Data was obtained from the curves in Figure 2; points were taken near the peaks and dips of the pulse waveforms: (———), a curve fit through the data using Equation 8.

observe fluorescence from the directly excited J' = 4.5, N' = 4 upper rotational level, the spectrometer is tuned to the $P_1(5)$ line. Fig. 2 illustrates typical experimental results. The peaks and dips of the laser pulse and the fluorescence pulse nearly coincide; this indicates that a quasi-equilibrium between the directly excited upper and lower rotational levels is established very quickly. When fluorescence power is plotted versus laser power, the fluorescence signal exhibits considerable saturation, as shown in Fig. 3.

The fluorescence spectrum is found to be markedly non-Boltzmann and sharply peaked at the directly excited level throughout the laser pulse. This is due to two effects: the competition between electronic quenching and rotational relaxation processes (4) and the short length of the laser pulse. Because the pulse is so short, steady state is not established throughout the upper rotational levels. The peaks of the fluorescence pulses from levels which are not directly excited by the laser lag the laser pulse peaks by one to four nanoseconds, depending on the energy gap between the given level and the directly excited level.

Using the frozen excitation model to analyze the data shown in Fig. 3, and calibrating the system via Rayleigh scattering (8), a total OH number density of 4×10^{16} cm^{-3} was calculated for an assumed flame temperature of 2000 K in the methane-air torch. N_T was not compared directly with the results of absorption studies; future flat flame burner studies will involve direct comparison of absorption and fluorescence.

Acknowledgements
We are indebted to Dr. Fred E. Lytle and his students for their advice and assistance, and to Dr. Alan C. Eckbreth of the United Technologies Research Center for his many helpful suggestions. This work was supported by DOE Contract ER-78-S-02-4939.

IV. Literature Cited
1. Piepmeier, E., Spectrochimica Acta, 1972, 27B, 431.
2. Baronovski, A.P.; McDonald, J.R., Applied Optics, 1977, 16, 1897.
3. Daily, J.W., Applied Optics, 1975, 15, 955.
4. Lucht, R.P.; Laurendeau, N.M., Applied Optics, 1979, 18, 856.
5. Berg, J.O.; Shackleford, W.L., Applied Optics, 1979, 18, 2093.
6. Crosley, D.R., Ed., "Laser Probes for Combustion Chemistry," American Chemical Society: Washington, D.C., 1979; p.
7. Daily, J.W. (Crosley, D.R., Ed.), "Laser Probes for Combustion Chemistry," American Chemical Society: Washington, D.C., 1979, p.
8. Eckbreth, A.C.; Bonczyk, P.A.; Verdieck, J.F., Applied Spectroscopy Reviews, 1978, 13, 15.
9. Harris, J.M.; Lytle, F.E.; McCain, T.C., Analytical Chemistry, 1976, 48, 2095.

RECEIVED February 1, 1980.

Nitric Oxide Detection in Flames by Laser Fluorescence

DANIEL R. GRIESER and RUSSELL H. BARNES

Battelle–Columbus Division, 505 King Avenue, Columbus, OH 43201

Laser-Fluorescence techniques for NO are of interest for studying the mechanisms of NO formation and its influence on chemical processes and pollutant formation in flames. In general, the optical fluorescence techniques provide very high detection sensitivities and good spatial resolution.

The method described here for detecting NO in flames is based on the use of a frequency-doubled tunable dye laser to excite transitions in the (0,0) γ-band of NO in the range of 2250 to 2270 Å. Fluorescence is observed at wavelengths associated with the bands involving the (0,0), (0,1), (0,2), and higher ground-state vibrational transitions of the γ-band system.

The experimental system used for the NO flame fluorescence measurements is shown in Figure 1. The frequency-doubled beam from the dye laser was focused into the high-temperature reaction zone in a flat flame on a 2-1/4-inch-diameter burner. Fluorescence from the flame was collected in a direction perpendicular to the face of the burner using a cassegrain collection optic and focused through the slits of the spectrometer. The collection optic was located at a distance of about 8 inches from the face of the burner and used with an effective aperature of about f/2. An EMI 6256 SA photomultiplier was used as a detector. The output signal from the detector was processed using an ORTEC gated photon counting system that was synchronized to the laser pulse.

Excitation spectra for NO recorded by setting the spectrometer at specific wavelengths and continuously scanning the output wavelength of the dye laser are presented in Figures 2 and 3. Figure 2 shows a fluorescence excitation spectrum for a trace of NO in nitrogen flowing from the face of the burner at room temperature and no flame present. Figure 2 shows the same excitation spectrum for NO in the high-temperature zone of a $CH_4/O_2/N_2$ flame. Both excitation spectra were obtained at atmospheric pressure at the same position above the face of the burner. In the case of Figure 2, the spectrometer was set at 2262.0 Å for the (0,0) γ-band, while in Figure 3 the spectrometer was set at 2368.8 Å for the (0,1) γ-band. The spectrometer slits were set at 3 mm which

0-8412-0570-1/80/47-134-153$05.00/0
© 1980 American Chemical Society

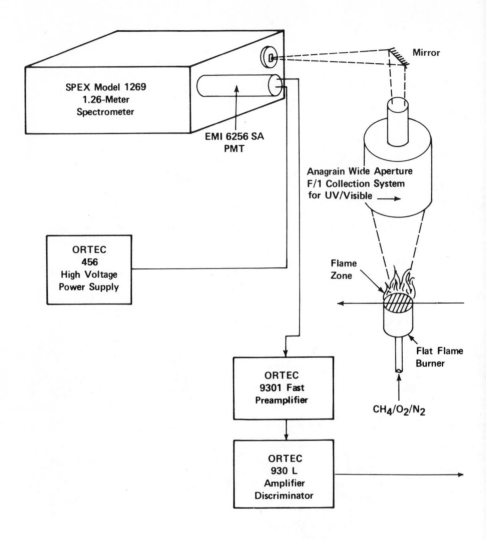

Figure 1. *Dye laser system for nitric oxide fluorescence measurements in CH$_4$–O$_2$–N$_2$ flame*

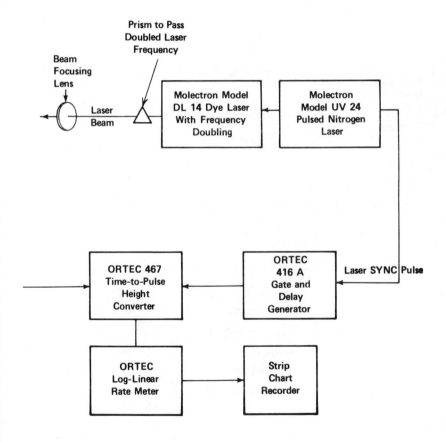

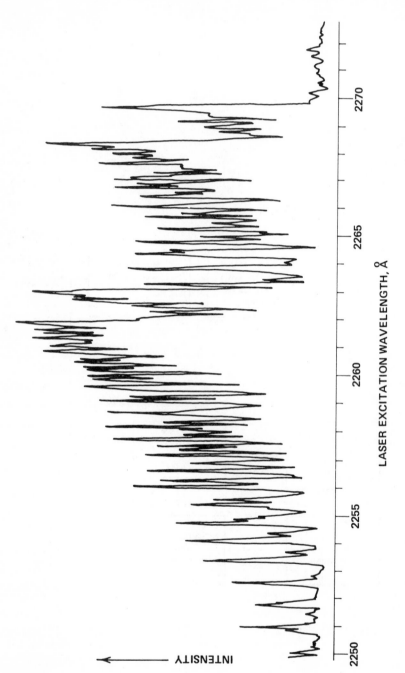

Figure 2. Laser-excitation spectrum for nitric oxide in N₂ flowing from burner at atmospheric pressure (spectrometer set for 0,0 γ-band at 2262.0 Å)

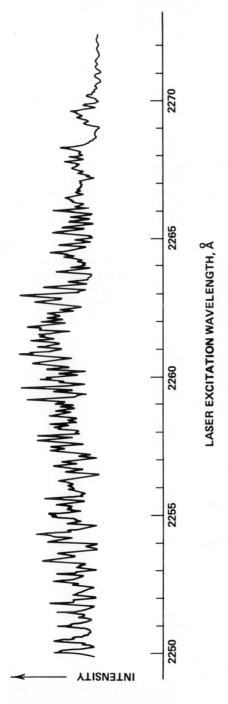

Figure 3. Laser-excitation spectrum for nitric oxide in CH_4–O_2–N_2 flame at atmospheric pressure (spectrometer set for 0,1 γ-band at 2368.8 Å)

is equivalent to a spectral bandpass of about 19.9 Å. In the case
of the flame spectrum, a trace of NO was added to the nitrogen to
enhance the intensity of the spectrum. It was, however, possible
to detect the natural level of NO in the flame which was measured
previously with the same burner by Merryman and Levy (1) using
a quartz probe and chemiluminescence analyzer and found to
be in the range of 20-30 ppm. In the case of Figure 3, it is
estimated that the level of NO in the doped flame is about 60 ppm.
Scan times of 60 minutes were required to record the excitation
spectra shown in Figures 2 and 3. The contribution of Rayleigh
scattering to the spectrum in Figure 2 would be expected to be
small because the cross sections for Rayleigh scattering are
several orders-of-magnitude smaller than fluorescence cross
sections. This was verified experimentally by comparing spectra
for the (0,0) γ-band transition, with those involving other lower-
electronic vibrational levels of the NO γ-bands. The spectral
lines in Figures 2 and 3 all correspond to identified rovibronic
transitions (2,3).

In summary, the results in this paper demonstrate that laser
fluorescence can be used to detect NO in atmospheric-pressure
flames. Detection sensitivities in the ppm range were observed
with laser pulse energies in the range of about 3 μJ. This
sensitivity can be increased significantly by using a higher
intensity laser.

This research was supported by the Division of Fossil Fuel
Utilization of the U. S. Department of Energy under contract no.
W-7405-Eng-92, Task 88.

Literature Cited

1. Merryman, E. L., and Levy, A., "Fifteenth Symposium (Inter-
    national) on Combustion", The Combustion Institute, 1974,
    p. 1073.
2. Zacharias, H., Anders, A., Halpern, J. B., and Welge, K. H.,
    Opt. Commun., 1976, 19, 116. Also see Errata: Opt. Commun.,
    1977, 20, 449.
3. Deezi, I., Acta Physica, 1958, 9, 125.

RECEIVED February 1, 1980.

Laser-Induced Fluorescence of Polycyclic Aromatic Hydrocarbons in a Flame

DONALD S. COE and JEFFREY I. STEINFELD

Department of Chemistry and Chemical Engineering and Center for Health Effects of Fossil Fuel Utilization, Massachusetts Institute of Technology, Cambridge, MA 02139

Laser-induced fluorescence spectroscopy (LIF) is being developed as an in-situ, real time diagnostic for polycyclic aromatic hydrocarbons (PCAH) in combustion systems. PCAH are known to be formed in sooting flames[1,2] and are of interest both for their carcinogenic properties and possible role in the soot formation process. Gas chromatography and mass spectrometry have provided probe measurements of PCAH in flames; however, there is a need for a real time, non-intrusive technique for measurement of PCAH in combustion systems. Probe measurements have indicated individual PCAH concentrations in the 10 ppb to 10 ppm range[3]. LIF has been shown to be capable of detection of flame radicals at these concentrations[4] and is expected to give similar limits for PCAH.

The individual PCAH of interest include naphthalene, pyrene, fluoranthene, phenanthrene, anthracene, benzpyrene, and others. In a combustion environment many PCAH will be present in varying concentrations, so that detection of an individual species requires deconvoluting complex spectra from the multicomponent mixture. This requires a detailed knowledge of excitation and fluorescence spectra for individual species under flame conditions. A literature search indicated that in most cases the available vapor phase spectra are insufficiently detailed for this purpose, even at near-room-temperature conditions.

Thus in the initial stages of the study, excitation and fluorescence spectra were measured for individual species in a cell (heated to approximately 100 C to provide sufficient vapor pressure) to determine their (near) room temperature spectra. Individual PCAH were then injected into a flame to determine the effects of flame temperatures on the spectra and to determine sensitivities. These spectra will then be used as a data base to attempt to deconvolute the complex spectra observed upon excitation of the flame itself.

0-8412-0570-1/80/47-134-159$05.00/0
© 1980 American Chemical Society

Experimental Apparatus

The equipment installed for this project consists of a
tunable dye laser and the fluorescence detection system. Both
the laser and the data acquisition are under computer control.
The apparatus schematic is shown in Figure 1. The laser is a
Phase-R Corporation DL-1400 flashlamp-pumped tunable dye laser.
Because of the high gain afforded by the coaxial flashlamp,
this unit has the capability of lasing directly throughout the
350-760 nm wavelength range. A KDP doubling crystal has been
installed, which can produce tunable u.v. radiation in the
290-350 nm range by frequency-doubling. Additional crystals
are available, which can extend this range down to 200 nm.
With the present arrangement, however, we are able to cover
the fluorescence excitation regions of most of the PCAH species
of interest.
        The laser produces a pulse approximately 300 nsec long
with peak powers of 1 Megawatt. Pulse repetition rates up to
10 per second are possible. The output bandwidth is narrowed
to 0.2 nm with a double prism oriented at Brewster's angle to
produce a horizontally polarized beam. By tuning the prisms
(i.e., changing the reflection angle -- see Figure 1), the
output wavelength can be scanned over 30-50 nm for a given dye
solution. The tuning of the prisms, frequency doubler (if used),
and etalons (if used) are all carried out by stepping motors
controlled by the on-line computer. The output wavelength is
further stabilized by a feedback system in the fluid pump line
which maintains a constant temperature difference between the
cooling water and the dye solution, thus minimizing turbulence
in the dye.
        The laser-induced fluorescence is observed perpendicular
to the beam axis with a low-dispersion monochromator and photomultiplier (PMT). The laser pulse energy is monitored with a
photodiode (PD) calibrated against a Scientech Model 3600 power
meter. The 300 nsec pulses from the PMT and the PD are
averaged with dual gated integrators on a pulse-to-pulse basis.
The integrator outputs are then read into the PDP-8/L computer,
through a pair of 10-bit A/D converters. The computer then
performs pulse-to-pulse averaging and normalization and
statistical analysis of the data, as well as scanning the
optical components of the laser itself. Typically, averaging
10 pulses yields statistically acceptable data.

Results

        Figure 2 shows fluorescence spectra for pyrene and fluor-
anthene in atmospheric pressure cells pumped at essentially
the same wavelength. These spectra are typical of the types
of profiles obtained for PCAH. The spectra are broad band with
no significant fine structure. Comparison of the two spectra

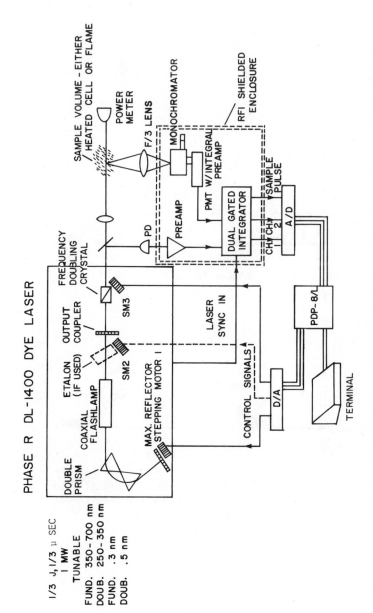

Figure 1. Schematic of laser-induced fluorescence instrumentation

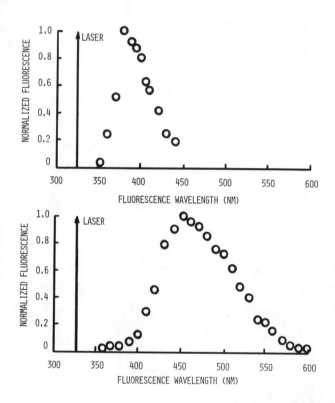

Figure 2. Fluorescence spectra—atmospheric pressure cell (air): upper figure, *pyrene at 78°C;* lower figure, *fluoranthene at 90°C; fluorescence-analyzing band-pass of 4 nm.*

shows that, at least in a low temperature (100 C) two component mixture of pyrene and fluoranthene, the individual concentrations could be ascertained by deconvoluting the fluorescence spectrum alone. Similar results hold for the excitation spectra.

Figure 3 shows the fluorescence spectra for pyrene and fluoranthene injected into the post-reaction zone of an ethylene-air flame. It was found that the fluorescence signal did not attenuate appreciably as the sample volume was moved downstream of the injector. This indicates that the species are not being consumeu in the flame. The temperature of the injected stream is not known; however, the injector flow rate was kept low to allow more efficient heating of the stream. This temperature will be monitored in future measurements. Comparison of these spectra with those in Fig. 2 indicates the effect of increased temperature. In both cases, the spectrum is broadened somewhat but the qualitative features are still distinguishable and can be attributed to the species injected. Thus, it appears that the effect of high temperatures on the spectra is not so gross that it precludes identification of individual species. Further, the spectra at flame temperatures can be measured by injection into flames, providing a means of calibrating the LIF method for both spectral signature and sensitivity for individual PCAH.

The detection limits of individual PCAH will be determined by injecting known amounts into a flame. An estimate of the limit for pyrene can be obtained from the atmospheric pressure cell measurements. Pyreene was detected in the cell at 50°C and 1 atm of air, where its vapor pressure is about 0.1 milli-torr, or about 0.1 ppm. While this limit can be improved by optimization of the optics and electronics, it is sufficient to detect pyrene in the concentrations expected from probe measure-ments.

Conclusion

PCAH have been observed in a flame using laser induced fluorescence spectroscopy by injecting individual species into the post-reaction zone. While the spectra are broadened by the elevated temperature, the spectra are qualitatively similar to low temperature (100 C) spectra and are indicative of the particular species injected. Thus, the injection procedure appears to be a feasible method of determining PCAH spectra at flame temperatures. These spectra will be used as a data base to determine individual PCAH concentrations in flames from their LIF spectra.

Acknowledgement

This work is supported by U.S. Public Health Service Grant No. 5P01-ESO1640.

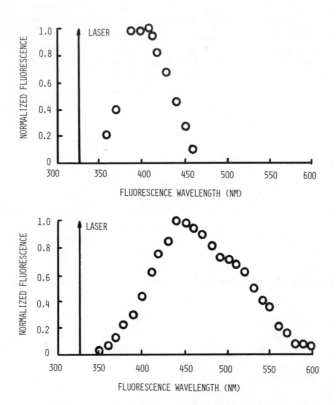

Figure 3. Fluorescence spectra—sample injected in an ethylene–air flame: upper figure, *pyrene 8 mm downstream of injection point;* lower figure, *fluoranthene 13 mm downstream of injection point; fuel equivalence ratio of 1.7; fluorescence-analyzing bandpass of 4 nm.*

References

1. Bittner, J.D. and Howard, J.B., "Role of Aromatics in Soot
    Formation" Alternative Hydrocarbon Fuels: Combustion and
    Chemical Kinetics (C.T. Bowan and J. Birkeland, eds.)
    Progress in Aeronautics and Astronautics, Vol. 62, American
    Institute of Aeronautics & Astronautics, New York, 1978
    p. 335-358.

2. Wagner, H.G., "Soot Formation in Combustion" Seventeenth
    Symposium (International) on Combustion 1979, p. 3.

3. Crittenden, B.D., and Long, R., "Formation of Polycyclic
    Aromatics in Rich, Premixed Acetylene and Ethylene
    Flames" Comb. and Flame, 20, 359-368, 1973.

4. Numerous examples may be found in the present volume.

RECEIVED February 5, 1980.

Flow Visualization in Supersonic Flows

N. L. RAPAGNANI and STEVEN J. DAVIS

Chemical Laser Branch, Air Force Weapons Laboratory, Kirtland Air Force Base, Albuquerque, NM 87117

Since the invention of flowing chemical lasers, a persistent problem has been the difficulty of understanding the mixing phenomena so that accurate modeling of these devices could be accomplished.[1] The system efficiencies are controlled in great part by the mixing and, if one hopes to build a bigger and better device, a degree of understanding of the mixing is certainly needed.

In recent years there has been a considerable amount of effort devoted to developing new nonintrusive techniques and also applying well-established techniques to interrogate these flow fields. Some of the methods which have been applied are chemiluminescence and Schlieren photography, Coherent Anti-Stokes Raman Scattering, and Laser Doppler Velocimetry.[2] All of the above techniques have been used in an attempt to understand the mixing process and construct a map of the flow field in the laser. While useful, these techniques have some inherent problems and difficulties. What was needed was a fast and efficient method for obtaining mixing efficiency on new nozzle concepts.

The recent introduction of laser induced fluorescence as a diagnostic has opened up a new field as a nonintrusive flow field diagnostic.[3] In this paper we describe how seeding of the flow field of these devices with I_2, which was made to fluoresce by an argon ion (Ar+) laser, gave us information on the mixing process.

Theory:

There is a fortuitous match between strong I_2 absorption and the 5145Å line of the Ar+ laser.[4] This absorption is from the v" = 0 level of the $X^1\Sigma$ state to the v' = 43 level of the $B^3\Pi o$ state. The absorption can be made specific to one or two rotational levels of the v" = 0 level by insertion of an intracavity etalon and forcing the Ar+ laser to oscillate on a single longitudinal mode. The absorption process then becomes very efficient and the resulting fluorescence becomes much brighter. The v' = 43 level of $B^3\Pi o$ state fluoresces to a multitude of v"

This chapter not subject to U.S. copyright.
Published 1980 American Chemical Society

levels giving a characteristic yellow emission. The resulting
fluorescence is extremely intense when produced this way. In a
typical run, the Ar+ laser was properly tuned by maximizing the
fluorescence in a sealed glass cell containing I_2 vapor. The
laser was stable enough that further tuning was not required.
        There are several advantages in using I_2 as the seed for
LIF studies. First, the vapor pressure of I_2 is sufficiently
high that I_2 vapor (not crystals) can be easily entrained in a
He gas flow and carried into the flow field as a true molecular
vapor. Thus, the I_2 can be injected through extremely small
orifices. Secondly, the I_2 fluorescence utilized in the present
work extends to much longer wavelengths than the excitation
source. Consequently, the LIF is easily isolated from any
scattered laser light by insertion of a long pass filter over
the viewing port. Thirdly, the radiative lifetime of the relevant
excited state in I_2 is long enough to allow excited molecules to
travel a significant distance (~1 cm) in the flow direction if
the flow is supersonic and pressure low enough. Thus, the
visible fluorescence will persist downstream from the excitation
source and one can track the flow field for a single excitation
point. The visible fluorescence was used to monitor the flow
field and both black and white and color photographs of the flow
field were obtained.
Experiment:
        A 2-liter stainless steel vessel, partially filled with I_2
crystals, was connected directly to the secondary He feed supply
line of a supersonic chemical laser. A complete description is
given in a separate paper.[2] The laser cavity had viewing windows
on top and bottom so that the nozzle array could be viewed in a
direction perpendicular to the optical axis. The pump laser was
a Spectra-Physics Model 170-03 Ar+ laser equipped with an intracavity etalon.
        The Ar+ beam was directed into the chemical laser cavity
along the optical axis by a focusing optical train. The spot
size in the cavity was a fraction of a millimeter, although
tighter focusing could have been done, thus increasing the
spatial resolution. The Ar+ beam could be translated in two
dimensions, up and down the nozzle face, at a single position in
the flow direction, and also downstream from the nozzle face.
Hence, the flow field could be visually mapped out.
        The collision free lifetime of the V'=43 level of the I_2 B
state is on the order of a few microseconds. The I_2 pressure in
the laser cavity was estimated to be only ~1m torr.[5] The He
pressure was less than 5 torr for all runs. At these conditions
neither self-quenching nor electronic quenching by He will
significantly alter the lifetime.[6] A lifetime of a few microseconds is nearly ideal to use as a tracer in our supersonic
flow fields in which velocities of 10^5 cm/sec are typical. This
means that the fluorescing I_2 would "light up" a region down-

stream of the excitation point of ~1 cm. It is important to
note that the region of excitation is defined by the spot size
of the laser which is variable as described above. Collisions
with He will, however, cause the originally excited V'=43 level
to be relaxed to lower V'. This causes the visible fluorescence
to become red shifted.

He gas with the I_2 vapor was injected into the flow field
through the secondary nozzles. A complete description of the
geometry is given in Figure 1. The Ar+ laser beam was positioned
in the flow field and photographs were taken. The majority of
photographs were taken with the laser operating in the cold flow
condition in which no F_2 was added in the combustor. Hot flow
conditions were imitated using N_2 and He gases. Hot flow experiments where I_2 reacts with F to produce IF with subsequent LIF
on IF will be completed in the coming year.

Results and Discussions:

Typical results of our effort are summarized in Figure 2.
The right portion of this photograph shows the visible fluorescence from I_2 (B-X) as viewed perpendicular to the optical axis
through the top of the cavity. The left portion (2b) of the
picture emphasizes the extremely narrow region (slit) that is
fluorescing in the optical axis direction. Figure 2b was
obtained by means of looking through a dichrocic mirror down the
optical axis of the chemical laser, collinear to the Ar+ beam.
The fluorescence downstream from the Ar+ beam is observed to be
well defined and remains approximately the width of the Ar+
beam. In all these photographs, the Ar+ laser excited a region
at the nozzle exit plane, one-half way up in the vertical direction. In both figures, the Ar+ pump beam can be seen intersecting the flow field with the fluorescence reflecting off the
nozzle face.

The first remark one can make about Photograph 2 is that of
flow nonuniformity. This particular nozzle bank has a severe
nonuniformity problem and actually some of the secondary nozzles
are not flowing. By redirecting the beam higher in the cavity,
the clogged nozzles were observed to flow much better. This
demonstrated the high degree of spatial resolution since the
nozzle was shown to have severe nonuniformities in flow over
small dimensions, <1.5 mm, in the vertical direction. By
directing the Ar+ beam up and down in the vertical direction and
in the downstream direction, the entire flow field was mapped
out. The nonuniformities of the flow field caused by the nozzle
seen at the nozzle plane in Figure 2 continued far downstream
~5 cm and the gradients in the vertical direction seen at the
nozzle plane also persisted.

Chemiluminescence photographs of the flow field taken
perpendicular to the optic axis indicated no such nonuniformities, because any such photos only give spatially integrated
particle density data. Consequently, vertical density gradients

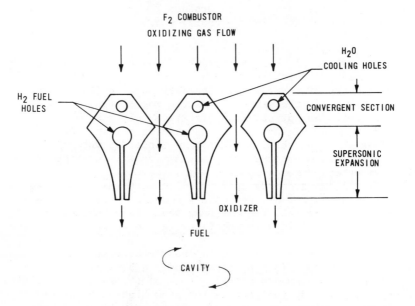

Figure 1.   Detailed description of the CL-II nozzle: H_2 slit width = .002 in.; F_2 throat width = .008 in.; F_2 exit width = .151 in.; F_2–F_2 nozzle width = .191 in.

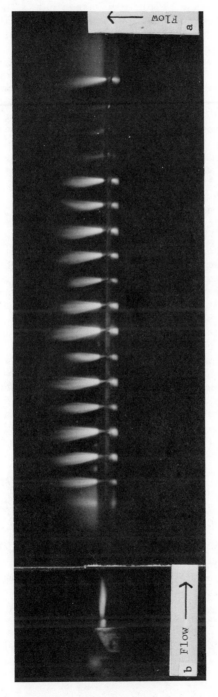

Figure 2.   Visible fluorescence from $I_2(B - X)$: (a) viewed perpendicular to the optical axis; (b) viewed along the optical axis.

caused by the nozzle are completely washed out. The amount of iodine actually injected into the flow field is only a small fraction of the He carrier gas. Thus, potential mixing problems caused by the massive I_2 injection are minimized. Visual examination of the fluorescence indicated no changes in the flow structure as the I_2 flow was reduced to zero; the only change was that the entire fluorescence intensity became weaker. As mentioned earlier, a pressure increased will cause the fluorescence to become red shifted. The characteristic bright yellow I_2 fluorescence becomes orange to red as the pressure increased. In our color photos this was easily discernable and was interpreted as being due to density gradients in the cavity caused by intersecting shock waves. These intersecting shocks were clearly visible on the color photos.

Conclusions:

These results demonstrate that LIFS can be applied successfully to study mixing zones in chemical laser cavities. Nozzle performance can be quickly and accurately evaluated. It is important to note that LIFS is not specific to iodine. Other species which can be made to fluoresce, such as sodium vapor, could be injected instead of I_2. Then, by use of a dye laser, visual fluorescence could be seen. By using several different species of different mass, one could study mixing processes with high accuracy and resolution. Different species could be injected into adjacent nozzles and the mixing phenomena could be studied in color. One of the least understood mixing methods used in CW chemical lasers is that of trip jets. These are jets of secondary gas injected into the primary and secondary nozzles near the nozzle exit plane and are used to enhance the mixing process. A fluorescent species seeded into the trip jets could be used to better understand to what extent this process enhances mixing. The spatial resolution of this technique is limited only by the spot size of the Ar+ laser which can be focused to diameters of a few microns.

References:
1. S.W. Zelazny, et al, "Modeling HF/DF Lasers: An Examination of Key Assumptions," AIAA J. 16 Pg 297, 1978.
2. C.W. Peterson, "A Survey of the Utilitarian Aspects of Advanced Flowfield Diagnostic Techniques," AIAA J. 18, Pg 1352, 1979.
3. K. L. Kompa and S.D. Smith, "Laser-Induced Processes in Molecules," Springer-Verlog, New York, 1979.
4. J. B. Koffend and R.W. Field, "CW Optically Pumped Molecular Iodine Laser," JAP, Vol 48 No. 11, 1977.
5. N. L. Rapagnani and S. J. Davis, "Laser Induced I_2 Fluorescence Measurements in a Chemical Laser Flow Field,"AIAA J. 18, Pg 1402, 1979.
6. G. A. Capelle and H. P. Broida, Lifetimes and Quenching Cross Sections for $I_2(B^3\Pi0_u+)$ J Chem Phys 58 4212, 1973.

RECEIVED March 31, 1980.

LASER-INDUCED FLUORESCENCE:
ATOMS

What Really Does Happen to Electronically Excited Atoms in Flames?

KERMIT C. SMYTH, PETER K. SCHENCK, and W. GARY MALLARD

National Bureau of Standards, Washington, D.C. 20234

In recent years numerous experiments have been reported on the fluorescence and energy transfer processes of electronically excited atoms. However, for flame studies the rates of many possible collision processes are not well known, and so the fate of these excited atoms is unclear. An interesting example concerns the ionization of alkali metals in flames. When the measured ionization rates are interpreted using simple kinetic theory, the derived ionization cross sections are orders of magnitude larger than gas kinetic (1,2,3). More detailed analyses (4,5) have yielded much lower ionization cross sections by invoking participation of highly excited electronic states. Evaluation of these models has been hampered by the lack of data on the ionization rate as a function of initial state for the alkali metals.

Opto-galvanic spectroscopy detects the absorption spectra of atoms (6) and some molecules (7) in a flame by measuring current changes induced by optical irradiation at a wavelength corresponding to an electronic transition. Two steps are involved:

$$A + h\nu \rightarrow A^*$$ photon absorption
$$A^* + M \rightarrow A^+ + e^- + M$$ collisional ionization.

Thus, the overall ionization starting from a given excited state is monitored. Since the efficiency of collisional ionization is highest when the energy required for ionization is lowest, this method is particularly sensitive for detecting high-lying states. We have found that two-photon transitions are readily observable for many atomic species. If one selects a particular atom and then monitors the laser-induced current changes for a series of electronic states, the observed signal magnitudes are sensitive to the competing ionization and quenching processes. In this paper we compare our experimental results on Na with model calculations which incorporate state-specific ionization and quenching rates.

Experimental Results Two-photon transitions in Na from the 3s ground state to high-lying s and d states have been observed

This chapter not subject to U.S. copyright.
Published 1980 American Chemical Society

using stepwise excitation (two lasers operating at different wavelengths). Thus, we have (1) 3s → 3p at λ_1 and (2) 3p → nd at λ_2 where n, the principal quantum number, is between 4 and 12. The transition probabilities for each step are well known (8).

Two dye lasers were excited simultaneously by a pulsed nitrogen laser (pulsewidth = 7 ns), and the two unfocussed beams irradiated the flame in a counter-propagating, collinear geometry. With 25% of the nitrogen laser pumping the dye laser operating at λ_1, typical laser energies were 35-70 μJ, and the first transition 3s → 3p was optically saturated. However, for the 3p → nd transitions care was taken to avoid optical saturation; typical laser energies at λ_2 were 1-300 μJ.

The flame was fuel rich H_2/air, burning on a 5-cm long slot burner; the estimated temperature was 2000 K (7). Aqueous Na solutions were aspirated into this flame, with concentrations of 1 ppm for measuring the λ_1 signal (3s → 3p) and 10 ppb for the signal using both dye lasers. The latter solution corresponds to a Na number density of ~10^9 cm^{-3} in the flame.

Figure 1 shows a portion of the data obtained for the stepwise excitation of Na. Figure 2 plots the observed signal enhancement (defined as signal with λ_1 + λ_2 divided by the signal with λ_1 only) versus the absorption coefficient for the stronger ($3p_{3/2}$ → nd) of the two components.

Modelling the Data   The present model extends an earlier version (9). Here we seek to evaluate the expected signal from a series of d states in Na by using a rate equation model which includes absorption, stimulated emission, collisional ionization, and quenching. The essential features are the following: (a) Only the so-called "n-manifold" states ($\ell \geq 2$) are considered, since the Na d states mix very rapidly with states of higher ℓ (k = 10^{10}-10^{11} s^{-1} (10)). This mixing is assumed to be complete before ionization and quenching occur. The s and p states are thus ignored. (b) Quenching is assumed to proceed via many small steps of $\Delta n = -1$; i.e. nd → (n-1)d → ... 3d → 3p (11, 12). (c) A state-specific ionization rate constant is calculated (see below) using the cross section for Na 3s ionization derived by Hollander (4). (d) N_2 is assumed to be the collider in the ionization and quenching processes (4, 10, 11). Although charged species may have large quenching cross sections for excited Na, their concentrations are orders of magnitude lower than that of N_2 and so their contribution is neglected. (e) The radial intensity distributions of the two lasers are described by a Gaussian function. (f) A Voigt line analysis is required for calculating the optical transition rates since the observed linewidths for the Na d states are several times the nominal 0.01 nm bandwidth of the laser. These large linewidths arise from the large ℓ-changing (10) and elastic (13) collision rates.

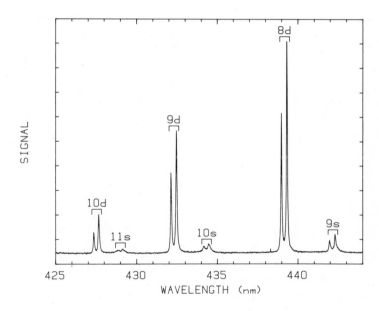

Figure 1. Optogalvanic signal for stepwise excitation of sodium (3s → 3p → nd, ns) in an H₂–air flame. Each transition is split into two components by the fast mixing of the fine structure states, $3p_{1/2} \longleftrightarrow 3p_{3/2}$. The data are not normalized for the variation of laser power with wavelength. At this level of sensitivity the one-photon signal (3s → 3p) is undetectable.

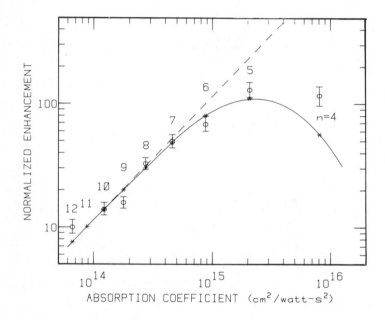

Figure 2. Comparison of the stepwise excitation results (○) with the model calculation (*). The enhancement (the two-photon signal divided by the one-photon signal) normalized for laser energy is plotted against the absorption coefficient for the 3p → nd transitions. For visual clarity a curve is drawn through the points of the model calculation and a dashed line of unit slope is drawn through the data at high principal quantum number, n.

The optical transition rates have been calculated using the experimentally measured laser energies. The next task is to derive overall global quenching and ionization rates for each Na n-manifold state. This is accomplished as follows. Humphrey et al. (11) have measured total loss cross sections from several n-manifold states, and these show a steady decrease with n. By detailed balancing one can estimate the fraction of energy transfer collisions which leave the Na atom in states of higher and lower n (this calculation neglects chemical reactions). Once these branching ratios are evaluated, an overall quenching rate can be calculated by summing all the downward rates for nd → (n-1)d → ... 3d → 3p. The rates at each step are given by $k = \sigma N \bar{v}$, where N is the number density and $\bar{v}$ is the average velocity. Table I lists the results of these calculations, the loss cross sections of Humphrey et al. (11), as well as the estimated loss cross sections for higher and lower n.

Similarly, the overall ionization rates can be evaluated. At 2000 K and 1 atm, Hollander's state-specific rate constant becomes $k_{ion} = 1.46 \times 10^{10} \exp(-\Delta E/kT)$ s^{-1}, where ΔE is the energy required for ionization. For each n-manifold state the fraction ionized by collisions is determined, as well as the fraction transferred to nearby n-manifold states in steps of $\Delta n = \pm 1$. Then the fractions ionized from these nearby n-manifold states are calculated. In this way a total overall ionization rate is evaluated for each photo-excited d state. The total ionization rate always exceeds the state-specific rate, since some of the Na atoms transferred by collisions to the nearby n-manifold states are subsequently ionized. Table I summarizes the values used for the state-specific cross sections and the derived overall ionization and quenching rate constants for each n-manifold state. The required optical transition, ionization, and quenching rates can now be incorporated in the rate equation model. Figure 2 compares the results of the model calculation with the experimental values.

Discussion   (a)   For high principal quantum numbers (n ≥ 7) ionization is much faster than quenching (see Table I), and so the details of the possible quenching processes are unimportant. Essentially all of the atoms excited to a given nd state are ionized, and the observed signal is simply proportional to the absorption probability of the second step (λ_2).
(b)   For n = 5 and 6 there is keen competition between ionization and quenching processes. It is here that the rate equation model is most sensitive to the actual state-specific rate constants employed. Using Hollander's values to derive ionization rates (4) and the results of Humphrey et al. for total loss rates (11), the agreement between theory and experiment is good. If ionization rates 100x larger are employed, the direct proportionality between the observed signal and the absorption probability is maintained down to n = 4. Clearly, this does not agree with the experimental

Table I. State-Specific Cross Sections and Global Rate Constants for $\ell \geq 2$ States.

Principal Quantum Number n	State-Specific Cross Sections (nm^2)		Global Rate Constants (s^{-1})	
	$\sigma(\text{ionization})$[1]	$\sigma(\text{loss})$[2]	k(ionization)	k(quenching to 3p)
3	3.2(-4)	0.54	2.1(6)	3.6(9)
4	0.015	0.50	1.2(8)	1.7(9)
5	0.091	0.43	9.0(8)	1.0(9)
6	0.24	0.36	2.5(9)	6.3(8)
7	0.43	0.32	4.2(9)	4.2(8)
8	0.63	0.18	5.2(9)	2.5(8)
9	0.81	0.15	6.3(9)	1.7(8)
10	0.98	0.12	7.3(9)	1.2(8)
11	1.13	0.10	8.2(9)	8.5(7)
12	1.25	0.09	9.0(9)	6.5(7)

The notation 2.1(6) represents 2.1×10^6. For 1 atm of N_2 at 2000 K a cross section of $0.148 \ nm^2$ corresponds to k (= $\sigma \bar{N} \bar{v}$) of $1 \times 10^9 s^{-1}$.

1. Ref. 4, $\sigma = 2.17 \ nm^2 \ \exp(-\Delta E/kT)$.

2. Values for n = 5-8 are from Ref. 11, for n = 4 and $\geq$ 9 are estimates, and the ratio of n = 3 to n = 4 is taken from Ref. 12.

results. However, further work is needed to ascertain just how
sensitive the model is to the state-specific ionization rates.
(c) For n = 4 the model predicts a smaller signal than actually
observed. Some of the model's assumptions are least tenable for
low values of n. At n = 4 the ℓ-changing collision rate constant
is ~$10^9 s^{-1}$ ($\underline{10}$) and thus is no longer much faster than quenching.
Also, the quenching cross sections have been assumed to be
independent of temperature, which may well be incorrect ($\underline{14}$).
Differences in the quenching cross sections would affect the
model estimates most strongly for low n values. Finally, although
quenching of the 4d state predominantly gives 3d ($\underline{12}$), it may be
necessary to consider energy transfer to other nearby states.

<u>Conclusions</u> (1) Hollander's "low" cross section for collisional
ionization of Na is sufficient to model the opto-galvanic signal
magnitudes as a function of excitation energy. Abnormally high
cross sections are not required.
(2) Essentially all (> 90%) of the Na atoms excited to n $\geq$ 7 are
ionized at a flame temperature of 2000 K. For n = 7 the energy
needed for ionization is 2249 cm^{-1}, which is approximately 2 kT
(kT = 1390 cm^{-1}).

Literature Cited

1. Tj. Hollander, P.J. Kalff, and C.T.J. Alkemade, J. Chem.
    Phys. <u>39</u>, 2558 (1963).
2. D.E. Jensen and P.J. Padley, Trans. Faraday Soc. <u>62</u>, 2140
    (1966); R. Kelly and P.J. Padley, Trans. Faraday Soc. <u>65</u>,
    355 (1969) and Proc. Roy. Soc. London A <u>327</u>, 345 (1972).
3. A.N. Hayhurst and N.R. Telford, J. Chem. Soc. Faraday I <u>68</u>,
    237 (1972); A.F. Ashton and A.N. Hayhurst, Comb. & Flame <u>21</u>,
    69 (1973).
4. Tj. Hollander, AIAA Journal <u>6</u>, 385 (1968).
5. G.N. Fowler and T.W. Preist, J. Chem. Phys. <u>56</u>, 1601 (1972).
6. G.C. Turk, J.C. Travis, J.R. DeVoe, and T.C. O'Haver, Anal.
    Chem. <u>51</u>, 1890 (1979).
7. P.K. Schenck, W.G. Mallard, J.C. Travis, and K.C. Smyth,
    J. Chem. Phys. <u>69</u>, 5147 (1978).
8. W.L. Wiese, M.W. Smith, and B.M. Miles, NSRDS-NBS <u>22</u> (1969).
9. J.C. Travis, P.K. Schenck, G.C. Turk, and W.G. Mallard,
    Anal. Chem. <u>51</u>, 1516 (1979).
10. T.F. Gallagher, R. E. Olson, W.E. Cooke, S.A. Edelstein,
    and R.M. Hill, Phys. Rev. A <u>16</u>, 441 (1977).
11. L.M. Humphrey, T.F. Gallagher, W.E. Cooke, and S.A.
    Edelstein, Phys. Rev. A <u>18</u>, 1383 (1978).
12. J.E. Allen, Jr., W.R. Anderson, D.R. Crosley and T.D.
    Fansler, 17th Comb. Symp. (International), 797 (1979).
13. A. Flusberg, R. Kachru, T. Mossberg, and S.R. Hartmann,
    Phys. Rev. A <u>19</u>, 1607 (1979).
14. N.S. Ham and P. Hannaford, J.Phys. B <u>12</u>, L199 (1979).

RECEIVED April 11, 1980.

Collisional Ionization of Sodium Atoms Excited by One- and Two-Photon Absorption in a Hydrogen-Oxygen-Argon Flame

C. A. VAN DIJK[1] and C. TH. J. ALKEMADE

Fysisch Laboratorium, Rijksuniversiteit Utrecht, The Netherlands

In the past, much work has been done on electrical properties of flames ([1],[2],[3],[4]). As in numerous other fields, the
laser has revolutionized the scene. It has been demonstrated
that the laser is capable of saturating optical transitions of
atoms present in flames ([5],[6]). When we think of an atom in
terms of a two-level system with equal degeneracies and an ionization continuum, saturation of the intermediate level means
that for half the atoms the energy necessary to ionize is dimin-
ished by the excitation energy provided by the laser. In other
words, the saturated level has been promoted to a quasi ground
level. For sodium atoms in a flame of 1800 K irradiated with
laser light tuned to e.g. the sodium D-lines, an increase is expected in the collisional ionization rate of $\exp(E_{3S}-E_{3P})/kT \approx 10^6$.
In the experiments which we devised to support the basic idea
described above, we used two thin iridium probes, which were sus-
pended into the flame and biased to 300 V. The negative probe
was fixed in position in the immediate neighborhood of the laser
excited volume in order to collect the ions; the positive probe
was movable and usually at a distance of 10 mm from the negative
probe. Laser beam and probes were well outside the flame front.
A resistor completed the circuit and signals were measured
across the resistor. The laser was a flashlamp pumped tunable
dye laser with a pulse duration of ≈ 1 μs and a peak power of
several kW; the bandwidth was 0.014 nm in the neighborhood of
589 nm. We used a stoichiometric H_2-O_2-Ar flame of 1800 K,
shielded with a mantle flame of identical composition. In the
inner flame a 2500 μg/ml NaCl solution was nebulized. An extensive description of the experiment can be found elsewhere ([7]).
Simultaneously with the ionization signals we monitored
fluorescence signals. Gated integrators processed both these
signals. We observed ionization signals with the laser tuned to

[1] Current address: Chemistry Department, Michigan State University, East Lansing,
MI 48824.

0-8412-0570-1/80/47-134-183$05.00/0
© 1980 American Chemical Society

the $3S_{1/2}-3P_{1/2}$, $3S_{1/2}-3P_{3/2}$, $3P_{1/2}-5S_{1/2}$, $3P_{3/2}-5S_{1/2}$ one-photon transitions and the $3S_{1/2}-3D_{3/2,5/2}$, $3S_{1/2}-5S_{1/2}$, $3S_{1/2}-4D_{3/2,5/2}$ two-photon transitions of sodium. With the first two transitions, the laser beam had a diameter of 3 mm and a power density of the order of 10^4 W/cm^2; for the remaining five transitions, the beam was focused to a diameter of 100 μm and the power density was of the order of 10^4 kW/cm^2. The ionization pulses as observed with an oscilloscope closely resembled the laser pulse, which puts a lower limit of several μs on the time-constants involved. The time-constants of the circuitry were negligible.

With the irradiated probe negative, the ionization signal decreased by 10% when the distance between the probes was increased from 2 to 10 mm. However, a decrease in ionization signal by a factor of 5 occurred in the same experiment with the polarity of the probes reversed (see Fig. 1). This observation shows that the current is limited by the sodium ions. Calculations, using literature values (8,9) for the mobility of sodium ions in flames, support this observation.

Scanning the laser across a sodium transition produces excitation profiles as shown in Fig. 2. Irrespective of whether a one- or two-photon transition was involved, we found the width of the resulting ionization profile to exceed that of the corresponding fluorescence excitation profile. We account for this fact by assuming that the process

$$Na^* + Z \underset{k_r}{\overset{k_i}{\rightleftarrows}} Na^+ + e + Z \tag{1}$$

is equilibrated and that the concentrations of the ions and electrons are equal. Here Na* denotes the excited atom, Z a flame gas atom or molecule, k_i the collisional ionization rate constant and k_r the recombination rate constant. As we did not observe significant saturation of the ionization signal as a function of probe voltage, the ionized region is electrically neutral to a good approximation. A simple calculation, using $[Na^+] = [e]$, shows that under these conditions the number density $[Na^+]$ of the ions is proportional to the square root of the excited state density [Na*] which is in semi-quantitative agreement with the observed behavior of the profiles. Saturation broadening is seen to occur (7), because the fluorescence excitation profile in Fig. 2 is much broader than either the laser profile (0.014 nm) or the fluorescence emission profile (0.0070 nm).

We obtained a measure of the degree of ionization as a function of the total sodium density in the flame by plotting the ratio of the ionization signal to sodium solution concentration versus the latter concentration on double logarithmic scales as shown in Fig. 3. For concentrations in excess of

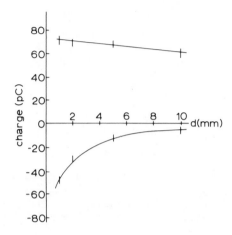

Combustion and Flame

Figure 1. Ionization charge as a function of the distance between the probes. The probe in contact with the laser beam ("beamprobe") is kept in a fixed position. Upper curve, beamprobe negative; lower curve, polarity of probes reversed. Laser tuned to $3S_{1/2} - 3P_{3/2}$ transition. Power density of the laser pulse is 7×10^4 W/cm²; diameter of laser beam is 3 mm.

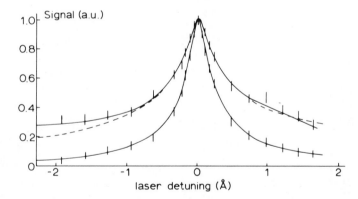

Combustion and Flame

Figure 2. Simultaneously recorded ionization profile (upper curve) and fluorescence excitation profile (lower curve) of the $3S_{1/2} - 3P_{3/2}$ transition. Fluorescence monitor: the $4P - 3S$ transition; (– – –), the square root of the fluorescence curve. Power density of the laser pulse is 5×10^4 W/cm²; diameter of laser beam is 3 mm.

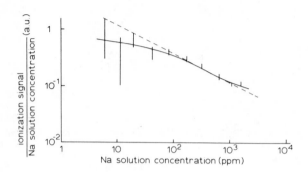

Combustion and Flame

Figure 3. Ratio of ionization signal to sodium solution concentration plotted against the sodium solution concentration on double-logarithmic scales: (– – –), the theoretically expected slope of −0.5 for large sodium solution concentrations, which indicates that the degree of ionization is small. The laser was tuned to the $3S_{1/2}$ − $3P_{3/2}$ transition. Power density of the laser pulse is 4×10^4 W/cm²; diameter of laser beam is 3 mm.

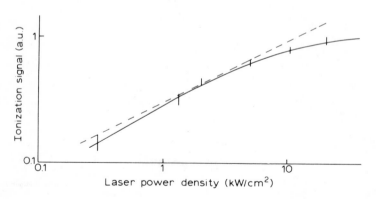

Combustion and Flame

Figure 4. Ionization signal as a function of the laser power density on double-logarithmic scales. The laser was tuned to the $3S_{1/2}$ − $3P_{3/2}$ transition; (– – –), a slope of 0.5; diameter of the laser beam is 3 mm.

200 µg/ml a limiting slope of -0.5 was observed for both the $3S_{1/2}-3P_{3/2}$ one-photon transition and the 3S-5S two-photon transition. This result can also be explained on the basis of eqn. (1) using $\beta = [Na^+]/[Na]_{total}$ as the definition of the degree of ionization, where $[Na]_{total} = [Na^0] + [Na^*] + [Na^+]$ and $[Na^0]$ is the concentration of the ground state. Because of the saturation $[Na^0] \simeq [Na^*]$, within a factor of the order of unity due to the degeneracies of the levels involved. Consequently, one obtains $\beta \sim 1/\sqrt{[Na]}_{total}$ for large sodium concentrations. The latter relationship supports the limiting behavior observed in Fig. 3.

A log-log plot of ionization signal versus laser power, as shown in Fig. 4, shows an initial slope of 0.5 and tends to level off at higher densities. This again can be explained by the equilibrated process (1), which leads to the square root relationship between ion-density and excited state density for low laser power where no saturation occurs. The initial slope of 0.5 also shows that multiphoton ionization is negligible since the latter process would give rise to a stronger dependency of ionization signal on the laser power.

Small quantities of excess oxygen had the effect of increasing the degree of ionization of sodium, whereas no such effect was observed when adding comparable quantities of N_2 to the flame. We ascribe this difference to the different electron affinities, which are 0.44 eV for O_2 and -1.5 eV for N_2.

Laser-enhanced ionization has also been observed by other workers and is sometimes called the 'opto-galvanic effect' (10).

In conclusion, we have presented three independent experimental results which can be explained on the basis of equilibrated collisional ionization using a laser-saturated level and electrical neutrality of the probed volume.

Possible applications of laser enhanced ionization in flame diagnostics are: 1. simultaneous observation of ionization and fluorescence signals from various levels might provide more information on the sequence of processes leading to and from the ionization continuum; 2. the measurement of ion mobilities, relating to cross-sections for elastic collisions between ions and flame particles; 3. measurement of ionization rate constants relating to cross-sections for inelastic collisions between excited atoms and other flame particles; 4. measurement of recombination rate constants, relating to cross-sections for inelastic collisions between ions, electrons and neutrals.

ACKNOWLEDGMENT

Some of the information contained in this chapter was adapted with permission from Combustion and Flame.

LITERATURE CITED

1. Wilson, H.A., Rev. Mod. Phys. 1931, 3, 156.
2. Calcote, H.F., King, I.R., 'Fifth Int. Symp. on Combustion', Reinhold Publ. Corp. 1955, p. 423.
3. Hollander, Tj., Kalff, P.J., Alkemade, C.Th.J., J. Chem. Phys. 1963, 39, 2558.
4. Kelly, R., Padley, P.J., Proc. Roy. Soc. Lond. 1972, A327, 345.
5. Omenetto, N., Benetti, P., Hart, L.P., Winefordner, J.D., Alkemade, C.Th.J., Spectrochimica Acta. 1973, 28B, 289.
6. Van Calcar, R.A., Van de Ven, M.J.M., Van Uitert, B.K., Biewenga, K.J., Hollander, Tj., Alkemade, C.Th.J., J.Q.S.R.T. 1979, 21, 11.
7. Van Dijk, C.A., Ph.D. Thesis, Ch. V, Rijksuniversiteit Utrecht, 1978; available on request.
8. Snelleman, W., Ph.D. Thesis, Rijksuniversiteit Utrecht, 1965.
9. Ashton, A.F., Hayhurst, A.N., Trans. Far. Soc. 1970, 66, 833.
10. Green, R.B., Keller, R.A., Schenck, P.K., Travis, J.C., Luther, C.G., J. Am. Chem. Soc. 1976, 98, 8517.
11. Van Dijk, C. A., Alkemade, C. Th. J., Combustion and Flame, submitted for publication.

RECEIVED February 1, 1980.

On Saturated Fluorescence of Alkali Metals in Flames

C. H. MULLER, III[1], MARTIN STEINBERG, and KEITH SCHOFIELD

Quantum Institute, University of California, Santa Barbara, CA 93106

The concept of saturated laser fluorescence appears attractive in that the fluorescence intensity is directly related to the particular species' concentration and becomes roughly independent of the laser intensity at saturation. Such a mode has been invoked already to monitor absolutely flame concentrations of Na ($\underline{1}$-$\underline{4}$), OH ($\underline{5}$), C_2 ($\underline{6},\underline{7}$), CH ($\underline{7},\underline{8}$), CN ($\underline{8}$), and MgO ($\underline{4}$). However, during a recent study of the behavior of Na and Li in flames ($\underline{9}$-$\underline{11}$), we have observed evidence for laser induced chemical reactions under saturated conditions which has significant implications for the quantitative exactness of such measurements.

Observations. Using a Chromatix CMX-4 flashlamp pumped laser, saturated excitation of the $Na(3^2P_{3/2})$ level at various points in a series of fuel rich $H_2/O_2/N_2$ flames produces fluorescence intensities which markedly change with downstream location, decreasing significantly in the one dimensional flow above a flat flame burner. Additional measurements indicate that the extent of this decrease closely correlates to the H-atom concentrations at these flame locations. This behavior is very different from non-saturating conditions for which the sodium atom concentration decreases only slightly downstream due to slight temperature and diffusion effects. Obviously additional loss processes are operative during the 1-2 μs laser pulse duration for saturated conditions and the assumption that sodium is present solely as atoms in such fuel rich conditions clearly becomes invalid.

Experiments with lithium also show a change in behavior. This is not as directly obvious due to the flame distribution of lithium between Li and LiOH which occurs via the controlling reaction $Li + H_2O = LiOH + H$ and which relates Li to H-atom concentrations under normal non-radiated conditions. However, it is apparent that a similar behavior to sodium is exhibited but is disguised to a large extent by the normal lithium flame chemistry.

These additional loss processes have been identified and characterized in this work. Their neglect will necessarily lead to

[1] Current address: General Atomic Company, P.O. Box 81608, San Diego, CA 92138.

0-8412-0570-1/80/47-134-189$05.00/0
© 1980 American Chemical Society

absolute concentration measurements that are too low by extents that depend on the specific flame and location.

Interpretation. The data for hydrogen rich flames is most satisfactorily explained by invoking laser induced reactions between the excited states of the alkali metal and H_2O and H_2 which constitute the major flame species,

$$Na(^2P_{1/2,3/2}) + H_2O \underset{k_{42}}{\overset{k_{24}}{\rightleftharpoons}} NaOH + H \quad \Delta H^O_{2000\ K} = -27.5\ kJ\ mol^{-1}$$

$$Na(^2P_{1/2,3/2}) + H_2 \underset{k_{52}}{\overset{k_{25}}{\rightleftharpoons}} NaH + H \quad = +40.6\ kJ\ mol^{-1}.$$

For sodium, such reactions become energetically favorable for the excited states. Moreover, they appear to be sufficiently fast kinetically to reach a steady state during the initial part of the laser pulse. Their overall effect is to drain off elemental sodium or lithium from the dynamic excitation/physical quenching cycle into these chemical sinks. The actual extent depends on the relative magnitudes of the production and loss fluxes.

Participation of both the hydroxide and hydride is found to be the case for both alkalis. The analysis is unsatisfactory with the inclusion of solely the hydroxide or the hydride. An extension of the usual 2 or 3-level atomic model (2,12-14) to incorporate these chemical schemes is indicated in Figs. 1 and 2 and has been analyzed in detail. Assuming sufficient time for the attainment of a steady state distribution, and rapid coupling of the $^2P_{1/2}$ and $^2P_{3/2}$ states, the model predicts that for sodium under saturated conditions, the fluorescence intensity, I_f, will vary as

$$\frac{n_o}{I_f} = \frac{4}{A_{21}} + \frac{3}{A_{21}} \left\{ \frac{k_{24}[H_2O]}{(k_{41}+k_{42}+k_{43})} + \frac{k_{25}[H_2]}{(k_{51}+k_{52}+k_{53})} \right\} \frac{1}{[H]},$$

where n_o is the total metal concentration. The subscript numbers refer to processes connecting levels; 1, 2 and 3 representing the ground and electronically excited states of the metal, and 4 and 5, the hydroxide and hydride species. A very similar expression can be derived for lithium. This relationship predicts the correct dependence on [H] and shows straight line dependence for points throughout each flame. Also, it indicates that in spite of the participation of chemistry, a linear curve of growth will still be evident at a fixed point in the flame. Using relative values of the fluorescence intensities and absolute measures of [H] derived by the Na/Li method (15), and equilibrium concentration values for the major products H_2O and H_2, it has been possible to derive, for sodium, values of $k_{24}/(k_{41}+k_{42}+k_{43})=0.016$ and $k_{25}/(k_{51}+k_{52}+k_{53})= 0.036$, considered accurate to within a factor of two.

These results are particularly interesting in that on closer examination they indicate by comparison with conventional quench cross section measurements (16), which refer to all loss channels, that the chemical reaction flux is quite minor. The interaction between excited Na or Li with either H_2O or H_2 proceeds predominately and quite efficiently via a physical non-adiabatic quench-

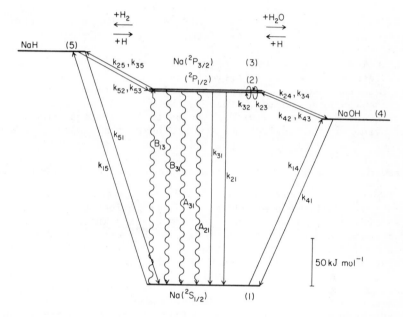

The Combustion Institute

Figure 1.   Energy-state diagram indicating the various radiative and quenching processes for the case of saturated laser excitation to the Na($^2P_{3/2}$) level, together with the chemical coupling to NaOH and NaH by reactions with H_2O and H_2, respectively

ing process. Nevertheless, even though the chemical contribution
is slight, amounting to about 2% and 0.5% of the total interaction
cross sections for sodium with H_2O and H_2, respectively, because
the overall fluxes are so large, these rates still are sufficient
to establish a steady state distribution over the hydroxide and
hydride states in a fraction of a µs.

The calculated steady state distributions in a typical case
are indicated in Figure 3. This more clearly summarizes the
drastic changes that occur on laser irradiation.

Laser Induced Ionization. Although laser induced ionization
effects now are well documented and form the basis for opto-galvanic analysis techniques (17,18), we find no indication for any
significant ionization in these flames on our concentration scales
with either sodium or lithium. This is supported by calculations
which indicate that although thermodynamically favored, kinetic
constraints are expected to prevent the ionization from proceeding
to an important extent on a µs time scale for either thermal or
alternate ionizing schemes. There appears to be no support for
Van Dijk and Alkemade's (19) contention that a state of ionization
equilibrium exists. Indications suggest the possible participation
of dimers that might be a complication in their study. Sequential
two-photon transitions at the D-line laser wavelengths are not
favored due to an absence of suitably located higher lying levels.

Implications. For sodium and lithium it is apparent that under
saturating conditions a steady state distribution is achieved over
the 5-level chemical model on a µs time scale. This is marginally
so for sodium and will not be the case with short pulse duration
lasers. Consequently, applications of quantitative saturated laser
fluorescence must make allowance for such potential complications.
This has been overlooked previously. Its consequences are evident
where such results have been compared with conventional absorption
measurements which all give consistently larger values. For example,
Pasternack et al. (4) found 5 and 7-fold discrepancies for Na and
MgO measurements and Bonczyk and Shirley (8) factors of 2 and 5 for
CH and CN, respectively. A reanalysis of the Na data (4) indicates
a need to increase their concentration and quenching rate data by
a factor of $[1+0.75[H]^{-1}\{k_{24}[H_2O]/(k_{41}+k_{42}+k_{43})+k_{25}[H_2]/(k_{51}+k_{52}+k_{53})\}]$ which we have found to be at least a factor of 2 or more.
Similar corrections are necessary to the work of Gelbwachs et al.
(1) and Smith et al. (2). Although this is not a large correction
in these particular cases there is no reason why it might not be
in other situations.

Summary. In saturated laser fluorescence studies of sodium
and lithium in a series of fuel rich $H_2/O_2/N_2$ flames there is
evidence for the involvement of laser induced chemical reactions
with H_2O and H_2. Although their reactive probabilities have been
shown to be small relative to the corresponding physical quenching
interactions they are still sufficient to establish significant

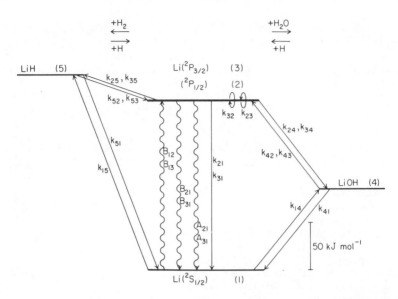

The Combustion Institute

Figure 2.   Corresponding energy-state diagram for lithium. However, this differs owing to the simultaneous excitation and monitoring of both the $^2P_{3/2, 1/2}$ levels because of their slight separation.

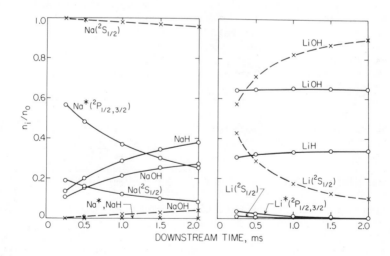

Figure 3.   Fractional distribution of sodium and lithium over the various atomic and molecular states in a $H_2-O_2-N_2$ flame of unburnt volume ratios 3:1:5 as a function of downstream location under normal or saturation laser radiation: (○), near saturation; (×), without laser excitation.

steady state concentrations of the corresponding hydroxide and hydride within the μs pulse duration. This explains the previously reported 'low' estimates resulting for such measurements. Because a simple general test of the participation of such chemical effects is not apparent, it will obviously be necessary to restrict applications of quantitative saturated mode measurements to systems utilizing nanosecond pulse lengths and should be of particular value at reduced pressure.

Acknowledgments. This research has been supported by the National Science Foundation under Contract Number CHE78-21458.

Literature Cited.
1. Gelbwachs, J.A., Klein, C.F., Wessel, J.E., Appl.Phys.Lett., 1977, 30, 489.
2. Smith, B., Winefordner, J.D., Omenetto, N., J. Appl. Phys., 1977, 48, 2676.
3. Daily, J.W., Chan, C., Combust. Flame, 1978, 33, 47.
4. Pasternack, L., Baronavski, A.P., McDonald, J.R., J. Chem. Phys., 1978, 69, 4830.
5. Mailander, M., J. Opt. Soc. Am., 1978, 68, 650.
6. Baronavski, A.P., McDonald, J.R., Appl. Opt., 1977, 16, 1897.
7. Mailander, M., J. Appl. Phys. 1978, 49, 1256.
8. Bonczyk, P.A., Shirley, J.A., Combust. Flame, 1979, 34, 253.
9. Muller, C.H.,III, Schofield, K., Steinberg, M., Chem. Phys. Lett., 1978, 57, 364; 1979, 61, 212.
10. Muller, C.H.,III, Schofield, K., Steinberg, M., 10th Materials Res. Symp. Characterization High Temp. Vapors and Gases, Nat. Bur. Stand. (U.S.) Spec. Publ. 561/2, 1979, p. 855.
11. Muller, C.H.,III, Schofield, K., Steinberg, M., J. Chem. Phys., (in press).
12. Daily, J.W., Appl. Opt., 1977, 16, 568.
13. Allen, J.E.,Jr., Anderson, W.R., Crosley, D.R., Fansler, T.D., Symp. (Int.) Combust., (Proc.), 1979, 17, 797.
14. Van Calcar, R.A., Van de Ven, M.J.M., Van Uitert, B.K., Biewenga, K.J., Hollander, Tj., Alkemade, C.Th.J., J. Quant. Spectrosc. Radiat. Transfer, 1979, 21, 11.
15. Bulewicz, E.M., James, C.G., Sugden, T.M., Proc. R. Soc. London, 1956, A235, 89.
16. Lijnse, P.L., "Review of Literature on Quenching,Excitation and Mixing Collision Cross Sections for the First Resonance Doublets of the Alkalis," Fysisch Laboratorium Report 398, Rijksuniversiteit Utrecht, The Netherlands, 1972.
17. Young, J.P., Hurst, G.S., Kramer, S.D., Payne, M.G., Anal. Chem., 1979, 51, 1050A.
18. Turk, G.C., Travis, J.C., De Voe, J.R., O'Haver, T.C., Anal. Chem., 1979, 51, 1890.
19. Van Dijk, C.A., Alkemade, C.Th.J., Present Symposium.

RECEIVED February 25, 1980.

Saturation Broadening in Flames and Plasmas As Obtained by Fluorescence Excitation Profiles

N. OMENETTO, J. BOWER, J. BRADSHAW, S. NIKDEL, and J. D. WINEFORDNER

Department of Chemistry, University of Florida, Gainesville, FL 32611

The high irradiance provided by pulsed tunable dye lasers is capable of saturating both single photon and is some cases 2-photon transitions of atoms in flames at atmospheric pressure. Besides the attainment of a saturation plateau as the source spectral irradiance is increased, it is well-known[1-3] that in a strong irradiation field, saturation broadening occurs, i.e., the broadening of the excitation line profile. A theoretical treatment on saturation broadening and experimental verification has recently been carried out by us[4]. Generally, atoms are considered to be dispersed as trace constituents in a gas at atmospheric pressure and characterized by a 2-level system. In our theoretical treatment coherence effects are neglected because dephasing, coherence interrupting collisions are considered to be fast in atmospheric pressure flames. In addition, several limiting cases will be presented here.

Line Source-Lorentzian Atom Profile. In this case, the laser is assumed to be monochromatic line source and the atomic absorption profile is assumed to be homogeneously broadened. In addition, the laser beam is assumed to be spatially uniform. The atom profile can be considered homogeneous if the following conditions hold: (1) Doppler broadening is negligible compared to collisional broadening; and (2) in the case of combined Doppler-collisioned broadening, velocity changing collisions are so fast that atoms cannot be considered to belong to any particular Doppler internal during the time of interaction with the laser beam. Assuming the steady state limit for the 2-level atom and introducing a Lorentzian dispersion function for the atomic absorption profile[4], the ratio of concentrations n_2/n_T (2=upper state, T=total population density, $n_T + n_1 + n_2$) can be evaluated. Evaluation of the fluorescence radiance, B_F, for the case where the atomic density is low and substitution for n_2 allows[4] evaluation of the FWHM of the excitation profile, $\delta\lambda_{exc}$,

$$\delta\lambda_{exc} = \delta\lambda_L \sqrt{1 + \frac{\rho(\lambda_\ell)}{\rho^s(\lambda_\ell)}} \qquad (1)$$

0-8412-0570-1/80/47-134-195$05.00/0
© 1980 American Chemical Society

where $\delta\lambda_L$ is the FWHM of the Lorentzian broadened absorption line profile, $\rho(\lambda_\ell)$ is the integrated spectral energy density of the laser at the laser excitation wavelength, λ_ℓ, and $\rho^s(\lambda_\ell)$ is the saturation energy density defined by

$$\rho^s(\lambda_\ell) = \frac{A_{21}\,\delta\lambda_{eff}}{\left(\dfrac{g_1 + g_2}{g_2}\right)B_{12}\,Y_{21}} \qquad (2)$$

where $\delta\lambda_{eff}$ is the effective width (FWHM) of the absorption profile, in m, the g's (dimensionless) are the statistical weights of the 2 levels involved in the excitation process, A_{21} is the Einstein emission probability, in s^{-1}, B_{12} is the induced absorption transition probability, in $J^{-1}m^3s^{-1}m$, and Y_{21} is the fluorescence quantum efficiency of the species. It is evident that the excitation profile width increases with $\sqrt{\rho(\lambda_\ell)}$ and with $\sqrt{Y_{21}}$ and a plot of $\delta\lambda_{exc}^2$ vs $\rho(\lambda_\ell)$ should give a straight line with a slope and an intercept which can be used to estimate $\rho^s(\lambda_\ell)$ and $\delta\lambda_L$, respectively. Of course, at the limit of $\rho(\lambda_\ell) \to 0$, $\delta\lambda_{exc} \to \delta\lambda_L$.

Gaussian Laser Profile-Voigt Atom Profile. This case turns out to be a better approximation of our experimental situation, i.e., the laser FWHM is fairly broad compared to the absorption line width and the absorption profile of atoms in an atmospheric combustion flame is described by a Voigt profile. Here the laser is assumed to have a Gaussian spectral profile as well as a Gaussian atomic absorption profile. In this case, convolution of two Gaussian functions is still a Gaussian function. Evaluation of the ratio n_2/n_T, and the fluorescence radiance, B_F, allows determination of the half width of the fluorescence excitation profile, $\delta\lambda_{exc}$ as

$$\delta\lambda_{exc} = \frac{\sqrt{\delta\lambda_\ell^2 + \delta\lambda_a^2}}{\sqrt{\ell n\,2}} \sqrt{\ell n(2 + \frac{\rho}{\rho^s})} \qquad (3)$$

where $\delta\lambda_\ell$ is the laser FWHM, $\delta\lambda_a$ is the atomic absorption line FWHM, ρ is the integrated energy density of the laser at the excitation peak, in $J\,m^{-3}m^{-1}$, and ρ^s in the saturation spectral density for the atomic process of concern, in $J\,m^{-3}m^{-1}$, and is given by

$$\rho^s = \frac{\sqrt{\delta\lambda_\ell^2 + \delta\lambda_a^2}\,\sqrt{\pi}}{\left(\dfrac{g_1 + g_2}{g_2}\right)B_{12}\,\tau\,2\sqrt{\ell n\,2}} \qquad (4)$$

where all terms have been defined above except τ which is the effective lifetime of the excited state, $(A_{21} + k_{21})^{-1}$. In this case, it is clear that $\delta\lambda_{exc}$ varies at a slower rate with ρ than in the previous limiting case. It is also clear that high quantum efficiency flames will be more sensitive for observation of broadening effects. At the limit, when $\rho \to 0$, $\delta\lambda_{exc} \to \sqrt{\delta\lambda_\ell^2 + \delta\lambda_a^2}$ which is to be expected. Finally if $\rho \gg \rho^s$, then

$$\delta\lambda_{exc} = \frac{\sqrt{(\delta\lambda_\ell)^2 + (\delta\lambda_a)^2}}{\sqrt{\ell n\ 2}} \sqrt{\ell n\ \frac{\rho}{\rho^s}}$$

and so by plotting $(\delta\lambda_{exc})^2$ vs $\ell n\rho$, a straight line results where the slope gives $\frac{1}{\sqrt{(\delta\lambda_\ell)^2 + (\delta\lambda_a)^2}}$ and the intercept gives ρ^s

Experimental. The experimental set up consisted of a N_2-pumped-dye laser (Molectron UV-14, DL-400), spatial filters to isolate the central part of the dye laser beam, a H_2-O_2-Ar or N_2 flame supported by a capillary burner with Ar or N_2 sheath, and a fluorescence detection system at right angles (a JY-H-10 monochromator, a photomultiplier, and a PAR 162-164 boxcar averager). All measurements were taken 1 cm above the burner top; the concentration of Ca, Sr, In, and Na was low (1 µg/ml). The fluorescence waveform was monitored with a 75 ps sampling head (PAR 163). The laser spectral bandwidth was also measured with a JY-HR-1000 monochromator ($\delta\lambda_s \cong 0.1$ Å).

Results. Experimental and theoretical fluorescence excitation halfwidths were obtained. In Table I, the ratio of the maximum value of $\delta\lambda_{exc}$ (corresponding to that obtained with full laser power) and the minimum value of $\delta\lambda_{exc}$ (corresponding to that obtained at $\rho<<\rho^s$, where $\delta\lambda_{exc} = \sqrt{(\delta\lambda_\ell)^2 + (\delta\lambda_a)^2}$. The saturation power was evaluated for each element from the experimental saturation curve. In Table II, a comparison is given between experimental values of the laser spectral bandwidth obtained by direct measurement (JY-HR1000), and by 2 values based on laser excited fluorescence. Apart from several unexplained discrepancies, the agreement between experiment and theory was excellent.

TABLE I

COMPARISON BETWEEN THE THEORETICAL AND THE EXPERIMENTAL VALUES OF THE FLUORESCENCE EXCITATION PROFILE HALFWIDTHS FOR THE $Ar/O_2/H_2$ FLAMES. [a]

| Element | $[(\delta\lambda_{exc})_{max}/(\delta\lambda_{exc})_{min}]$ [b] | | | |
| | Theoretical [a] | | Experimental | |
	$Ar/O_2/H_2$	$N_2/O_2/H_2$	$Ar/O_2/H_2$	$N_2/O_2/H_2$
Ca	2.5	2.4	3.0	2.2
Sr	2.8	2.8	2.9	5.4
Na	2.4	2.0	4.3	1.6
In	1.7	1.4	2.3	1.3

[a] values are considered to be within ± 10%.

[b] $(\delta\lambda)_{max}$ refers to the value obtained with the laser at full power while $(\delta\lambda)_{min}$ refers to that obtained when the laser is attenuated with neutral density filters until the fluorescence signal is linearly related to the laser irradiance (see values reported in Table II).

[c] calculated according to Equation 3 in the text.

TABLE II

COMPARISON BETWEEN THE VALUES OF THE LASER SPECTRAL BANDWIDTH AS OBTAINED BY DIFFERENT METHODS. [a]

Element	Direct [b] Measurement	Fluorescence [c] Excitation Profile		Saturation [d] Broadening	
		$Ar/O_2/H_2$	$N_2/O_2/H_2$	$Ar/O_2/H_2$	$N_2/O_2/H_2$
Ca	0.23	0.20	0.20	0.26	0.21
Sr	0.23	0.42	0.23	0.46	0.51
Na	0.36	0.23	0.40	0.30	0.35
In	0.24	0.19	0.28	0.26	0.21

[a] values are within $\pm$ 10%.

[b] values obtained by scanning the laser beam through a 1-m grating monochromator ($\Delta\lambda$ resolution = 0.12 Å). Values are not corrected for the instrumental profile.

[c] values obtained by scanning the attenuated laser beam through the atomic vapor in the flame.

[d] values calculated from the slope of the plot obtained from Equation 3 in the text.

References:

1. C. A. Van Dijk, Ph.D. Dissertation, Utrect, The Netherlands, 1978.
2. J. W. Hosch and E. H. Piepmeier, Appl. Spectrosc., 32, 444, (1978).
3. S. Ezekial and F. Y. Wu, in Multiphoton Processes, J. H. Eberly, and P. Lambropoulos, Editors, John Wiley, New York, 1978.
4. N. Omenetto, J. Bower, J. Bradshaw, C. D. Van Dijk, and J. D. Winefordner, J. Quant. Spectrosc. Radiat. Transfer, Submitted.

Research was supported by grants from AFOSR F44620-76-C-0005 and WPAFB F33615-78-C-2038.

RECEIVED February 11, 1980.

Determination of Flame and Plasma Temperatures and Density Profiles by Means of Laser-Excited Fluorescence

J. BRADSHAW, S. NIKDEL, R. REEVES, J. BOWER,
N. OMENETTO, and J. D. WINEFORDNER

Department of Chemistry, University of Florida, Gainesville, FL 32611

The fluorescence technique, like other methods based on scatter (elastic or inelastic), has been shown by us[1-3] and others to be a reliable unperturbing method of measuring spatial/ temporal flame temperatures and species concentrations. To avoid the dependency of the fluorescence signal on the environment of the emitting species, it has been shown by several workers that optical saturation of the fluorescence process (i.e., the condition occurring when the photoinduced rates of absorption and emission dominate over the spontaneous emission and collisional quenching rates) is necessary. Pulsed dye lasers have sufficient spectral irradiances to saturate many transitions. Our work has so far been concerned with atomic transitions of probes (such as In, Pb, or Tl) aspirated into combustion flames and plasmas.

Concepts and Methods

The temperature of a flame, plasma, or hot gas can be estimated by using the steady state fluorescence expressions derived by Boutilier, et al[1] for spectral continuum excitation. Several unique methods which can be used to measure spatial temperatures (volumes < 10 mm^3) have been developed by us and will be reported in detail in a paper to be submitted for publication.[4] The methods are generally based upon the introduction of inorganic 3-level probes (Tl or Pb) into a flame and measuring the ratio of fluorescence signals resulting between levels 3 and 2 and 3 and 1 following excitation of level 3 via levels 1 or 2. Because of the restrictions regarding overall length of this report and because of the future availability of the published paper[4] concerning these new methods, we will here only give the approaches and several flame temperatures measured by the described methods.

Method 1. Linear 2-Line Method[5,6] In this method, the ratio of fluorescence signals, $B_{F_{3\to2}^{1\to3}}$ and $B_{F_{3\to1}^{2\to3}}$ (the upper subscripts represents the measured fluorescence transition and the lower subscripts represents the excitation transitions), is

0-8412-0570-1/80/47-134-199$05.00/0
© 1980 American Chemical Society

measured. By calibration of the spectrometric system and measuring the ratio of fluorescence and excitation intensity ratios, the flame temperature can be determined from a simple expression. This method requires calibration, linear behavior of the fluorescence intensity with excitation intensity and excitation beam matching. In addition, efficient quenching species in some flames (hydrocarbon fuels) and pre- and post-filter effects lead to deterioration of the signal-to-noise ratios. Laser excitation is advantageous for spatial measurements and improved signal-to-noise ratios.

Method 2. Saturation Method for Sequential Pumping. In this method, atomic fluorescence of the inorganic probe is produced at 3→1 and at 3→2 after excitation at 1→3 and/or 2→3 respectively. However, in this case, it is necessary to "saturate" the excited level, 3, in order to use the method[1,2] In addition, in order for the flame temperature to be evaluated it is necessary for the mixing first order rate constant, k_{21}, between the metastable, 2, and ground state, 1, to be much greater (> 20X) than the sum of the total deactivation rate constants between levels 3 and 1 and also between 3 and 2. This method also requires calibration of the spectrometric measurement system, saturation of level 3, corrections or minimization of scatter and post filter effects, and beam matching of 2 dye laser beams are needed for the excitation process.

Method 3. Saturation Method With Peak Detection. In this method, developed by Omenetto and Winefordner[2,3], it is necessary to excite fluorescence 3→1 with 1→3 and a short time later (< 1 μs) excite 3→1 with 2→3. In this case the atomic system effectively acts on a 2-level atom since excitation and measurement of fluorescence is done at the peak of the excitation profil prior to relaxation of the system to a 3-level steady state process[6] The temperature here is related simply to the ratio $B_{F_{3→1}}^{1→3}/B_{F_{3→1}}^{2→3}$ and statistical weights of the levels and is independent of non-radiational rate constants as in the preceeding case and of calibration as in the two preceeding cases. On the other hand, this method requires the use of a fast rising laser pulse to perturb the inorganic probe to reach a 2-level steady state and the use of fast electronics to measure the fluorescence prior to relaxation of the system to a steady state involving all three levels (1,2,3). This method also requires 2 spatially and geometrically matched dye laser beams which will cause the probe to be rapidly saturated.

Methods 4 and 5. Two other novel methods for flame temperature measurement will be reported upon in the full paper to be published,[1] but no results will be given here. One of these methods (Method 4) involves saturation of level 3 via simultaneous pumping of both 1→3 and 2→3 and taking the ratio of the resulting

3→1 fluorescence and the 3→1 fluorescence resulting when exciting 2→3. This method has most of the same potential difficulties of Method 2. Method 5 involves the use of one laser beam and linear or saturation behavior: in this case, the ratio of the probe fluorescence resulting at 3→1 with 2→3 excitation and the laser induced emission resulting at i→1 (i>3) with 2→3 excitation. This method has a number of advantages: (i) saturation is not necessary; (ii) only one laser wavelength is needed; (iii) no need to spatially match laser beams: (iv) calibration of the fluorescence spectrometer is still needed but there is no need to calibrate the excitation intensity: post filter and scatter effects are minimal, and (v) temporal (single pulse) measurements of temperatures are feasible.

In Method 1-4, by measuring the fluorescence signals close together (say 1 μs), then temperatures corresponding to nearly "frozen" flame conditions are obtainable. We are currently in the process of making such temporal temperature measurements; these results will be published at a later date. By beam expansion of the laser beam(s) and isolation of the central homogeneous section, it is also possible to resolve spatially small flame volumes, e.g., depending upon the spectrometer entrance slit or slit aperature, < 10 mm^3.

Spatial density profiles of atomic (and molecular) species can also be made via saturation fluorescence approaches. For a "2-level" atom, like Sr, a plot of $1/B_F$ vs $1/E_\lambda$ (B_F is the fluorescence radiance, in J s^{-1}m^{-2}sn^{-1}, and E_λ is the excitation spectral irradiance, in J s^{-1}m^{-2}nm^{-1}) allows estimation of the quantum efficiency, Y of the fluorescence process (and thsu estimation of "radiationless" rate constants) and the total number density n_T, of the species of interest by means of

$$\frac{1}{B_F} = (\frac{1}{C})(\frac{g_1 + g_2}{g_2}) + (\frac{1}{C})(\frac{g_1}{g_2})(\frac{8\pi hc^2 \cdot 10^{-7}}{\lambda^5 Y})(\frac{1}{E_\lambda}); \quad C = (\frac{\ell}{4\pi})h\nu An_T$$

where: ℓ is the fluorescence path length, $h\nu$ is the fluorescence (or excitation photon energy) in J, A is the emission probability, in s^{-1}, the g's are the statistical weights of the 2 levels, h is Planck's constant, and c is the speed of light. The plot of $1/B_F$ vs $1/E_\lambda$ has a slope which includes $(n_T Y)^{-1}$ and an intercept which includes n_T^{-1}. If both B_F and E_λ are measured in absolute units, then n_T and Y can be obtained. Even if B_F is measured in relative units Y can be determined by multiplying through by C and then calibrating ordinate in units of $(g_1 + g_2)/g_2$. If the atom is a 3 (or multi) level system, then the radiationless rate constants must be known and included in the expression for $1/B_F$ as a function of $1/E_\lambda$ unless the fortunate circumstances arises where two of the 3 levels essentially coalesce into a single level and once again we have essentially a 2-level atom. In this case absolute measurement of both B_F and E_λ is necessary.

Table I. Measured Flame Temperatures

	$H_2/O_2/Ar$[a,b,c]	$H_2/O_2,N_2$[a,b,c]
Linear 2 line Method (Source Spectral Radiance was 5 to 10^3 W/cm^2 nm) Ratio taken was $B_{F_{3\to2/1\to3}}$ over $\overline{B_{F_{3\ 1/2\ 3}}}$	2200 ± 30 K[d]	1980 ± 30 K[d]
Saturation Method-Sequential Pumping (λ_{f1} = same in both cases) (Source Spectral Radiance was 1×10^7 W/cm^2 nm) Ratio taken was $B_{F_{3\to1/1\to3}}$ over $\overline{B_{F_{3\to1/2\to3}}}$	2120 ± 30 K[3]	1990 ± 30 K[d]

a. Lijnse and Elsenaar (P.L. Lijnse and R.J. Elsenaar, J. Quant. Spectrosc. Radiat. Transfer, 12 (1972) 1115.) obtained temperatures of 2136 K for $H_2/O_2/Ar$, 2/1/4 and 1970 K for $H_2/O_2/N_2$ 1.9/0.95/4.

b. Hoomayers (H.P. Hoomayers, Ph.D. Thesis, University of Utrecht, 1966.) obtained temperatures of 2350 K for $H_2/O_2/Ar$, 1.72/0.85/3.45 and 2160 K for $H_2/O_2/N_2$, 1.72/0.85/3.45. The source of the systematic errors in the values measured by Lijnse and Elsenaar, by Hoomayers, and by us is not known.

c. Line reversal temperature measurements by us for the same flames were 50-100 K higher than our fluorescence values.

d. The random error of $\pm$ 30 K was due to shot noise on the signal.

EXPERIMENTAL SYSTEM

Out experimental system consisted of a N_2-pumped dual dye laser (Molectron UV-14 with Molectron DL-400 and Lambda-Physik FL-2000), operated at 20Hz, spatial filters to isolate the central portion of the dye laser beams, H_2-O_2-Ar and H_2-O_2-N_2 flames supported on a Meker type flame shielded flame with Ar or N_2 outer sheaths, and a fluorescence detection system consisting of a 0.1 m grating monochromator, a Hamamatsu R928 photomultiplier tube, and a Tektronix 151 sampling oscilloscope with 0.5 s averaging time constant. All measurements were taken 1.5 cm above the burner top (previous studies indicated the flame to be nearly constant in temperature with height, 1-3 cm, and with width. The flames studied in this report included: H_2/O_2/Ar, 2/1/4 and H_2/O_2/N_2, 2/1/4: the flows are relative volume ratios at standard temperature and pressure for the unburnt gases.

RESULTS AND DISCUSSION

Flame temperatures were determined by both the linear 2-line method and by the saturation method with sequential pumping for both flames. The measured values are given in Table I.

Assuming the sum of the radiationless and radiational rate constants between levels 3 and 2 for inorganic probe, like Tl, are much less than the non-radiational mixing constant k_{21}, then the 2-level expression relating $1/B_F$ to $1/E_\lambda$ applies. Using this relationship and the following measured parameters: slit area, 0.5 x 1.5 mm^2: solid angle, 0.26 sr; fluorescence depth, ℓ = 0.4 cm; 100 ppm Tl aspirated; $B_{F3\to1/1\to3}^{max}$ (H_2/O_2/N_2) = 1.1 x 10^{-1} W/cm^2 sr and $B_{F3\to1/1\to3}^{max}$ H_2/O_2/Ar) = 1.8 x 10^{-1} W/cm^2 sr, and using A_{31} = 0.41 x 10^8 s^{-1} (taken from Wade, et al[8]), then n_T = 1.0 x 10^{11} cm^{-3} for the H_2/O_2/Ar flame. These values compare favorably with population densities measured for similar flames and similar aspiration conditions.

REFERENCES

1. G.D. Boutilier, M.B. Blackburn, J.M. Mermet, S.J. Weeks, H. Haraguchi, J.D. Winefordner, and N. Omenetto, Appl. Optics, 17 (1978) 2291.
2. N. Omenetto and J.D. Winefordner, Prog. Anal. Atomic. Spectrosc., 2 (1979)1.
3. N. Omenetto and J.D. Winefordner, Chapter 4 in Analytical Laser Spectroscopy, N. Omenetto, ed., Wiley, New York, 1979.
4. J.D. Bradshae, N. Omenetto, J.N. Bower, and J.D. Winefordner, Appl. Optics, (to be submitted).
5. N. Omenetto, R.F. Browner, J.D. Winefordner, G. Rossi, and P. Benetti, Anal. Chem., 44 (1972) 1683.
6. H. Haraguchi, B. Smith, S. Weeks, D.H. Johnson, and J.D. Winefordner, Appl. Spectrosc., 31 (1977) 156.
7. J.N. Bower, J.D. Bradshaw, N. Omenetto, and J.D. Winefordner, Appl. Optics, (to be submitted).
8. M.K. Wade, M. Czajkowski, and L. Krause, Canad. J. Phys., 56, (1978) 891.

RECEIVED February 11, 1980.

SPONTANEOUS RAMAN SCATTERING

Raman-Scattering Measurements of Combustion Properties

MARSHALL LAPP

General Electric Corporate Research & Development, P.O. Box 8,
Schenectady, NY 12301

Laser light-scattering techniques for measuring flame gas
properties have advanced to the stage where they now can be employed to determine key flow and combustion field variables.
These include temperature, major constituent densities, gas velocity, and correlations of these properties. We discuss here the
advantages and limitations of various potential light-scattering
probes, and illustrate these with recent results for vibrational
Raman scattering flame diagnostics.

Light scattering is often viewed as a desirable probe of gas-
phase processes because it can be utilized to obtain three-
dimensional spatial resolution as well as well-defined, short-
duration temporal resolution. The former is achieved by triangu-
lation between the incident (laser) source beam and the field of
view of the optical detection apparatus (See Fig. 1), while the
latter can be obtained through use of a pulsed laser source.
Additionally, probes based upon light scattering are nonimmersed;
they do not require placement inconveniently near hostile environments, and are usually nonperturbing over a wide range of
experimental conditions.

A wide variety of information can be obtained from scattering probes, including fluid velocity, gas temperature, total gas
density, and constituency (i.e., species densities) (1-6). A
convenient overview of these diagnostic methods is provided by
grouping them into categories of elastic, or unshifted scattering and inelastic, or shifted, scattering. In Table I, this
grouping reveals that the elastic processes are focused mainly
upon velocity data, which require observations of light scattered
from particles either seeded into the flow or naturally present,
and upon total gas density. Temperature measurements, while
possible, are difficult to obtain in this fashion. (Note that
these "elastic" methods are actually slightly inelastic - as, in
fact, is implied by the usual name "laser Doppler velocimetry"

0-8412-0570-1/80/47-134-207$06.00/0
© 1980 American Chemical Society

Table I: Information Available from Elastic and Inelastic Molecular Light Scattering

Observation	Scatterer	Scattering Process	Information	Comments
Elastic unshifted scattering	Particles	Tyndall (Mie)	Characterization of particle distribution	Information can be difficult to interpret for non-ideal particle systems
			Velocimetry	Particles must be small enough to follow flow fluctuations
	Gas	Rayleigh	Total density	Few particles Favorable configuration Major composition known
			Temperature	Few particles Favorable configuration Major composition known Equilibrium Difficult to instrument
Inelastic (shifted) scattering	Gas	Raman Fluorescence Nonlinear processes	Temperature and component densities	Nonequilibrium OK Wide range of signal strengths and complexities to obtain detailed data from various systems, including hostile environments

(LV) for velocity measurement - but the dominant character of the scattered signature is that it is concentrated about the incident frequency.)

The inelastic processes - spontaneous Raman scattering (usually simply called Raman scattering), nonlinear Raman process-es, and fluorescence - permit determination of <u>species</u> densities as well as temperature, and also allow one, in principle, to de-termine the temperature for particular species whether or not in thermal equilibrium. In Table II, we categorize these inelastic processes by the type of the information that they yield, and indicate the types of combustion sources that can be probed as well as an estimate of the status of the method. The work that we concentrate upon here is that indicated in these first two categories, viz., temperature and major species densities deter-mined from vibrational Raman scattering data. The other methods - fluorescence and nonlinear processes such as coherent anti-Stokes Raman spectroscopy - are discussed in detail elsewhere (5).

Further breakdown of vibrational Raman scattering (RS) methods for temperature and density measurement can be achieved by classification according to the character of the laser source used (2). In Table III, we show such a classification according to the experimental capabilities provided through use of these various sources. Here, we view these capabilities in terms of those features that can be provided by optical scattering diagnostics, that are of most value to flame analyses. These involve the ability to determine precise temporally- and spatially-resolved data leading to probability density functions (pdf's, or histograms of flame properties, which give the proba-bility that any particular instantaneous measurements of a state variable within a large ensemble of measurements will be found between specified limits), frequency spectra, and spatial mapping. We note that at this time, no laser source (i.e., no single RS method) can provide simultaneously information of high quality for pdf's, complete frequency spectra, and spatial gradients.

The work to be described further in the next section is that indicated in the first entry of Table III - RS data for pdf's (and potentially, for spatial gradients) obtained through use of short time duration, energetic laser pulses. This work has been motivated by an examination of the questions: What are the key fluid quantities necessary to model turbulent flames? And what methods are well enough developed to produce the required data with confidence and without substantial additional proof-of-prin-ciple experiments? These questions lead one to recognize that, in <u>increasing</u> degrees of measurement difficulty, we are concerned with flow field quantities, combustion field quantities, and finally, pullutant field quantities (19). In terms of the most basic needs of combustion modelers, this listing is also organized in <u>decreasing</u> importance for determining the fundamental qualities of flames. (Of course, if one's goal is to model the nitric

Table II: Space- and Time-Resolved Measurements from Inelastic Light Scattering. All methods are suitable for nonequilibrium conditions. Here, RS refers to Raman scattering, CARS to coherent anti-Stokes Raman spectroscopy, and RIKES to Raman-induced Kerr effect.

Information	Environment	Method	Status	Comments
Temperature	Clean flame zones	RS	Accomplished	Don't need a priori composition Too weak for luminous systems
Major species densities	Clean flame zones	RS	Accomplished	Don't need a priori temp. for low temps Too weak for luminous systems
Temperature	Bright and/or particle-laden flame zones	CARS	Accomplished	Strong signal Tolerates particle loading and strong flame luminosity More difficult to instrument and interpret than RS
Major and intermediate species densities	Bright and/or particle-laden flame zones	CARS	Possible	Strong signal Tolerates particle loading More difficult to instrument and interpret than CARS for temp.
		Raman gain	Possible	Developmental alternative to CARS
Minor species densities	Bright and/or particle-laden flame zones	Fluorescence	Semi-quant.	Strong signal
		Saturated fluor.	Probably quant. for some species	Reduces dependence of fluorescence on collisional quenching
		RIKES, other nonlinear processes	Possible	Developmental High experimental demands for increased quality of data

Table III: Comparison of Vibrational Raman Scattering Fluctuation Measurement Capabilities for Different Laser Source Characteristics.

Character of Laser Source	pdf	Frequency Spectra	Spatial Gradients	Comments/[References]
Short energetic laser pulses, low rep rate (ex. dye, Nd:YAG, Q-switched ruby lasers)	Yes	No	Yes	Can use variable wavelength laser. [This work - pdf's. (2,7,8)] [Avg. values, stand. deviations, correlations. (3,4)] [Spatial gradients in jets (9)] [Spatial gradients in flame. (10,11)]
Long strongly energetic laser pulses (ex. free-running ruby laser, intracavity exp.)	Yes	Mid (kHz) to high (~50 kHz) freq.	Difficult	Can probe somewhat luminous particulate flows. [Time history for ~300 µs. (12,13)]
cw laser (operated either cw or chopped), with time domain analysis using: –Fourier transform –Autocorrelation function –Moments of photon count distribution	Yes	Low (Hz) to mid (kHz) freq.	Difficult	Restricted to low luminosity flows. pdf's obtained from finite sequence of photon count moments. [Spectral density. (14,15)] [Autocorrel. fct. & pdf's (16,17)] [pdf's. (18)]

oxide emission of a particular combustion source, then determination of the pollutant field quantities is of prime importance.
However, in determining the overall characteristics of flames, the
above ordering of quantities represents a reasonable priority list
for modelers.)

In Table IV, we see that established techniques for velocity
measurement allow us to determine the average momentum flux,
average velocity, turbulent intensities, and shear stress. Next
on the list, to complete the flow field description, is the fluctuation mass flux, and first on the combustion field list is the
temperature and major species densities of the flame gases.
These are the quantities to which we are giving our attention.
Vibrational Raman scattering is being used for the temperature and
density data, and, when taken simultaneously with velocity data
from coupled LV instrumentation (8), provides also the fluctuation
mass flux through use of fast chemistry assumptions and the ideal
gas law for atmospheric pressure flames.

Raman Scattering Diagnostics

The fundamentals of the Raman effect can be understood by
consideration of a classical model, in which an incident beam of
radiation (i.e., laser beam, for all practical purposes, in flame
diagnostics) passes through an ensemble of molecules. The resultant laser beam electric field distorts the electronic cloud
distribution of each molecule, causing oscillating dipoles; these
induced dipoles are related to the incident laser beam electric
field by the molecular polarizability. The dipoles, in turn,
produce a secondary radiating field at essentially the same frequency as that for the incident beam. This radiation is termed
Rayleigh scattering.

Since different orientations of the rotating and vibrating
molecules produce different polarizabilities, these molecular
motions modulate the polarizability at the rotational and vibrational frequencies. Sidebands are thereby created which are
displaced from the incident laser beam frequency by shifts
corresponding to the molecular vibrational and rotational frequencies. These shifts can be understood from this type of classical argument; the scattering intensities and selection rules for
the appearance of specific bands, however, can come only from the
introduction of quantum mechanical considerations. Thus, for
example, the understanding of molecular vibrational and rotational
energy level structure leads to an understanding of how Raman
scattering band analyses are related to the determination of
molecular vibrational and rotational "population" temperatures,
i.e., to the determination of well-defined gas temperatures even
in the absence of equilibrium between molecular internal modes
and translation. (However, equilibrium within each internal mode
is necessary in order to ascribe a meaningful population temperature to that mode.)

Table IV: Ordering of Predictive Needs for Combustion
Modeling with Measurement Capabilities (In Increasing
Experimental Difficulty).

Here, Instantaneous Value $\tilde{X}$ = Mean Value X + Fluctuation Value X';
LV denotes laser velocimetry; RS, Raman scattering, RayS, Rayleigh scattering; and pdf, probability density function.

Flow Field Quantities		Measurement Techniques
Average Momentum Flux	ρu^2	Pitot Tube
Average Velocity	u	LV, Pitot Tube
Turbulent Intensity	$<u'^2>$	LV
Shear Stress	$<u'v'>$	LV
Fluctuation Mass Flux	$<\rho'u'>$	LV + RS or RayS

Combustion Field
   Quantities

Temperature $\tilde{T}$ and Major Species Mass Fractions $\tilde{M}_i$	Mean and Variance of $\tilde{T}$, $\tilde{M}_i$	RS; RayS for $\tilde{T}$
	pdf for $\tilde{T}$, $\tilde{M}_i$	RS; RayS for $\tilde{T}$
		[Thermocouples (mean)]
		[Gas Sampling (mean)]
		[Absorp./Emission Spectroscopy (not 3-dimensional)]
Density	ρ	RayS, knowing composition and cross sections; RS using ΣM_i, or T with fast chem. and ideal gas law

Additional properties of Raman scattering related to its
fundamental character that are of significant importance to its
use in flame diagnostics include a linear, species-specific dependence on molecular number density, independent of other species present (of critical importance for density measurements);
nonperturbing nature for incident laser beam intensities sufficiently strong to produce useful combustion measurements; sensitivity to a wide range of molecular species, including homo-
nuclear molecules (such as N_2 and O_2) that are important for
flame diagnostics and which do not possess infrared spectra; and,
as was mentioned in the Introduction, effectively instantaneous
time response (limited only by the incident laser pulse width,
often in the microsecond range, but achievable in the nanosecond
range or less), excellent three-dimensional spatial resolution
(demonstrated to less than 0.1 mm^3, although often in the 1 mm^3
range), and remote, in situ capability. Useful though these
properties are, one must keep clearly in mind that the spontaneous
Raman effect is weak, and that its use is therefore limited to
measurements on major and intermediate density flame species.
Some perspective on the relative strength of Raman processes
compared with more familiar scattering effects can be gained from
inspection of Table V. That this inherent weakness can be overcome by current experimental methods can be seen by reference to
Table II, which suggests the main diagnostic applications of
these processes and the present degree of accomplishment.

In Fig. 2 we see a schematic of the primary molecular scattering processes present when a laser beam impinges upon nitrogen
test gas. The pure vibrational Raman scattering is shifted from
the incident laser line (and therefore, from the Rayleigh scattering and from the center of the pure rotational Raman spectrum,
shown here as an envelope of individual lines) by 2331 cm^{-1} for
nitrogen. These sharp bands, termed Q-branches to designate no
change of rotational quantum number, correspond to $\Delta v = +1$, where
v is the vibrational quantum number (Stokes band), and to $\Delta v = -1$
(anti-Stokes band). Since the Stokes bands arise from 'lower"
vibrational energy levels, while the anti-Stokes bands arise from
"upper" levels, the ratio of these intensities can be a sensitive
indicator of temperature (1-5). This can be seen in Fig. 2 from
their corresponding intensities at different temperatures. In
addition, the detailed contours of these bands, shown at higher
resolution in the inset diagrams in Fig. 2, present further oppor-
tunities for temperature determination. Here, the successive
peaks in the 1500 K Stokes profile correspond to successive bands
- the ground state band (v = 0 → v = 1) and the upper state bands
(v = 1 → v = 2, v = 2 → v = 3, ...). A similar profile exists on
the anti-Stokes side. Thus, temperature can be determined from
contour fits, using a spectrometer, ratios of contour peak height
intensities using a spectrometer, polychromator, or spectral
filters, etc. (20)

In a similar fashion, pure rotational Raman scattering can

Table V: Typical Cross Section Values for Scattering Processes

Scattering Process	Differential Cross Section (cm^2/sr)
Particle Scattering (Mie) - 10 μm diam.	10^{-7}
Particle Scattering (Mie) - 0.1 μm diam.	10^{-13}
Atomic Fluorescence - Strong, Visible	$10^{-13} - 10^{-18}$
Molecular Fluorescence - Simple Molecules	$10^{-19} - 10^{-24}$
Rayleigh Scattering - N_2 (488 nm)	10^{-27}
Rotational Raman Scattering - N_2 (All Lines, 488 nm)	10^{-29}
- N_2 (Strong Line, Including Fractional Population Factor, 488 nm)	6×10^{-31}
Vibrational Raman Scattering - N_2 (Stokes Q-Branch, 488 nm)	$5. \times 10^{-31}$

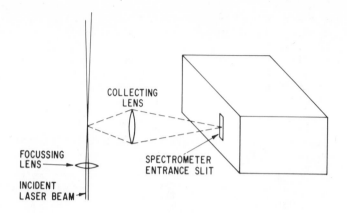

Figure 1.   Geometry for typical Raman-scattering measurement

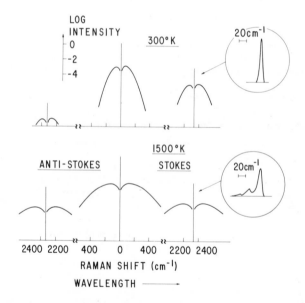

Figure 2.   Raman and Rayleigh scattering from N_2 at ambient (300 K) and ele-vated (1500 K) temperatures for an exciting laser line in the midvisible. The central unshifted peak corresponds to Rayleigh scattering, which is flanked by rotational Raman scattering represented here by wing envelopes of the rotational line peak intensities. The vibrational O-branches on the Stokes and anti-Stokes sides are shown at the charac-teristic Raman shifts for N_2 of 2331 cm^{-1}. These Q-branches are surrounded by weaker vibrational bands called the O- and S-branches, shown also by wing envelopes. Note that relative intensities are drawn on a logarithmic scale and that large breaks occur along the wave number and wavelength axes. The spectral contours of the Q-branches shown in the two inset diagrams are presented on a linear scale and have been calculated using a triangular spectrometer slit function with 6 cm^{-1} ($\sim$ 0.18 nm) FWHM.

be used as a sensitive temperature indicator (20) with the advantage of stronger signal intensities (See Table V), but with the
disadvantage that the spectral signatures of many of the important flame molecules all fall in the same general spectral region,
thus complicating both density and temperature determinations.
However, careful application of this technique has resulted in
significant applications to flame analyses (21,22).

Recent Results for pdf Data

Following the discussion in the Introduction of the various
possible approaches to flame measurements - both from the points
of view of the needs of combustion modelers as well as from the
capabilities of flame spectroscopists - we describe here one such
avenue of research. In this approach, we utilize a high energy-
per-pulse laser source to produce Raman data leading to probability density functions (where each datum that contributes to the
pdf is a statistically-independent measurement resulting from a
single laser pulse) of the most important state variables, viz.,
temperature, and major flame species densities. These data are
combined with concommitant measurements of flow velocity using
laser volocimetry apparatus to form near-simultaneous data sets
or correlations of significant pairs of variables. They are of
prime importance in contributing to the understanding of both the
fluid mechanic and flame chemistry portions of combustion systems,
since turbulent fluctuations can occur over very small spatial
scales and over short time durations, and since chemical reactions are strongly dependent upon the instantaneous values of
temperature as well as various species densities.
      The overall combustor/optical layout is shown in Fig. 3,
which illustrates temperature measurement by the Stokes/anti-
Stokes method. Typical results for temperature pdf's at four
radial positions (2,7)- near the centerline to near the flame
boundary - and at an axial distance 50 fuel-tip diameters downstream of the fuel line tip are shown in Fig. 4. The shaded parts
of the pdf contours (from 300 to 800°K), which increase in area
near the flame boundary, correspond substantially to scattering
from ambient temperature air, and therefore provide a measure of
flow intermittency. The upper limit of these bins was chosen to
be 800°K because the accuracy possible for the Stokes/anti-Stokes
temperature measurement method degrades rapidly at temperatures
below roughly that value (2,7). Thus, treating the fluctuation
temperature data for T < 800°K in any greater detail was un-
warranted.
      The data shown in Fig. 4 give an estimate of the spatial
variation of the temperature pdf's, and therefore of the average
value and of higher moments, but not of instantaneous values of
the gradients. Data acquired simultaneously in space (as well as
in time) is required for the determination of instantaneous
gradients; information yielding such results has been obtained by

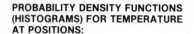

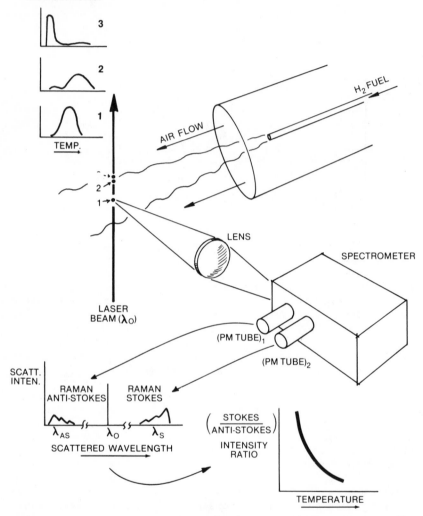

Figure 3. Schematic of turbulent combustor geometry and optical data acquisition system for vibrational Raman-scattering temperature measurements using SAS intensity ratios. Also shown are sketches of the expected Raman contours viewed by each of the photomultiplier detectors, the temperature calibration curve, and several expected pdf's of temperature at different flame radial positions. The actual SAS temperature calibration curve was calculated theoretically to within a constant factor. This constant, which accounted for the optical and electronic system sensitivities, was determined experimentally by means of SAS measurements made on a premixed laminar flame of known temperature. Measurements of N_2 concentration were made also with this apparatus, based on the integrated Stokes vibrational Q-branch intensities. These signals were related to N_2 gas densities by calibration against ambient air signals.

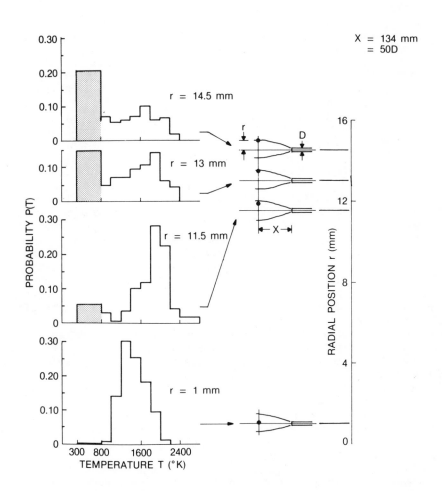

Figure 4. Probability density functions of temperature for H_2–air turbulent diffusion flame determined at various radial positions 134 mm downstream of the fuel line tip according to procedures indicated in Figure 3. The measurement positions are drawn schematically in the center of the figure to correspond to the radial positions r on the scale at the RHS.

Hartley (9) and Bridoux et al (10,11). Time-averaged data demonstrating the feasibility of instantaneous gradient data acquisition have also been presented by Black and Chang (23). At the
same axial position as that corresponding to Fig. 4, we show in
Fig. 5 the joint pdf for instantaneous values of temperature
T x velocity u, i.e., the correlation <T·u>. Further details of
the latter measurement are presented in Ref. 4.

Also described in Ref. 4 is a new optical layout for LV data
acquisition which permits a significant increase in the overlap
between the Raman and LV probe test volumes. The worth of the
various correlations of density and temperature with velocity is
critically dependent upon the accuracy of this overlap at all
flame measurement positions. Thus, one must either lock the
Raman and LV probes together in a precise but movable fashion -
a rather difficult procedure for the precision required for
"bench scale" laboratory flames - or else translate the flame.
We have chosen the latter approach, and show in Fig. 6 a sketch
of a movable fan-induced co-flowing turbulent jet combustion
tunnel. The working section is a 15 cm x 15 cm square pipe with
large glass windows giving clear optical access to the turbulent
diffusion flame produced on a 3-mm-diameter fuel tube.

The accuracy of the temperature pdf data obtained with the
Raman Stokes/anti-Stokes technique has been assessed by tests
made on a known and well-calibrated laminar premixed flame source,
viz., a porous plug burner (20). These data, which were checked
by analytical calculations based upon the optical and electronic
properties of our detection system, showed a roughly 5-7%
standard deviation, which has been considered acceptable for
present measurement purposes (2,7). However, additional problems,
not considered in this type of test, can exist. For example:
Does our turbulent flame test volume (an approximate cylinder,
0.7 mm high, with a volume less than 0.1 mm^3) correspond essentially to isothermal conditions at all times? Are the flame
gases in the test volume at chemical equilibrium? Are assumptions
such as Lewis number Le (ratio of thermal diffusivity to mass
diffusivity) equal to one valid in modeling the flame gases? And
so forth.

In order to probe some of these questions - an essential
endeavor in forming a clear interpretation of our results - we
wish to compare our experimentally-determined data with predictions from a simple model. The experimental data available (See
Fig. 3) are instantaneous values of flame temperature from the N_2
Stokes/anti-Stokes intensity ratio (plotted as histograms in Fig.
4) and simultaneously-obtained values of N_2 density (determined
from the absolute value of the N_2 Stokes intensity
calibrated against the value obtained for N_2 in ambient air).
Accordingly, we have produced "comparison" plots using the following scheme (24): If we calculate flame gas density and temperature as a function of flame stoichiometry (i.e., as a function of the fuel/air equivalence ratio ϕ; see Fig.7), then we can

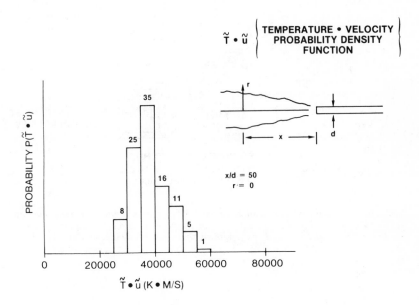

Figure 5. *Probability density function (pdf or histogram) for temperature* X *velocity for turbulent diffusion flame. These data correspond to a test zone along the axis, 50 fuel-tip diameters downstream from the fuel line tip.*

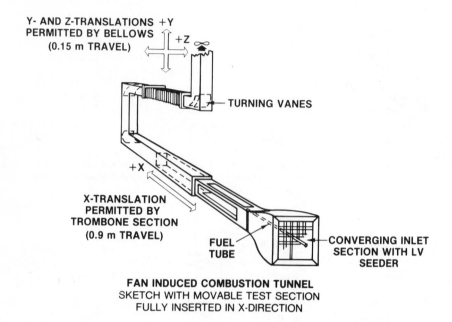

Y- AND Z-TRANSLATIONS +Y
PERMITTED BY BELLOWS
(0.15 m TRAVEL)
+Z

TURNING VANES

+X

X-TRANSLATION
PERMITTED BY
TROMBONE SECTION
(0.9 m TRAVEL)

FUEL
TUBE

CONVERGING INLET
SECTION WITH LV
SEEDER

FAN INDUCED COMBUSTION TUNNEL
SKETCH WITH MOVABLE TEST SECTION
FULLY INSERTED IN X-DIRECTION

Figure 6. Fan-induced combustion tunnel. This sketch shows the movable test section fully inserted in the x-direction. For purposes of scale, the square test section is 0.15 m × 0.15 m and the length of the optical viewing windows is 0.9 m.

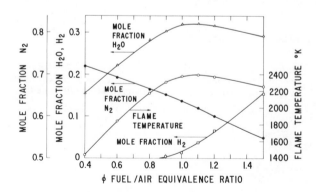

Figure 7. Plots of major flame species and temperature for H_2–air flame as a function of flame stoichiometry (i.e., fuel–air equivalence ratio ϕ) for adiabatic conditions

cross-plot the density vs. temperature for any particular species, with ϕ as a parameter along the curve. This is shown for nitrogen in Fig. 8, where the solid curve corresponds to calculated values of nitrogen concentration plotted as a function of the flame temperature for the assumptions suggested in the preceeding paragraph, i.e., isothermal test volume, adiabatic flame conditions with chemical equilibrium, Le = 1, etc. Here, the upper branch of the curve (i.e., that part of the curve corresponding to values of N_2 concentration greater than that for ϕ = 1 - the stoichiometric point) corresponds to fuel-lean conditions, and the lower branch to fuel-rich conditions. Thus, any experimental datum for nitrogen concentration and temperature (shown as a box symbol in Fig. 8) should, to a degree according to its satisfaction of the ideal assumptions, fall along or nearby the theoretical curve - i.e., correspond, to within the optical/electronic experimental accuracy, to some value of stoichiometry. Departures from the theoretical curve then indicate either spread in the experimental data - caused by random or systematic errors - or a failure in our simple adiabatic model to account for the observed data.

The data plotted in Fig. 8 were taken near the flame boundary, 50 fuel-tip diameters downstream, with no optical background corrections made to the vibrational Raman raw data. With such corrections, the data appear as in Fig. 9. (It is these data that are plotted in Fig. 4 in the top histogram, corresponding to r = 14.5 mm.) Similarly, Figs. 10 and 11 show data taken near the flame axis; the data in Fig. 11 appear in Fig. 4 in the bottom histogram, corresponding to r = 1 mm. What do we learn from these plots?

Firstly, the rough agreement of the N_2 concentration-vs-temperature data with adiabatic calculations implies that gross errors are unlikely in the Raman data acquisition, and that the basic analytical and experimental assumptions are reasonable. Secondly, the scatter of the data in these preliminary runs appears to exceed that expected from the photon statistics, and indicates, most likely, that greater control over experimental "calibration" parameters (such as ambient air N_2 Stokes signals) is required, as are greater amounts of data to define better the experimental results. Furthermore, inspection of these pairs of curves does not permit one to decide clearly between the procedures of utilizing or not utilizing background radiation corrections (the clear measurement of which is difficult); alternately, one can interpret this result as a rough indication that precise optical background measurements are not inordinately critical to the interpretation of the data.

Finally, we note what may turn out to be a significant departure in the fuel-rich data. In Figs. 10 and 11, the N_2 concentrations appear to deviate from the theoretical curve increasingly as ϕ increases. No detailed explanation for this behavior has emerged yet, but possible departure of Le from unity for a H_2-rich flame may lead to an explanation in terms of "non-ideal" behavior of mass and heat transport.

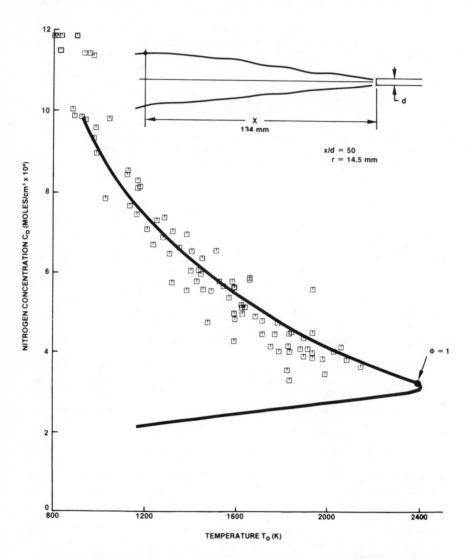

Figure 8. Nitrogen concentration vs. temperature, determined from Raman data at position shown in H₂–air turbulent diffusion flame. The solid theoretical curve, corresponding to adiabatic conditions, was obtained by replotting the information in Figure 7. The theoretical point for stoichiometric combustion ($\phi = 1$) is shown on this curve as a filled-in circle. These Raman data were not corrected for optical background at the Raman spectral band position.

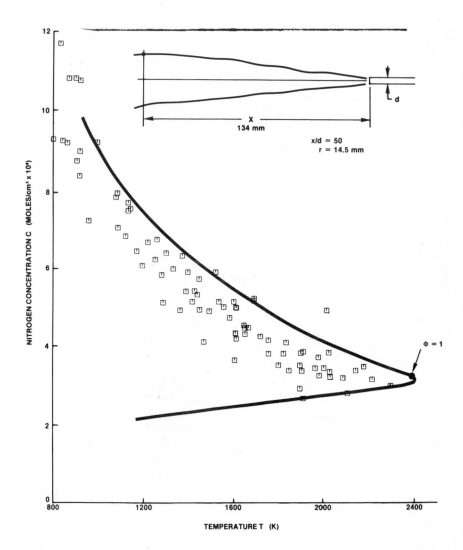

Figure 9. Nitrogen concentration vs. temperature, determined from Raman data at position shown in H_2–air turbulent diffusion flame. These Raman data were corrected approximately for optical background at the Raman spectral band position.

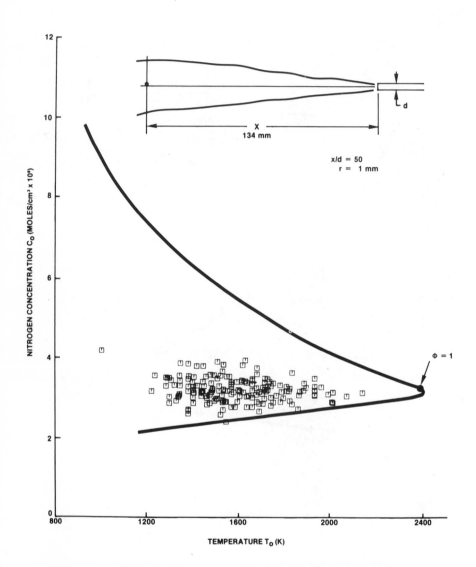

Figure 10.   Nitrogen concentration vs. temperature, determined from Raman data at position shown in H₂–air turbulent diffusion flame. These Raman data were not corrected for optical background at the Raman spectral band position.

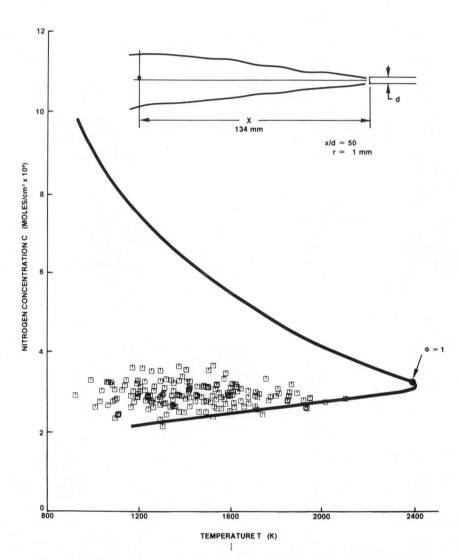

Figure 11.   Nitrogen concentration vs. temperature, determined from Raman data at position shown in H_2–air turbulent diffusion flame. These data were corrected approximately for optical background at the Raman spectral band position.

The emphasis in this work has been on the acquisition of simultaneously-obtained instantaneous values of temperature and concentration, with as high a spatial resolution as practical for such experiments. The temporal and spatial resolution requirements result from the necessity to probe within (if at all possible) characteristic turbulence time and length scales. The accuracy of our experiments (which, in any case, utimately depends upon a trade-off with resolution (1)), is considered to be adequate to achieve the diagnostic goal of providing data of value to flame modelers; this can be seen by comparison of the fluctuation temperature measurement uncertainty (characterized by a 5-7% standard deviation) with the broad temperature spread of the measured pdf's (extending, in Fig. 4, from values near ambient temperature to values in the vicinity of the adiabatic flame temperature).

Conclusion

Time- and space-resolved fluctuation data for flame gas temperature and major species densities have been obtained from Raman scattering and from stronger inelastic scattering processes. When combined with information about velocity from laser velocimetry, these data and their correlations provide key new information for flow field and combustion field modeling.

Acknowledgement

The author is grateful to his colleagues, M. C. Drake, C. M. Penney, and S. Warshaw, with whom he has collaborated in all phases of this work, and to B. Gerhold and R. M. C. So for valuable discussions on analyses of flames. He also acknowledges the generous support of the Office of Naval Research (Project SQUID), the Air Force Office of Scientific Research, and the U. S. Department of Energy for portions of this research effort.

Literature Cited

1. Lapp, M.; Penney, C. M. in "Advances in Infrared and Raman Spectroscopy"; Clark, R. J. H.; Hester, R. E., Ed., Vol. 3; Heyden; London, 1977; Chap. 6.

2. Lapp, M.; Penney, C. M. in "Proceedings of the Dynamic Flow Conference 1978 on Dynamic Measurements in Unsteady Flows"; Proceedings of the Dynamic Flow Conference 1978; P. O. Box 121, DK-2740 Skovlunde, Denmark, 1979; p. 665.

3. Lederman, S. Prog. Energy Combust. Sci.,1977, 3, 1.

4. Lederman, S.; Celentano, A.; Glaser, J. Phys. Fluids, 1979, 22, 1065.

5. Eckbreth, A. C.; Bonczyk, P. A.; Verdieck, J. F. Appl.
    Spectrosc. Rev., 1978, 13, 15.

6. Rambach, G. D.; Dibble, R. W.; Hollenbach, R. E. WSS Paper
    No. 79-51, 1979 Fall Western States Section Combustion Meet-
    ing; Western States Section/The Combustion Institute, Pitts-
    burgh, PA.

7. Lapp, M. in "Proceedings of the Sixth International Conference
    on Raman Spectroscopy"; Schmid, E. D.; Krishnan, R. S.;
    Kiefer, W.; and Schrötter, H. W., Ed., Vol. 1; Heyden: London,
    1978; p. 219.

8. Warshaw, S.; Lapp, M.; Penney, C. M.; Drake, M. in this
    volume.

9. Hartley, D. in "Laser Raman Gas Diagnostics"; Lapp, M.; Penney
    C. M., Ed., Plenum Press: New York, 1974; p. 311.

10. Bridoux, M.; Crunelle-Cras, M.; Grase, F.; Sochet, L. R. in
    "Proceedings of the Sixth International Conference on Raman
    Spectroscopy"; Schmid, E. D.; Krishnan, R. S.; Kiefer, W.;
    Schrötter, H. W., Ed., Vol. 2; Heyden: London, 1978; p. 256.

11. Bridoux, M.; Crunelle-Cras, M.; Grase, F.; Sochet, L. R.
    C. R. Acad. Sc. Paris, 1978, 286, 573.

12. Pealat, M.; Bailly, R.; Taran, J. P. E. Opt. Comm., 1977, 22,
    91.

13. Bailly, R.; Pealat, M.; Taran, J. P. E. in "Proceedings of
    the Sixth International Conference on Raman Spectroscopy";
    Schmid, E. D.; Krishnan, R. S.; Kiefer, W.; Schrötter, H. W.,
    Ed., Vol. 2; Heyden: London, 1978; p. 258.

14. Chabay, I.; Rosasco, G. J.; Kashiwagi, T. in "Proceedings of
    the Sixth International Conference on Raman Spectroscopy";
    Schmid, E. D.; Krishnan, R. S.; Kiefer, W.; Schrötter, H. W.,
    Ed., Vol. 2; Heyden: London, 1978; p. 516.

15. Chabay, I.; Rosasco, G. J.; Kashiwagi, T. J. Chem. Phys.,
    1979, 70, 4149.

16. Birch, A. D.; Brown, D. R.; Dodson, M. G.; Thomas, J. R.
    J. Fluid Mech., 1978, 88, 431.

17. Birch, A. D.; Brown, D. R.; Dodson, M. G.; Thomas, J. R.
    J. Phys. D: Appl. Phys., 1975, 8, L167.

18. Penney, C. M.; Warshaw, S.; Lapp, M.; Drake, M. in this volume.

19. Lapp, M.; So, R. M. C. to appear in "Proceedings of the AGARD
    Specialists Meeting on Testing and Measurement Techniques in
    Heat Transfer and Combustion"; Brussels, May 5-7, 1980.

20. Lapp, M. in "Laser Raman Gas Diagnostics"; Lapp, M.; Penney,
    C. M., Ed.; Plenam Press: New York, 1974; p. 107.

21. Drake, M.; Rosenblatt, G. M. Chem. Phys. Lett., 1976, 44, 313.

22. Williams, W. D.; Power, H. M.; McGuire, R. L.; Jones, J. H.;
    Price, L. L.; Lewis, J. W. L. AIAA Paper 77-211, 1977.

23. Black, P. C.; Chang, R. K. AIAA J., 1978, 16, 295.

24. Drake, M.; Lapp, M.; Penney, C. M.; Warshaw, S., submitted for
    publication.

RECEIVED March 31, 1980.

Temperature from Rotational and Vibrational Raman Scattering: Effects of Vibrational-Rotational Interactions and Other Corrections

MICHAEL C. DRAKE

General Electric Corporate Research & Development, P.O. Box 8, Schenectady, NY 12301

CHAMNONG ASAWAROENGCHAI and GERD M. ROSENBLATT

Department of Chemistry, The Pennsylvania State University, University Park, PA 16802

Raman spectroscopy and the closely related technique of coherent anti-Stokes Raman spectroscopy are becoming increasingly important techniques for measuring temperatures in combustion and other high temperature reactive environments. The determination of temperatures from rotational or vibrational Raman spectra requires comparison with theoretical relative peak intensities or band profiles. In this paper we examine theoretical factors which enter into calculations of relative Raman intensities in order to assess the thermometric accuracy and useful temperature ranges of rotational and vibrational Raman scattering from N_2, O_2 and H_2. For pure rotational Raman scattering the factors considered are intensity corrections for centrifugal distortion (f_{oo}) and for rotational scattering from vibrationally excited molecules (η^{oo}). For vibrational Raman scattering, the factors considered are O and S-branch scattering, anisotropic Q branch scattering, and intensity corrections arising from vibrational-rotational interactions (f_{o1}).

Rotational Raman Scattering

Analysis of experimental rotational Raman scattering from N_2, O_2, and H_2 has been used to determine temperatures in premixed laboratory flames (1,2). Temperatures based upon rotational Raman scattering from N_2 and O_2 had lower uncertainties (1-4%) than those based upon vibrational Raman scattering (3-9%) because rotational Raman scattering is generally more intense and gives rise to many more transitions. However, careful application of Raman intensity theory is required.

The theory of rotational and vibrational Raman intensities is discussed in detail elsewhere (e.g., References 1-6). Relative rotational Raman intensities are proportional to Raman line strength factors (S'). For rigid rotator, harmonic oscillator diatomic molecules $S'(J_i,J_f) = 3(J_i+1)(J_i+2)/(2(2J_i+3))$ where J is a rotational quantum number. However, real molecules are not rigid rotators and S' must be

0-8412-0570-1/80/47-134-231$05.00/0
© 1980 American Chemical Society

multiplied by a correction factor f_{00} to account for centrifugal-distortion. From James and Klemperer [7]

$$f(J)_{00} = [\, 1+(4/\chi\,)(B_e/\omega_e)^2(J^2+3J+3)\,]^2$$

where $\chi = (\alpha_{//}\, -\alpha_{\perp})_e / \{r_e\,[\partial\,(\alpha_{//}\, -\alpha_{\perp})/\partial r]_e\}$

Experimentally determined values of χ for $H_2(0.38\pm0.01)$, $N_2(0.45\pm0.09)$, and $O_2(0.23\pm0.07)$ have recently been reported [8] and are used here to calculate f_{00} values for H_2 and N_2 shown in Figure 1. The f_{00} values for N_2 (and O_2) are small but are much larger for H_2 because B_e/ω_e is about an order of magnitude larger for H_2. The inclusion of the f_{00} factor lowers temperatures calculated from N_2 rotational spectra by 1% and temperature calculated from H_2 rotational spectra by 7% for temperatures near 2000K [1,2]. In addition, because of the very large values of f_{00} for H_2 transitions with $J > 5$ there is some question whether this first order linestrength correction factor is sufficiently accurate for quantitative intensity analysis. Further experiments are in progress to elucidate this potential limitation of H_2 rotational Raman intensity analysis.

Temperatures measured from N_2 (and O_2) rotational Raman spectra are considerably more precise than those from H_2 because of the larger, more uncertain f_{00} correction for H_2 and because many fewer lines are measured with H_2. However, when excited vibrational states are sufficiently populated ($T \gtrsim 1000K$ for N_2 or O_2), an additional correction factor η is needed to correct measured rotational Raman peak intensities for the contribution from pure rotational Raman scattering from molecules in vibrationally excited states. This correction is necessary because transitions involving given rotational levels in different vibrational levels overlap to a substantial degree at low J but are displaced at high J. The displacement $\delta\nu$ is easily determined from known molecular energy levels from

$$\delta\nu = 4(J_{lower} - 3/2)\,\alpha_e v$$

where α_e is the usual spectroscopic rotational vibration interaction constant. Although the effect of the vibrationally excited molecules (η correction) is important for N_2 and O_2, it is not required for H_2 spectral analysis since α_e for H_2 is sufficiently large that rotational transitions in different vibrational levels are easily resolved by conventional spectrometers. Since the η correction factor for N_2 and O_2 is a strong function of temperature (vibrational population) and of spectrometer resolution (experimental slit function), it introduces additional complexities and uncertainties into N_2 and O_2 rotational temperature calculations. The η corrections raises computed temperatures over calculations which neglect it – by 8% for N_2 rotational spectra at 2000K for triangular spectrometer slit widths of 2.6 cm^{-1} fwhm.

The effect of the η correction is demonstrated in Figure 2 in which an analysis of experimentally measured rotational Raman intensities from O_2 in a H_2-O_2 premixed flame is presented. The calculated

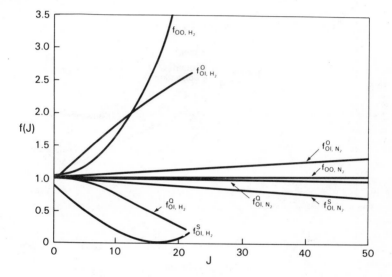

Figure 1. Rotational–vibrational line strength correction factors for pure rotational Raman scattering $(f_{oo})_o$ and for O-, S-, and Q-branch vibrational Raman scattering $(f_{ol}, f_{ol}{}^S, and f_{ol}{}^Q)$. The value J is the rotational quantum number of the initial level; ($\bigcirc$), Stokes; ($\triangle$), anti-Stokes.

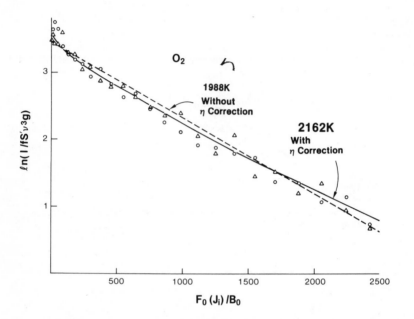

Figure 2.   Temperature-analysis plot for rotational Raman scattering from O_2 in an H_2–O_2 premixed flame. The experimental spectrum is in Figure 5 of Ref. 1. All data are corrected for centrifugal distortion: (– – –), analysis without consideration of rotational transitions from vibrationally excited O_2 molecules (η correction); (——), results from the more accurate analysis including the η correction.

temperature is 1988K without the η correction and 2162K with the η correction. Not withstanding the scatter in the data, the fit is clearly better using the η correction.

Vibrational Raman Scattering

Vibrational Raman scattering from diatomic molecules is dominated by $Q(\Delta J=0)$ branch transitions although weaker $O(\Delta J=-2)$ and $S(\Delta J=+2)$ side branches do occur. The Q branch line strength $[S'(J_i,J_f)=\alpha_o(2J+1)+J(J+1)(J+2)/(2(2J-1)(2J+3))]$ includes an isotropic part (the α_o term) and a smaller anisotropic contribution. Vibrational Raman band profiles for N_2 and H_2 that were calculated previously in order to determine flame temperatures included only the isotropic part of the Q branch (see, for example, References 5 and 6). Here additional factors of O- and S-branch scattering, anisotropic Q branch scattering, and vibrational–rotational interaction intensity corrections (f_{o1}) have all been included. Figure 3 shows the effect of these factors on N_2 vibrational Raman spectra at 2000K. In the Q branch region the two curves are very similar but depart measurably for the high vibrational level transitions. The temperature errors which would result from not including the factors discussed in this paper for N_2 at 2000K would equal zero for temperatures determined from Stokes-antiStokes intensity ratios, $<5K$ for temperatures determined from Stokes intensity ratios for $v = 1 \to 2$ and $v = 0 \to 1$, and $< 20K$ for temperatures determined from a complete Stokes vibrational band profile fit. Thus these correction factors have little effect on temperatures determined from N_2 or O_2 vibrational Raman spectra.

However, those same correction factors are important in H_2 vibrational Raman intensity analyses primarily because of the large values of $f_{o1}(H_2)$ given in Figure 1. For a Stokes vibrational Raman spectrum of H_2 at 2000K the calculated temperature would be 1935K if only the isotropic part of the Q branch were included and 1989K if the line strength correction factor f_{o1} were added.

Summary

Temperature corrections arising from higher orders effects in Raman intensity analysis are summarized in Table 1. At 300K all of the corrections are negligible but at elevated temperatures the effects can be large. Particularly important for rotational Raman spectra from N_2 and O_2 are corrections (η) for vibrationally excited molecules. Because these corrections are strongly dependent on the spectrometer slit function, they are difficult to determine reliably and may limit the applicability of spontaneous rotational Raman spectra for these molecules to $<2200K$. Above these temperatures vibrational Raman spectra from N_2 and O_2 probably would provide more accurate temperatures. The much higher spectral resolution available using rotational CARS (coherent antiStokes Raman spectroscopy) may make η

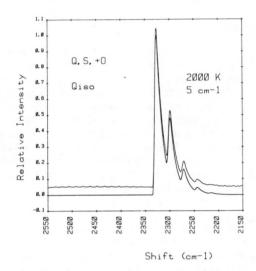

Figure 3.   Calculated band profiles of Stokes vibrational Raman scattering from N_2 at 2000 K assuming a triangular slit function with FWHM = 5.0 cm^{-1}. The bottom curve includes the isotropic part of the Q-branch only. The top curve is a more exact calculation including O- and S-branch scattering, the anisotropic part of the Q-branch and line-strength corrections owing to centrifugal distortion. The base lines have been shifted vertically for clarity.

Table I.   Temperature Corrections Caused by Higher-Order Effects

	η		f		O,S, and Q_{ani}	
	300K	2000K	300K	2000K	300K	2000K
Rotational Raman						
N_2 and O_2	0	+(8-10)%	∼0	-1%	—	—
H_2	0	0	-1%	-7%	—	—
Vibrational Raman						
N_2 and O_2	—	—	0	∼0	0	∼0
H_2	—	—	∼0	+3%	∼0	+½%

corrections much smaller for this technique and extend its useful upper temperature range. Temperatures from H_2 rotational (or vibrational) Raman spectra may be in error because of the large and possibly inadequate f_{QQ} (and f_{Q1}^Q) corrections, particularly for transitions involving $J > 5$. Finally, temperatures determined from vibrational Raman spectra from N_2 and O_2 are not strongly influenced by the factors considered here (<20 K error at T=2000 K).

Literature Cited

1. Drake, M.C.; Rosenblatt, G.M. Comb. Flame, 1978, 33, 179.

2. Drake, M.C.; Rosenblatt, G.M. in National Bureau of Standards Special Publication 561, "Proceedings of 10th Materials Symp. on Characterization of High Temperature Vapors and Gases"; Hastie, J.W., Ed.; U.S. Govt. Printing Office, Washington, D.C., 1979; p. 609.

3. Drake, M.C.; Grabner, L.H.; Hastie, J.W. ibid p. 1105.

4. Weber, A. in "The Raman Effect"; Anderson, A., Ed.; Marcel Dekker, New York, 1973; Vol. 2; p. 543.

5. Lederman, S. Prog. Energy Combust. Sci., 1977, 3, 1.

6. Lapp, M.; Penney, C.M. in "Advances in Infrared and Raman Spectroscopy"; Clark, R.J.H., Hester, R.E., Ed.; Heyden and Sons, London, 1977; Vol. 3; Chap. 6.

7. James, T.C.; Klemperer, W. J. Chem. Phys. 1959, 31, 130.

8. Asawaroengchai, C.; Rosenblatt, G.M. J. Chem. Phys. in press.

RECEIVED February 1, 1980.

Temperature-Velocity Correlation Measurements for Turbulent Diffusion Flames from Vibrational Raman-Scattering Data

S. WARSHAW, MARSHALL LAPP, C. M. PENNEY, and MICHAEL C. DRAKE

General Electric Corporate Research & Development, P.O. Box 8, Schenectady, NY 12301

Raman scattering flame diagnostic methods have been developed to provide improved test probe capability for hostile flame environments (1, 2, 3, 4). One long-term goal of such efforts is to contribute to a better understanding of the interplay between turbulent fluid mechanics and flame chemistry through application of laboratory measurements to flame modeling (5, 6). Virtually simultaneous measurements (from at least a fluid mechanic point of view) of a range of key flame properties can produce substantially increased insight over measurements of single properties (7). We have focused upon the simultaneous determination of fluctuations in flame temperature and gas velocity for our initial study.

We present here preliminary results for the (temperature x velocity) probability density function shown in this paper as $\langle T \cdot u \rangle$, where the quantities within the average brackets are instantaneous values. These data have been obtained from a co-ordinated experimental program utilizing pulsed laser vibrational Raman scattering and cw real fringe laser velocimetry (LV). These instantaneous temperature and velocity values can be related to values of the average fluctuating mass flux $\langle \rho' u' \rangle$ for our experimental conditions, utilizing assumptions of the ideal gas law and fast flame chemistry. Here ρ' and u' are fluctuation values of density and velocity, respectively, Knowledge of flame properties such as $\langle \rho' u' \rangle$ provides key data needed for developing improved combustion models.

For these experiments, a well-defined H_2-air diffusion flame was produced in a co-flowing jet combustor (8). The axisymmetric configuration was chosen and implemented with care in order to produce well-defined flame conditions suitable for testing analytical modeling concepts. The 3-mm-diameter fuel tube was centered within a 100-mm-diameter glass pipe test section, through which air was driven by a fan. Air flow speeds were roughly 10 m/s. Using a fuel-to-air speed ratio of 11:1, 0.5-m-long diffusion flames were produced.

0-8412-0570-1/80/47-134-239$05.00/0

© 1980 American Chemical Society

The flame gas temperature was determined utilizing vibrational Raman scattering from nitrogen. A 1J, 1μs duration tunable
dye laser was used as the probe source. Multiple photomultipliers mounted in a polychromator monitored nitrogen Stokes and
anti-Stokes Raman signals, as well as signals from other major
flame constituents, viz., water vapor and hydrogen. The temperature was found from the ratio R of nitrogen vibrational Stokes
scattered intensity to the corresponding anti-Stokes signal:
R = K exp (Q/T), where the constant K incorporates spectroscopic
and optical system constants. The value of K was experimentally
determined by calibration against a premixed laminar flame with
a known temperature produced on a porous plug burner. The
characteristic vibrational temperature Q = hcω/k = 3374°K for
nitrogen. Here, h is Planck's constant, c is the speed of light,
ω is the vibrational constant, k is Boltzmann's constant, and T
is the temperature in Kelvin.

The statistical nature of the turbulent flame required the
analysis of many temperature and density data points from separate pulses for accurate results. Thus, an overall computer
system was used to control the various components of the combustion probe apparatus, and to collect and interpret the resultant data in an accurate and timely fashion. This system
produced a block of data for each laser shot that included information about the Raman signals, LV readings, and ancillary
data such as an identifying shot number and corresponding dye
laser pulse energy. Typical current operation permits about
twenty experimental run conditions daily, with up to several
hundred shots per run.

In Fig. 1 we show the system control flow chart. The laser
and computer system, after recharging from the previous laser
shot, was armed and readied for a seed particle to flow through
the LV probe region. Either 1 μm diameter alumina or 0.25 μm
titanium dioxide was used to seed the flame. The particle count
rate was continuously adjustable from 0.2 to 1000 valid velocity
measurements per second. For initial development purposes, an
average 0.5 s particle interarrival delay was chosen.

The virtually simultaneous LV and Raman measurements do not
materially interfere with each other. When a particle enters the
probe region, the light from real fringes is Mie-scattered. The
scattered light intensity appears as a sine wave with a Gaussian
envelope representing the finite beam overlap region. In Fig. 2
the particle flight time through a preset number of fringes of
known spacing is shown. If the measurement meets a minimum
accuracy criterion, based on comparison of particle flight time
through 5 and 8 fringes, then that velocity value is held and the
dye laser source for Raman scattering is triggered. The velocimeter instrumentation required a minimum 4 μs delay after the
required number of fringe crossings to validate the velocity
reading. An additional delay occurs between validation and the
dye laser pulse. This adjustable time delay is sufficient to
sweep the particle out of the volume so that the particle does

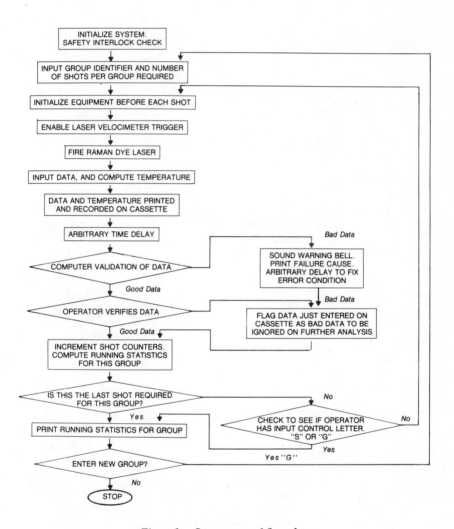

Figure 1. System control flow chart

not scatter the pulsed dye laser light. A 10 μs delay was
actually used in this experiment; with flame gas velocities
corresponding to our measurement positions for this experiment (8),
this delay produced data well within the spatial resolution goal
of a cube of 1 mm dimension. Note that the right-angle Mie
scattering of a 1 W cw argon laser beam from a 1 μm particle is
not strong enought to significantly alter the background level of
the photomultiplier detectors in the spectrometer focal plane at
other than the positions corresponding to 488 or 514 nm. If the
seeding is too heavy, it is possible that the dye laser pulse
could occur during the passage of a second seed particle in the
region, thus causing a non-damaging saturation of the Raman
photomultipliers, the signal from which is automatically rejected
by the computer.

Electronic signal conditioning circuitry was developed to
capture and hold transient signals produced in the Raman scattering process because the analog baseline signal was observed to
fluctuate with ground noise and temperature drift. In initial
experiments, photographic recordings were made of ocilloscope
records of the photomultiplier analog signals, which were then
reduced manually. In its current operational form, dual sample-
and-hold (S/H) circuits capture the analog signal from each
photomultiplier before and after the nominal 1-μs-duration dye
laser pulse. The delayed S/H captures its signal an adjustable
10 μs after the dye laser trigger pulse - long enough for the dye
laser noise to have abated. All of the channels are acquired in
parallel but are sampled sequentially by the computer.

Probability density functions, or histograms, of the product
of instantaneous temperature x velocity were obtained through use
of this combustion probe system for a variety of downsteam and
radial flame test positions. A typical histogram is shown in
Fig. 3, while Fig. 4 displays the same data (as well as data for
a test position further downsteam) in a "scattergram" format;
i.e., in a plot of velocity vs. temperature. Here, each datum
corresponds to a specific shot, while the histogram bins correspond to integrated results from numbers of shots.

The data presented in this paper were taken on a less than
optimum burner-optical geometry setup. The flame, Raman source
laser beam, and angle bisector of the laser velocimeter probe
beams were orthogonal to each other, permitting only roughly 20%
spatial overlap between the regions probed by the Raman and LV
methods. Thus, the statistical correlation of turbulent velocity
with temperature could well be masked by the large non-overlap
region. Additionally, separate collection optics caused difficulties in the simultaneous alignment of the LV and Raman probe
regions. These sources of error were corrected in our current
optical layout geometry where the LV beams' bisector is colinear
with the illuminating laser for Raman scattering, allowing for
more than 90% overlap. (See Fig. 5.) Also, we can now look at
the scattered LV light through the Raman collection optics. This

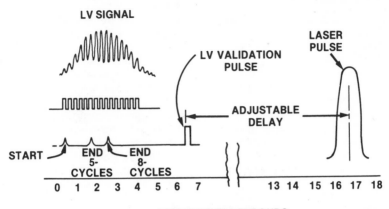

LV SIGNAL

LASER PULSE

LV VALIDATION PULSE

ADJUSTABLE DELAY

START **END 5- CYCLES** **END 8- CYCLES**

0 1 2 3 4 5 6 7   13 14 15 16 17 18

TIME IN MICROSECONDS

Figure 2.   LV–Raman scattering timing sequence. The LV signal, both raw and with conditioning and the LV validation pulse, are shown on the time scale 0–7 μsec. The resultant laser pulse occurs after an additional adjustable delay, which was set for these experiments at about 10 μsec in order to allow slow seed particles to escape from the test volume.

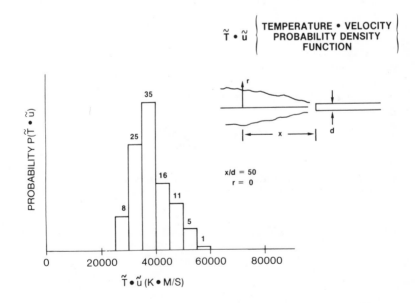

$\tilde{T} \cdot \tilde{u}$ | **TEMPERATURE • VELOCITY PROBABILITY DENSITY FUNCTION**

PROBABILITY $P(\tilde{T} \cdot \tilde{u})$

35

25

16

11

8

5

1

x/d = 50
r = 0

0   20000   40000   60000   80000

$\tilde{T} \cdot \tilde{u}$ (K • M/S)

INSTANTANEOUS VALUE $\tilde{X}$ = MEAN VALUE X + FLUCTUATION VALUE X′

Figure 3.   Probability density function (pdf or histogram) for temperature × velocity for turbulent diffusion flame. These data correspond to a test zone along the axis, 50 fuel-tip-diameters downstream from the fuel-line tip.

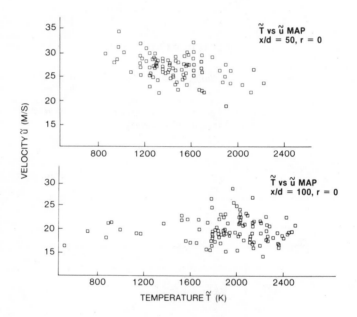

Figure 4.   Scattergram of temperature and velocity for same measurement position in the turbulent diffusion flame corresponding to Figure 3 and for a position twice as far downstream

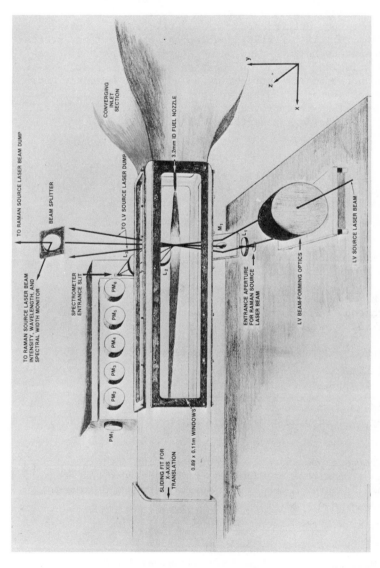

Figure 5.   Current overall optical layout for laser velocimetry and Raman scattering diagnostics, shown here on new fan-induced square-cross-section movable combustion tunnel. Note the co-linear Raman and LV probe laser source axes and the colinear detection optics.

clearly defines the cylindrical regions for the Raman and LV probes, utilizing the same axes and apertures.

The basic limitations to the overall accuracy of the data presented here lie in the Raman measurement process - inherently weak, but possessing sufficient intensity as utilized here to produce, for example, only 5-7% standard deviations for instantaneous temperature determinations in a "calibrated" premixed laminar flame (9).   Further development of this light scattering measurement technique - including better treatment of background radiation and optical calibration problems, greatly increased amounts of data contributing to pdf's and moments, and, when possible, use of fixed-bed optical probes and movable combustors - will lead to improved accuracy, and to increased utility for combustion modeling using these data.

Acknowledgement

The authors are grateful to the Office of Naval Research (Project SQUID), the Air Force Office of Scientific Research, and the US Department of Energy for sponsorship of portions of this work.

Literature Cited

1. Lapp, M.; Penney, C. M., Ed. "Laser Raman Gas Diagnostics"; Plenum Press: New York, 1974.

2. Lapp, M.; Penney, C. M. in "Advances in Infrared and Raman Spectroscopy"; Clark, R. J. H.; Hester, R. E., Ed., Vol. 3; Heyden and Sons: London, 1977; Chap. 6.

3. Eckbreth, A. C.; Bonczyk, P. A.; Verdieck, J. F. Appl. Spectrosc. Rev., 1978, 13, 15.

4. Lederman, S. Prog. Energy Combust. Sci., 1977, 3, 1.

5. Hartley, D.; Lapp, M.; Hardesty, D. Physics Today, 1975, 28, (12), 36.

6. "High Temperature Science: Future Needs and Anticipated Developments"; Committee on High Temperature Science and Technology, National Research Council; National Academy of Sciences: Washington, D.C., 1979; Chap. 3.

7. Bilger, R. W. Prog. Energy Combust. Sci., 1976, 1, 87.

8. Wang, J. C. F.; Gerhold, B. W. AIAA Paper 77-48, 1977.

9. Lapp, M.; Penney, C. M. in "Proceedings of the Dynamic Flow Conference 1978 on Dynamic Measurements in Unsteady Flows"; Proceedings of the Dynamic Flow Conference 1978: P. O. Box 121, DK-2740 Skovlunde, Denmark, 1979; p. 665.

RECEIVED March 31, 1980.

Observations of Fast Turbulent Mixing in Gases Using a Continuous-Wave Laser

C. M. PENNEY, S. WARSHAW, MARSHALL LAPP,
and MICHAEL DRAKE

General Electric Corporate Research and Development, Schenectady, NY 12301

Time- and space-resolved major component concentrations and temperature in a turbulent gas flow can be obtained by observation of Raman scattering from the gas. (1,2) However, a continuous record of the fluctuations of these quantities is available only in those most favorable cases wherein high Raman scattering rate and/or slow rate of time variation of the gas allow many scattered photons ($\gtrsim$ 100) to be detected during a time resolution period which is sufficiently short to resolve the turbulent fluctuations.(2,3) Fortunately, in other cases, time-resolved information still can be obtained in the forms of spectral densities, autocorrelation functions and probability density functions.(4,5)

Spectral densities and autocorrelation functions are Fourier transform pairs, and thus formally equivalent, although in a practical sense one or the other may be easier to measure in the range of interest. A probability density function (PDF) carries independent statistical information which shows the fraction of time during which the fluctuating quantity lies within each of a number of incremental ranges spanning the extent of its variation. Birch et al (6) have discussed the calculation of PDF's from observations of scattering from a continuous laser beam by the fluctuating flame gases. In this paper, we show that useful information about a PDF describing the fluctuations of the instantaneous concentration of a gas constituent can be obtained even in cases where an average of only one, or a few photons are detected per resolution period. Since the number of detected photons is proportional to the resolution period, the focus of our work is on the limits of time resolution of this technique for any specified experimental configuration.

The calculation of a PDF begins with experimental data in the form of a photon count distribution $F(j)$, defined in Table 1 along with the other functions discussed here.

The factorial moments derived from the count distribution are equal to the zero moments, Z_m, of the PDF, and simply related to the moments about its average value, C_m. (Table 1). The moments alone provide significant information about the concen-

0-8412-0570-1/80/47-134-247$05.00/0
© 1980 American Chemical Society

Table 1. Basic Quantities in Analyses of CW Laser Scattering for Probability Density Function. In Eq. 1 within the table, F(J) is the photon count distribution obtained over a large number of consecutive short periods. For example, F(3) expresses the fraction of periods during which three photons are detected. The PDF, P(x), characterizes the statistical behavior of a fluctuating concentration. Eq. 1 describes the relationship between Fj and P(x) provided that the effects of dead time and detector imperfections such as multiple pulsing can be neglected. In order to simplify notation, the concentration is expressed in terms of the equivalent average number of counts per period, x. The normalized factorial moments and zero moments of the PDF can be shown to be equal by substitution of Eq.1 into Eq.2. The relationship between central and zero moments is established by expansion of $(x-a)^m$ in Eq.(4). The trial PDF [Eq.(5)] is composed of a sum of k discrete concentration components of amplitude A_k at density x_k. [The functions $\delta(x-x_k)$ are delta functions.]

Relationship between photon count distribution and concentration PDF

$$F(j) = \frac{1}{j!} \int_0^\infty dx \ e^{-x} \times x^j P(x) \tag{1}$$

Factorial moments of count distribution

$$Z_m = a^{-m} \sum_{j=m}^{\infty} \frac{j!}{(j-m)!} F(j) \tag{2}$$

Moments from zero of PDF

$$Z_m = a^{-m} \int_0^\infty x^m P(x) \ dx \tag{3}$$

Moments about average of PDF

$$C_m = a^{-m} \int_0^\infty (x-a)^m P(x) \ dx \tag{4}$$

$$C_2 = Z_2 - 1, \ C_3 = Z_3 - 3Z_2 + 2, \ \text{etc.}$$

Trial function for PDF

$$P(x) = \sum_{k=1}^{K} A_k \ \delta(x - x_k) \tag{5}$$

tration fluctuations. For example, if the concentration is constant, then P(x) takes the form of a delta function, and all the zero moments equal unity. The second central moment yields the mean square deviation of the PDF, and all the odd central moments are zero if the PDF is symmetric about its average. Furthermore, the experimental moments can be used to correct for moderate dead time effects, which can be significant when fast time resolution is required. The details of this correction will be presented in a subsequent paper.

Although these and other characteristics provided by the moments are useful, a primary objective is to calculate the actual shape of the PDF. A straightforward approach is a standard least mean squares (LMS) fit ($\underline{8}$) of adjustable parameters in a trial function for the PDF, such as the one defined in Table 1. However, this kind of fit often does not produce a physically meaningful PDF because some of the coefficients A_k derived in the LMS fit turn out to be negative. One cause of negative coefficients is unavoidable statistical fluctuations in the count distribution F(J), which appear as deviations from the ideal distribution that would be obtained for the actual PDF in an unlimited data acquisition time. Another cause, which can be shown to produce negative coefficients even for an ideal count distribution, is an inopportune choice of trial components x_k. Even when all the A_k coefficients are positive for two different choices of the x_k, the actual fits may show considerably different shapes. Thus, if the choice of K and the x_k for a trial function is regarded as an initial bias, even physically acceptable final fits are bias-dependent. Furthermore, although linear optimization routines ($\underline{9}$) can provide a LMS fit with coefficients constrained to be positive, these routines are not always successful, and they also are bias-dependent.

We have found an alternative procedure which provides fits with positive coefficients which show consistent stability well within useful accuracy bounds when tested over a wide range of actual and simulated experiments. The procedure involves a large number (100-400) of LMS fits of a trial function of the form of Eq.(5) in Table 1. Several different values of K are used, typically ranging from four to eight, and the x_k are varied randomly over a limited range between each fit. Typically about one quarter of these fits produce count distributions within the expected statistical fluctuations of the original data, and contain no significant negative coefficients. All of these successful fits are combined to produce a composite distribution for the PDF which can be expressed, for example, as a histogram. The stability of this type of fit for repeated experiments with the same PDF, and repeated analysis of the same data with independent random choices for the x_k, suggest that this process, in effect on average over bias, reduces its influence to a low level.

The fitting procedure has been tested on computer simulated

data, time-varying signals generated by a light-emitting diode
(LED) and Raman scattering from an oxygen jet. The LED results
are described here because they combine a source whose time
variation could be verified directly, and a photomultiplier
detector viewing light at typical Raman scattering levels. The
experimental arrangement used with the LED is shown in Fig. 1.

In operation, the stop in the light path is adjusted so
that the count rate with no ND filter is sufficient (about 2.5
MHz) to provide an accurate time record of the LED source variation on the oscilloscope. At lower count rates obtained using
various ND filters, the gated counter (Tennelec Model TC592P) is
used to record the number of counts detected during consecutive
400 μsec periods. At the end of each period the counter provides
a voltage pulse whose height is proportional to the number of
counts recorded during that period. The distribution of these
voltages, accumulated over a large number of periods on the
multichannel analyzer, gives the pulse count distribution $F(J)$.
(Periods shorter than 400 μsec were not used for this demonstration because the counter analog output contained transients
which prevented reliable operation for shorter periods).

In Fig. 2 we show a quasi-sinusoidal LED output as measured
from the unattenuated signal on the oscilloscope, and the pulse
count distributions obtained with ND1 and ND3 filters (average
counts per 400 μsec channel of 100 and 1, respectively). Also
shown in this figure are two typical PDF fits to the latter data
(average count per cycle = 1), compared to the PDF calculated
from the LED signal variation displayed on the oscilloscope.
The figure illustrates that the 100 count per period count distribution reproduces the shape of the PDF fairly well, whereas
the 1 count per period distribution displays no obvious resem-
blance to the PDF. However, the typical fits calculated from
two independent sets of 1 count per period data, shown in Figs.
(2D) and (2E), reproduce the PDF to an accuracy which approaches
that of the 100 count per period result. The unoptimized For-
tran program which produced these fits requires approximately 20
seconds per run on a Honeywell DPS-2 computer. Our experience
with this and other shapes for the PDF leads to a conclusion
that the overall shape of a widely distributed PDF can be obtained reliably, though some ambiguity is found in the finer details
at average counts as low as one per period. As the average
count is increased to two or four per period, the resolution
improves steadily.

A significant advantage of the technique is that the equipment requirements are relatively modest and share a strong com-
monality with equipment and data analysis required for correlation or spectral density measurements, and laser velocimetry.
In particular, if the number of photons detected during each
period is recorded as a sequential record (instead of the simpler data recording mode utilized for this work) then autocorrelation functions and spectral densities of concentration

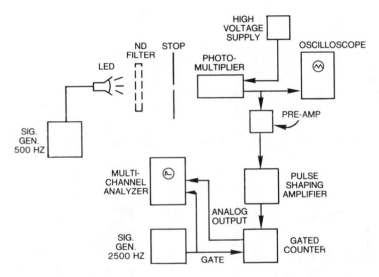

Figure 1. Experimental configuration to obtain detected photon-count distributions using a LED source

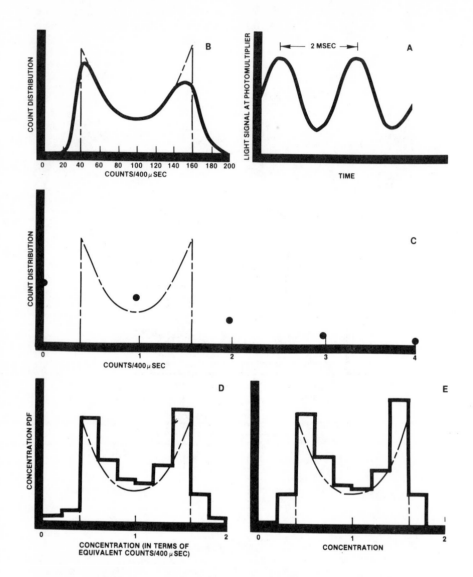

Figure 2.   *Experimental results obtained using configuration shown in Figure 1. An oscillograph tracing of the photomultiplier signal from the unattenuated LED is shown in A. The (slightly smoothed) pdf calculated from this signal is shown in B as a dotted curve, along with the photon count distribution obtained from the LED attenuated by a ND-1 filter, giving an average count of 100 detected photons per period. The calculated pdf is reproduced in C, D, and E. Also shown in C is the photon-count distribution out to F(4), for the ND-3 filter, giving an average count of 1 detected photon per period. This data actually extends out to 11 counts per period, usually observed a few times out of a total of 2,000,000 periods. In D and E, two typical pdf fits to independent sets of 1 cps, 2,000,000 total period data are shown. These fits were obtained using the multiple least-mean-squares technique described in the text.*

fluctuations, as well as PDF's, can be calculated from the same data set.

In comparison to the pulsed laser Raman technique which has been used for PDF measurements in flames (1,2) the CW laser technique is more sensitive to background light and probably less accurate when extremely fast time resolution (∿ μsec) is required. However, its relative ease of application, and the other advantages we have discussed, make this technique a good candidate for concentration PDF measurements in non-luminous turbulent flows, and low luminosity flames. For highly luminous subjects, a variation of this technique using a high frequency pulsed laser and gated detector can be used to provide strong background discrimination. This alternative may be advantageous for example, in measurements within sooting regions of a flame, where single pulse Raman measurements are questionable because the necessarily large laser pulse strongly heats the soot particles. (10)

Acknowledgements - We wish to thank Robert Dibble at Sandia Livermore for suggesting the test using the LED source, Marcus Alden at Chalmers University for his help during the earlier experimental phases of this work, and Donald R. White of this laboratory, who suggested the use of a high frequency pulsed laser for luminous gas diagnostics.

Literature Cited

1. Lederman, S., AIAA Paper No. 76-21, 1976.
2. Lapp, M. and Penney, C.M., in "Proceedings of the Dynamic Flow Conference 1978 on Dynamic Measurements in Unsteady Flows", p.665, Proceedings of the Dynamic Flow Conference, Skovlunde, Denmark, 1979.
3. Bailly, R., Pealot, M. and Taran, J.P.E., in "Proceedings of the Sixth International Conference on Raman Spectro scopy", Vol.2, (ed. by E.D. Schmid, R.S. Krishnan, W. Keifer, and H.W. Schrotter), pp.256-7, Heyden and Son, Ltd., 1978.
4. Birch, A.D., Brown, D.R., Dodson, M.G. and Thomas, J.R., J. Phys. D: $\underline{Appl.\ Phys.}$, 1975, $\underline{8}$, L167-L170.
5. Chabay, I., Rosasco, G.J. and Kashiwagi, T., J. Chem. Phys., 1979, $\underline{70}$, 4149.
6. Birch, A.D., Brown, D.R., Dodson, M.G. and Thomas, J.R., J. Fluid Mech., 1978, $\underline{88}$, part 3, 431-449.
7. Mandel, L. and Meltzer, D., IEEE J. Quantum Electronics, 1970, $\underline{QE-6}$, 661.
8. Margenau, H. and Murphy, G.M., "The Mathematics of Physics and Chemistry", Van Nostrand, Princeton, N.J., 1956.
9. Hillier, F.S. and Lieberman, G.J., "Introduction to Opera tions Research", Holden-Day, Inc., San Francisco (1967).
10. Eckbreth, A.C., Bonczyk, P.A. and Verdieck, J.F., $\underline{Appl.}$ Spect. Rev., 1978, $\underline{13}$, 15.

RECEIVED March 31, 1980.

A Nd:YAG Laser Multipass Cell for Pulsed Raman-Scattering Diagnostics

DOMENIC A. SANTAVICCA

Department of Mechanical and Aerospace Engineering, Princeton University, Princeton, NJ 08544

Spontaneous Raman scattering is an attractive diagnostic technique for measuring gas temperature and species concentration because it is a linear, non-resonant process and because of the unique spectral location of the vibrational Raman spectra of different molecules. It is unfortunately a very inefficient process which to date has limited its application to temperature and major species concentration measurements in relatively noise free environments.(1) Signal enhancement of measurements made in a steady environment can be achieved by signal averaging; however, signal averaging is not applicable to unsteady environments or to turbulent environments where the average Raman signal depends not only on the average temperature and average density but also on usually unknown density and temperature correlations.(2,3).

The effect of signal averaging is to increase the number of detected Raman photons which after background subtraction results in increased effective signal-to-noise ratio. In order to increase the photon yield without averaging over time intervals greater than the characteristic time scale of the phenomenon under study, it is proposed to use a pulsed multipass configuration, whereby the same pulse is repetitively passed through the scattering volume using an optical multipass cell.

Optical multipass cells have been used for the enhancement of CW Raman scattering(4); however, these cells are typically not well-suited for use with high power, pulsed lasers. A new multipass cell for use with a pulsed Nd:YAG laser is proposed whereby the 1.06 micron laser output is admitted into a multipass cell cavity where it is partially converted to 532nm with a Brewster's angle cut second harmonic generating crystal   The 532nm pulse is trapped in the mirrored cavity while the 1.06 micron pulse is dumped. This multipass cell concept has been demonstrated with the experimental set-up shown in figure 1. The pump laser is a Quanta-Ray Nd:YAG Model DCR-1A with an 8 nsec, 700 mJ (max), 1.06 micron output. The multipass cell cavity is bounded by the normal incidence harmonic beamsplitter (>99.5%

0-8412-0570-1/80/47-134-255$05.00/0
© 1980 American Chemical Society

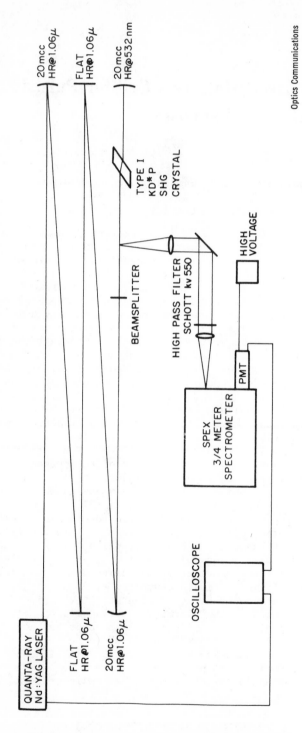

Optics Communications

Figure 1. Far-field isolator and 532-nm pulsed multipass cell experiment (5)

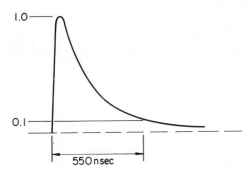

Optics Communications

Figure 2.   Nitrogen Stokes vibrational Raman signal from pulsed multipass cell (5)

reflectance at 532nm,   > 85% transmittance at 1.06 micron)
and the 20 meter radius of curvature mirror (> 99.7% reflectance at 532nm). The cavity mirror separation is 1.68 meters.
The first four mirrors (> 99.7% reflectance at 1.06 micron)
act as a far field isolator which locates the multipass
cavity 15 meters away from the laser and effectively isolates
the laser from the potentially damaging retroreflected 1.06 micron radiation from the normal incidence beam splitter. The
multipass cavity is aligned by monitoring the retroreflected
1.06 micron pulse which is found to emerge from the Nd:YAG laser
cavity, 120 nsec after the original pulse, when optimum alignment
is achieved.

The performance of this pulsed multipass cell is shown in
figure 2 where it is seen that the nitrogen vibration Raman
multipass signal decays to 10% of its initial strength in
550 nsec or 100 passes. This corresponds to a multipass cell
efficiency of 97.7% and a gain of 42.

The reader is directed to reference 5 for additional information on this work.

The author would like to acknowledge the financial support
of AFOSR Grant 76-3052, DOE contract EF-77-S-01-2762, and NSF
Grant ENG-77-12941.

Literature Cited

1. Lederman, S., Prog. Energy Comb. Sci., 1977, 3, 1.
2. Eckbreth, A. C., Comb. and Flame, 1978, 31, 231.
3. Setchell, R. E., AIAA Paper No. 76-28, 1976.
4. Hill, R. A., Mulac, A. J., and Hacket, C. E., Appl. Opt.,
    1977, 16, 7, 2004.
5. Santavicca, D. A., Opt. Comm., 1979, 30, 423.

RECEIVED February 11, 1980.

Time-Resolved Raman Spectroscopy in a Stratified-Charge Engine

J. RAY SMITH

Sandia Laboratories, Livermore, CA 94550

The objectives of this research were to develop techniques to measure both the mean and fluctuating nitrogen density and temperature in a combusting stratified charge engine. Such data is necessary for analytical engine model verification. The method chosen to achieve these measurements was spontaneous vibrational Raman scattering by a pulsed frequency doubled YAG laser to get a time resolution of 10 nsec. The nitrogen density was determined from the Stokes signal and the temperature was determined from the ratio of the anti-Stokes to Stokes signal. Setchell demonstrated the feasibility of using time-averaged Raman scattering in a combusting homogeneous charge engine.[1] A stratified charge engine with good optical access was recently developed, and its precombustion fuel-air distributions were determined by time-averaged Raman spectroscopy.[2] The latter engine's precombustion velocity and turbulence fields were measured by laser Doppler velocimetry and its performance and emissions were quantified by conventional methods.[3] The same engine design was used in the present study.

The short duration of the laser pulse at 5321 Angstroms precluded any movement of the spectrometer grating to allow spectral details to be resolved nor were there sufficient photons to use a multichannel detector. Therefore the spectrometer grating was fixed and the entire nitrogen Stokes band was integrated by a photomultiplier tube (PMT). Similarly, the anti-Stokes spectrum was integrated by a second photomultiplier. The major problem that must be solved in making Stokes nitrogen density measurements in a turbulent flow was pointed out by Setchell[4] to be the temperature dependence of the Raman scattered Stokes intensity. Because the transition probability is proportional to $(v + 1)$ where v is the initial vibrational state, the integrated Raman scattered intensity is not a unique function of number density. However, a theoretical analysis of the

0-8412-0570-1/80/47-134-259$05.00/0
© 1980 American Chemical Society

Stokes vibrational Raman spectrum of nitrogen has led to a method of making nitrogen number density measurements that are essentially independent of temperature. A variety of spectrometer slit convolutions and center wavelength settings were studied to determine their influence on the integrated Stokes Raman intensity versus temperature. As pointed out by Leonard[5] the basic approach is to select spectrometer settings which balance the increased transition probability of higher vibrational states against the decrease in population of the ground state as the temperature rises. Figure 1 is an example of the results for a trapezoidal slit of 10 by 50 Angstroms with the center wavelength varied from 6070 to 6073 Angstroms. By setting the spectrometer at 6072 Angstroms, it was possible to have the Stokes intensity vary by only ±2% while the temperature varied from 300 to 1970 Kelvins. The significance of this result is that it is not necessary to make simultaneous temperature measurements in a turbulent flow field in order to make density measurements.

A similar analysis of the anti-Stokes to Stokes intensity ratio expected from nitrogen as a function of bandwidth center position of the anti-Stokes spectra is shown in Figure 2. It appears extremely difficult to use spontaneous vibrational Raman scattering for determining temperatures of less than 1000 K due to the small anti-Stokes signal. Although temperatures in an engine are well above this level after the flame passes through the scattering volume, the present detection system does not have sufficient background rejection to make temperature measurements with good signal to noise ratios unless the stratification is small and the equivalence ratio less than 0.8.

The engine in this work was designed to simplify the fluid mechanics for modeling purposes. The intake and exhaust valves, fuel injector, spark plug and laser input/output windows were located in the cylinder side walls in the clearance volume above the piston. The head contained a 70 mm diameter, clear aperture window. The laser beam was passed through the small windows and scattered light was collected at right angles to the beam through the large window in the top of the engine. The intake valve was shrouded which caused the air flow to swirl. Propane at 3.35 MPa (485 psi) and 375 K was injected radially toward the center of the cylinder. The overall equivalence ratio of the data presented was 0.7 and the engine speed was 900 rpm.

The experimental arrangement is shown in Figure 3. A shaft encoder was used to synchronize the laser pulse to the chosen crank angle. The Raman scattered light was imaged onto the slit of a 3/4 meter, single spectrometer and detected by the cooled PMT's. Only the first six stages of the PMT were used in order to maintain linearity over a wide dynamic range. The PMT output charge was integrated by a preamplifier, pulse shaped by a spectroscopy amplifier, digitized, and collected by a mini-computer. The Raman signal was normalized by the laser pulse energy, and by accounting for the amplifier gain, the number of

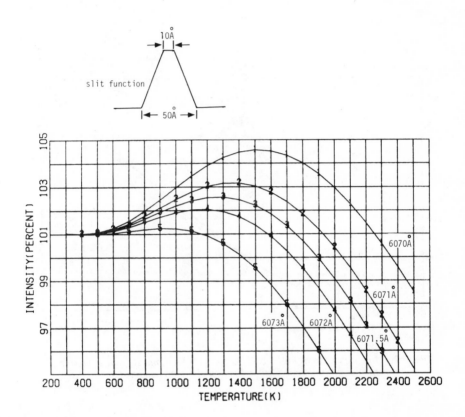

*Figure 1.   Relative Stokes vibrational Raman intensity for nitrogen for a trape-
zoidal slit function and various center positions*

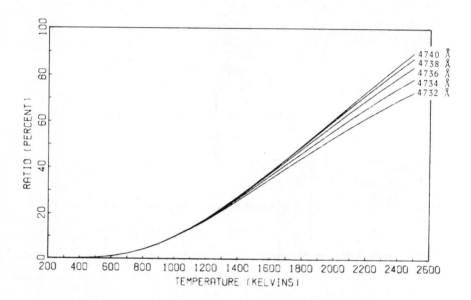

Figure 2.   Intensity ratio of anti-Stokes to Stokes vibrational Raman scattering for a trapezoidal slit function. Center position of Stokes bandpass at 6072 Å.

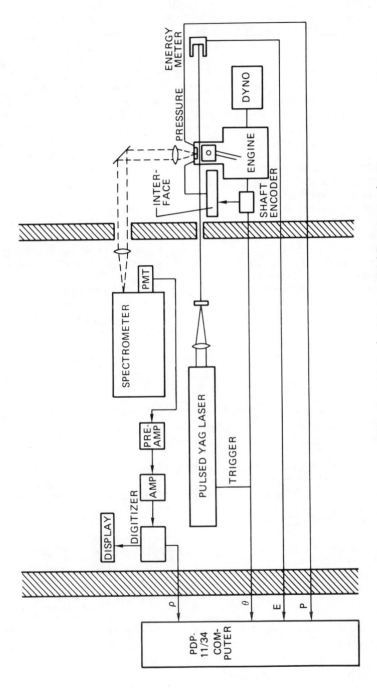

Figure 3. Experimental arrangements of time-resolved Raman experiment

photoelectrons was recorded. One thousand measurements were taken during successive engine cycles at each selected crank angle. Checks for possible sensitivity changes (due to window transmission) were made during the course of the data acquisition. The maximum sensitivity variation observed over a one hour period of lean engine operation was 2.4 percent change in the mean value.

Sub-microsecond time resolution was achieved by using a Quanta-Ray DCR-1A laser having a pulse width of 10 nsec. Although this laser was capable of producing in excess of 250 millijoules per pulse, only 50 millijoules were used in the experiment. Above this energy level the chances of window damage are greatly increased. All of the data presented were gathered with less than 20 MW/cm^2 power densities on the input/output windows. Gas breakdown was avoided by tilting the focusing lens relative to the beam axis thus introducing a large degree of astigmatism. This gave a scattering volume of 0.5 mm diameter by 1.25 mm length. The length was determined by the 4x magnification of the collection optics and the spectrometer entrance slit height of 5 mm. A check of the linearity of the Raman signal versus both nitrogen density and laser beam energy well beyond the ranges of the experiment was within two percent.

The Raman scattering process is very weak and the number of photoelectrons produced per pulse will obey Poisson statistics. If more than 100 photoelectrons are produced in each event, the uncertainty (one standard deviation) in the actual number of photoelectrons, N, may be approximated by $\sqrt{N}$ and the fractional uncertainty is $\sigma_p = \sqrt{N}$. Since nitrogen density fluctuations, σ_f, are not related to the photoelectron fluctuations, σ_p, they will combine randomly to give a signal fluctuation, $\sigma_s = (\sigma_f^2 + \sigma_p^2)^{1/2}$. Therefore the fractional fluctuations may be assessed by taking a sufficiently large sample of measurements to compute σ_s and using the mean N value of the signal to compute σ_p. This technique works well provided the actual fluctuations are at least half the size of the Poisson statistical fluctuations. Below this level the uncertainty in σ_p begins to dominate the computed σ_f value. In this experiment 200 to 800 photoelectrons were collected from each laser pulse which gave statistical uncertainties from 3.4 to 7 percent.

The mean values of the relative nitrogen density versus crank angle at the center of the combustion chamber are shown in Figure 4. Also shown (dashed line) for comparison is the nitrogen density expected from the piston motion. The density is relative to the nitrogen density in air at atmospheric pressure. Ignition occurred at 358 crank angle degrees and the flame arrived at the scattering volume at about 382 degrees. One observes the compression of the unburned gases ahead of the flame front after ignition.

The relative fluctuations in nitrogen density versus crank angle are indicated by the solid curve shown in Figure 5. Prior to ignition, the fluctuation level was quite low. The fluctua-

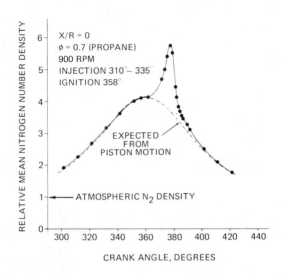

Figure 4. Relative mean nitrogen number density vs. crank angle

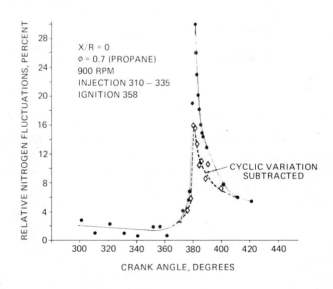

Figure 5. One standard deviation of fluctuations in nitrogen number density

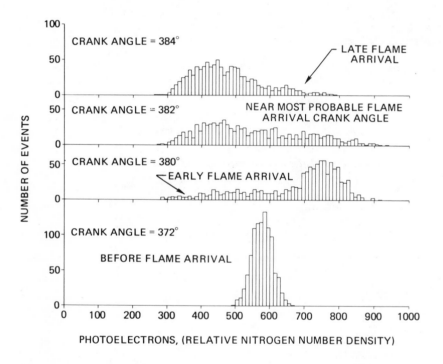

PHOTOELECTRONS, (RELATIVE NITROGEN NUMBER DENSITY)

Figure 6.   Histograms of nitrogen number density near time of flame arrival

tions reached a maximum near 382 degrees due to arrival of the flame. The fluctuations fell slowly in the post flame gases. The large peak value is due to the random variations in the arrival time of the flame front. Near 382 degrees some measurements were made just ahead of the flame front where high densities prevail and other measurements were made just behind the flame front where low densities prevail due to high temperatures. Clearly these large fluctuations are due to cyclic variations not turbulent fluctuations. The dashed curve is an attempt to remove this cyclic variation effect by using the most probable density value as the mean value of a normal distribution. The standard deviation of the distribution is determined from fitting the data to the side of the new mean that has not been distorted by flame arrival. The reduction of the apparent fluctuations near the flame arrival crank angle is dramatic. Both curves of Figure 5 have had the Poisson statistical fluctuations subtracted.

The histograms of Figure 6 represent the number of measurements versus the number of photoelectrons at crank angles near flame arrival in the scattering volume. The bin width is ten photoelectrons and the total number of events is 1,000 for each histogram. The normal shape of the histogram at 372 degrees is typical of those from 300 to 372 and 390 to 420 degrees. At 380 degrees early flame arrival caused the long tail in the distribution below the most probable density. Similarly, the distortion of the distribution to higher density values at 384 degrees was due to late flame arrivals. This explanation of the effects of cyclic variation on the distribution justifies the attempt to separate them from the real density fluctuations.

The Stokes signal-to-noise ratios were of the order of thirty to one even when the flame was in the scattering volume. It is likely that the increase in fluctuations immediately behind the flame front was flame induced turbulence. Improved fluctuation measurements are expected by using the temperature derived from the anti-Stokes channel for conditional sampling of the density data. However, this method cannot be used until the 5 microsecond integration time of present detection electronics is shortened to reduce the background luminosity signal on the anti-Stokes channel.

Acknowledgment
     This work supported by DOE and Motor Vehicle Manufacturers Association.

References

1. Setchell, R. E., 18th Annual Rocky Mountain Spectroscopy Conference, University of Denver, Aug. 2-3, 1976.
2. Johnston, S. C., SAE paper 790433, FEB. 1979.

3. Johnston, S. C., Robinson, C. W., Rorke, W. S., Smith, J.
    R., and Witze, P. O.. SAE paper 790092, Feb. 1979.
4. Setchell, R. E., 17th Aerospace Sciences Meeting, New
    Orleans, LA, Jan. 1979.
5. Leonard, D. A., Project SQUID, Tech. Rep. AVCO-1-PU, 1972.

RECEIVED February 1, 1980.

COHERENT RAMAN SPECTROSCOPY

Spatially Precise Laser Diagnostics for Practical Combustor Probing

ALAN C. ECKBRETH

United Technologies Research Center, East Hartford, CT 06108

With the increasing availability of laser sources, light scattering and wave mixing spectroscopic techniques are being increasingly employed in a broad spectrum of physical, chemical and biological investigations. The application of laser spectroscopy to the hostile, yet sensitive, environments characteristic of combustion, is particularly promising. Laser diagnostic techniques should facilitate improved understanding of a variety of combustion processes which should lead ultimately to enhanced efficiences and cleanliness in energy, propulsion and waste disposal systems. Spontaneous Raman scattering has received much attention for the remote, point probing of flames (1,2) but, due to its weak signal strength and incoherent character, is generally limited to investigations of major species and relatively clean flames (3). As soot levels increase, laser-induced interferences (4,5) can mask detection of the Raman signals, often by several orders of magnitude. Many practical flames, e.g. hydrocarbon-fueled diffusion flames, may consequently be beyond its applicability. With increasing emphasis on alternative and generally, less clean fuels, stronger diagnostic techniques need to be developed and refined.

Two techniques, which appear well suited to the diagnostic probing of practical flames with good spatial and temporal resolution, are coherent anti-Stokes Raman spectroscopy (CARS) and saturated laser fluorescence. The two techniques are complementary in regard to their measurement capabilities. CARS appears most appropriate for thermometry and major species concentration measurements, saturated laser fluorescence to trace radical concentrations. With electronic resonant enhancement (6), CARS may be potentially useful for the latter as well. Fluorescence thermometry is also possible (7,8) but generally, is more tedious to use than CARS. In this paper, recent research investi-

0-8412-0570-1/80/47-134-271$07.75/0
© 1980 American Chemical Society

gations into the practical feasibility of CARS and saturated
laser fluorescence at our laboratory will be reviewed. Tutorial
material will purposely be kept brief to minimize redundancy with
the earlier papers in this volume.

Coherent Anti-Stokes Raman Spectroscopy (CARS)

Coherent anti-Stokes Raman spectroscopy (CARS) (9,10,11,12)
is capable of the diagnostic probing of high interference environments due to its high signal conversion efficiency and coherent
signal behavior. CARS signal levels are often orders of magnitude
stronger than those produced by spontaneous Raman scattering. Its
coherent character means that all of the generated signal can be
collected, and over such a small solid angle that collection of
interferences is greatly minimized. CARS thus offers signal to
interference ratio improvements of many orders of magnitude over
spontaneous Raman scattering and appears capable of probing practical combustion environments over a broad operational range. In
experiments at UTRC (13), CARS has been successfully demonstrated
in a 50-cm dia. research scale combustion tunnel located in a jet
burner test stand. Measurements were made in the primary zone of
a highly swirled, coannular burner and in the exhaust of a JT-12
combustor can, both fueled with Jet A. CARS measurement demonstrations in combustion tunnels have also been performed recently
at Wright-Patterson AFB (14,15) and at ONERA. In England, CARS
measurements have been demonstrated in a gasoline-fired internal
combustion engine (16). With these "real world" demonstrations,
CARS is anticipated to see widespread utilization in practical
environments in the coming years. There is interest within NASA
to employ CARS for scramjet diagnostics and in the Army for bal-
listics studies. In this section, CARS will be briefly described
and its application to a variety of flames and molecular species
will be illustrated.

Theory. The theory and application of CARS are well explained
in several very good reviews which have appeared recently
(9,10,11,12). Briefly, incident laser beams at frequencies ω_1 and ω_2
(often termed the pump and Stokes beams respectively) interact
through the third order nonlinear susceptibility of the medium,
$\chi_{ijkl}^{(3)}$ $(-\omega_3,\omega_1,\omega_1,-\omega_2)$, to generate a polarization field which
produces coherent radiation at frequency $\omega_3=2\omega_1-\omega_2$. It is for
this reason that CARS is often referred to as "three wave mixing".
When the frequency difference $(\omega_1-\omega_2)$ is close to the frequency of
a Raman active resonance, ω_v, the magnitude of the radiation at

ω_3, then at the anti-Stokes frequency relative to ω_1, i.e. at $\omega_1 + \omega_v$, can become very large. Large enough, for example, that with the experimental arrangement described herein, the CARS signals from room air N_2 or O_2 are readily visible. By third order is meant that the polarization exhibits a cubic dependence on the optical electric field strength. In isotropic media such as gases, the third order susceptibility is actually the lowest order nonlinearity exhibited. The third order nonlinear suscep-tibility tensor is of fourth rank. The subscripts denote the polarization orientation of the four fields in the order listed parenthetically. In isotropic media, the tensor must be invari-ant to all spatial symmetry transformations and the 81 tensor elements reduce to three independent components, χ_{xyyx}, χ_{xyxy} and χ_{xxyy} where $\chi_{xxxx} = \chi_{xyyx} + \chi_{xyxy} + \chi_{xxyy}$. In CARS, which is frequency degenerate, $\chi_{xyxy} = \chi_{xxyy}$ and there are only two independent elements.

For efficient signal generation, the incident beams must be so aligned that the three wave mixing process is properly phased. The general phase-matching diagram for three wave mixing requires that $2\bar{k}_1 = \bar{k}_2 + \bar{k}_3$. $\bar{k}_i$ is the wave vector at frequency ω_i with absolute magnitude equal to $\omega_i n_i / c$, where c is the speed of light, and n_i, the refractive index at frequency ω_i. Since gases are virtually dispersionless, i.e., the refractive index is nearly invariant with frequency, the photon energy conservation condition $\omega_3 = 2\omega_1 - \omega_2$ indicates that phase matching occurs when the input laser beams are aligned parallel or collinear to each other. In many diagnostic circumstances, collinear phase matching leads to poor and ambiguous spatial resolution because the CARS radiation undergoes an integrative growth process. This difficulty is cir-cumvented by employing crossed-beam phase matching, such as BOXCARS (17), or a variation thereof (18,19). In these approaches, the pump beam is split into two components which, together with the Stokes beam, are crossed at a point to generate the CARS signal. CARS generation occurs only where all three beams intersect and very high spatial precision is possible.

Measurements of medium properties are performed from the shape of the spectral signature and/or intensity of the CARS radiation. CARS spectra are more complicated than spontaneous Raman spectra which are an incoherent addition of a multiplicity of transitions. CARS spectra can exhibit constructive and de-structive interference effects. Constructive interferences occur from contributions made from neighboring resonances, the strength of the coupling being dependent on the energy separation of the adjacent resonances and on the Raman linewidth which together determine the degree of overlap. Destructive interferences occur

when resonant transitions interfere with each other or with the
nonresonant background signal contributions of electrons and
remote resonances. For most molecules of combustion interest,
these effects can only be handled numerically. At UTRC, CARS
computer codes have been developed and validated experimentally
for the diatomic molecules, N_2, H_2,CO and O_2 (20) and one triatomic H_2O (21). Computer codes are extremely useful for studying
the parametric behavior of CARS spectra and, when validated, for
actual data reduction.

Experimental Approach. The CARS spectrum can be generated
in either one of two ways. The conventional approach is to employ
a narrowband Stokes source which is scanned to generate the CARS
spectrum piecewise, This approach provides high spectral resolution and strong signals and eliminates the need for a spectrometer. However, for nonstationary and turbulent combustion
diagnostics, it is not appropriate due to the nonlinear behavior
of CARS on temperature and density. Generating the spectrum
piecewise in the presence of large density and temperature
fluctuations leads to distorted signatures weighted toward the
high density, low temperature excursions from which true medium
averages cannot be obtained. The alternate approach (22) used here
is to employ a broadband Stokes source. This leads to weaker
signals but generates the entire CARS spectrum with each pulse
permitting instantaneous measurements of medium properties.
Repeating these measurements a statistically significant number
of times permits determination of the probability density function
(pdf) from which true medium averages and the magnitude of
turbulent fluctuations can be ascertained.
      Although CARS has no threshold per se and can be generated
with cw laser sources, high intensity pulsed laser sources are
required for most gas phase and flame diagnostics to generate
CARS signals well in excess of the various sources of interference and with good photon statistics, particularly with broadband
generation and detection. In the CARS work to be reported, a
frequency-doubled neodymium laser provides the pump beam and
drives the broadband Stokes dye laser as well as seen in Figure 1.
The laser actually emits two beams at the neodymium second
harmonic by sequentially doubling the primary and residual 1.06μ
from the first frequency doubler. The primary beam, 2ω, is
typically about 2W, i.e. 200 mJ pulses, 10^{-8} sec pulse duration,
at 10Hz, and the secondary, $2\omega'$, about an order of magnitude lower.
Various dyes and concentrations flowing through spectrophotometer
cells are employed to generate Stokes wavelengths appropriate to
the molecule being probed. Crossed-beam phase matching (BOXCARS)

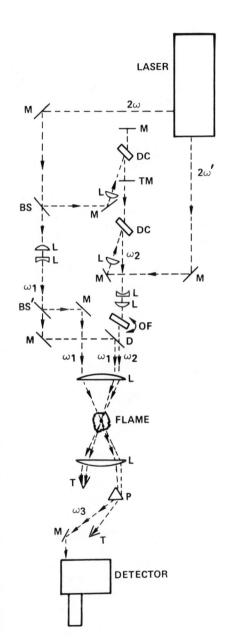

Figure 1. Schematic of BOXCARS experimental arrangement: BS, beamsplitter; L, lens; D, dichroic; OF, optical flat; P, prism; F, filter; DC, dye cell; T, trap; TM, partially transmitting mirror (24).

is used to ensure good spatial precision. The CARS signatures
are dispersed in a 0.6 or 1-m spectrograph and detected with an
optical multichannel analyzer (OMA) which permits capture of the
entire CARS spectrum in a single pulse. In laminar flames and
situations where fluctuation magnitudes are small, the CARS
spectrum can be averaged on the OMA or scanned with the monochromator using a boxcar averager. Greater detail about the
apparatus and procedures employed may be found in (23,24).

Thermometry. Nitrogen is the dominant constituent in airfed
combustion processes and is present in large concentrations despite
the extent of chemical reaction. Performing temperature measurements from N_2 provides information on the location of combustion
heat release and the extent of chemical reaction. Consequently,
considerable attention has been afforded N_2 thermometry (20). The
accuracy of CARS N_2 thermometry has been examined in premixed flat
flames by comparison with radiation-corrected, coated, fine wire
thermocouples. Employing constant Raman linewidth computer codes,
CARS gave temperatures slightly higher ($\sim$40K) than the thermocouples in the 1600-2100°K range. In the review by McDonald in
this volume, computer-synthesized N_2 CARS spectra (20,24) are
displayed as a function of temperature from 300 to 2400°K. At
high temperature, both the ground state band (v=0 to 1) and a
hot band (v=1 to 2) with rotational fine structure are apparent.
In Figure 2, the CARS spectrum from N_2 in a 2110°K flame is displayed together with the best visual computer fit which occurred
at 2150°K. The capability of CARS for measurements in highly
sooting flames has been demonstrated (25,26) and CARS has been
employed to perform detailed axial and radial temperature surveys
in sooting, laminar propane diffusion flames (24) Figure 3 presents
a comparison of single-pulse (10^{-8}sec) and averaged CARS signatures
in such a sooting flame at a spatial resolution of 0.3 x 1mm. The
spectra are interference free and of high quality. The single
pulse CARS spectrum is possible at laser energies an order of
magnitude lower than those typically employed for single pulse
spontaneous Raman scattering. Computer fitting these spectra
permits temperature to be determined and Figure 4 presents radial
profiles of temperature at five different heights in the sooting
flames. The axial temperature variation is displayed in the
review of McDonald in this volume. From the radial profiles, the
diffusive character of the flame is clearly evident. Recently,
as mentioned earlier, the feasibility of CARS for measurements in
practical combustion systems has been demonstrated. In tests at
UTRC (13),crossed-beam CARS thermometry has been performed in two

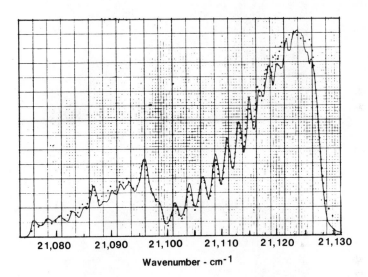

Combustion and Flame

Figure 2.   BOXCARS spectrum of N_2 over a 2.5-cm-diameter hexagonal flat flame burner operating on CH_4–air at 2110 K and 1-cm^{-1} spectral resolution. Dotted curve is the best computer fit at 2150 K, 0.8 cm^{-1} slit and 0.1 cm^{-1} Raman linewidths (20).

(a) SINGLE PULSE

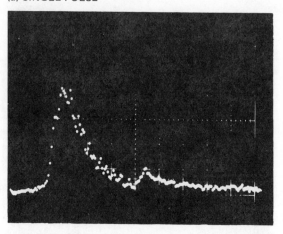

(b) AVERAGE

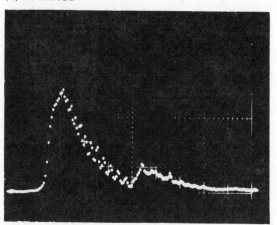

Combustion and Flame

Figure 3. Comparison of single-pulse and averaged CARS spectra of N_2 in a laminar propane diffusion flame recorded on optical multichannel analyzer (24)

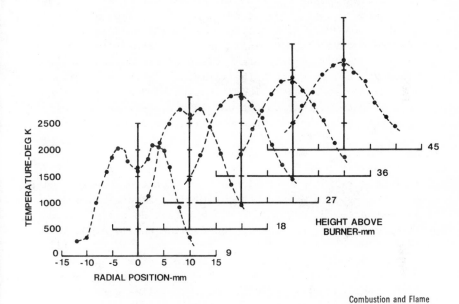

Combustion and Flame

Figure 4. Radial temperature profiles in a laminar propane diffusion flame (24)

different, liquid-fueled combustors housed in a 50-cm dia. combustion tunnel. Delicate instrumentation was housed in a control room adjacent to the burner test cell and the CARS signals were piped out employing 20 m long, 60µ dia. fiber optic guides (27). In Figure 5 are shown CARS signatures of N_2 at two different axial locations in the primary zone of a Jet A fueled swirl burner. At the upstream, x=8cm,location, CARS measurements were made through the fuel spray and the temperature was found to be about 900°K for an overall equivalence ratio of 0.8. At the downstream location, where the flame was visually very luminous, the temperature increased to 1500°K. In Figure 6 is shown a comparison of a single pulse and averaged CARS spectrum in the exhaust of a Jet A fueled JT-12 combustor can at cruise conditions. The temperature is approximately 1000°K. As seen there is little difference between the two, indicative of high single shot measurement quality. In Figure 7, the radial temperature profile of the JT-12 exhaust determined by CARS is shown. The variation in the centerline data values is due to small operational changes over a period of time. CARS measurements further downstream in the JT-12 exhaust were in good agreement with thermocouple probe values at several stoichiometries.

H_2 is ideal for combustion thermometry because of the simplicity of its spectrum as seen in the computer calculations (28) in Figure 8 . The vibration-rotation H_2 CARS spectrum consists of a series of well-spaced Q branch transitions. Figure 9 displays CARS signatures of H_2 obtained through a flat H_2-air diffusion flame (28). Such spectra, when reduced using simple algorithms deduced from computer calculations, permit temperature profiling of the flame as shown in Figure 10. There CARS measurements of temperature from H_2 and O_2 spectra show fairly good agreement with each other and with radiation-corrected fine wire thermocouple measurements. The O_2 spectra resemble those from N_2 except for the rotational fine structure.

Water vapor is the major product of hydrogen combustion and often the dominant product species of hydrocarbon-fueled combustion. Its measurement is an important gauge of the extent of chemical reaction and of overall combustion efficiency. Figure 11 presents a comparison of the experimentally-scanned and computer-synthesized spectrum of H_2O in the postflame region of a premixed CH_4-air flame (21). The gas temperature was approximately 1675°K. The very rich rotational structure of H_2O makes it a very promising candidate for thermometry. In the computer calculations, nearly eight hundred rotational transitions are accounted for. The various peaks represent groupings of transitions interfering constructively with one another. The H_2O CARS spectrum is quite

X = 6 CM   T = 900°K

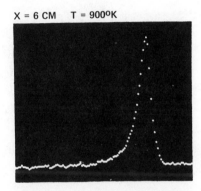

X = 39 CM   T = 1500°K

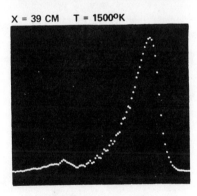

FREQUENCY ——————▶
0.586 CM⁻¹/DOT

Combustion and Flame

Figure 5.   Spatial variation of temperature from averaged CARS spectra of N_2 in swirl burner with Jet A fuel, an air flow of 0.15 lb/sec and an overall equivalence ratio of 0.8 (13)

SINGLE PULSE (10⁻⁸ SEC)

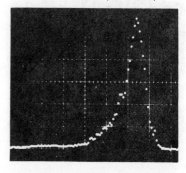

100 PULSE AVERAGE (10 SEC)

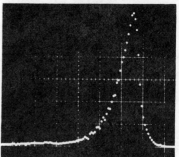

Combustion and Flame

Figure 6.   Comparison of averaged and single pulse N_2 CARS spectra in shrouded JT-12 combustor exhaust at cruise conditions (13)

FREQUENCY ———▶
0.574 CM⁻¹/DOT

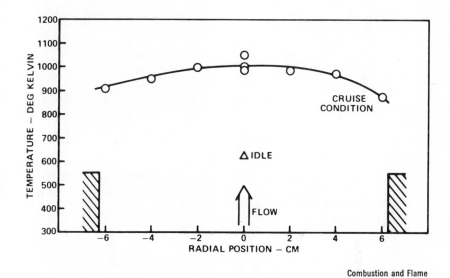

Combustion and Flame

Figure 7.   CARS temperature profile of shrouded JT-12 combustor can exhaust 13 cm downstream of exit plane at cruise conditions (13)

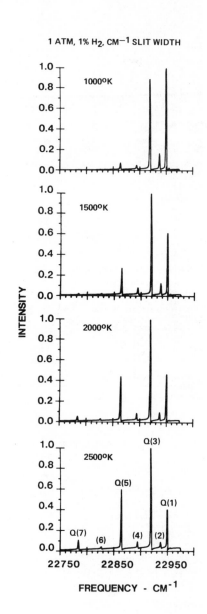

Figure 8. Computed temperature sensitivity of the CARS spectrum of H_2 at 1% concentration. Q-branch transitions are identified.

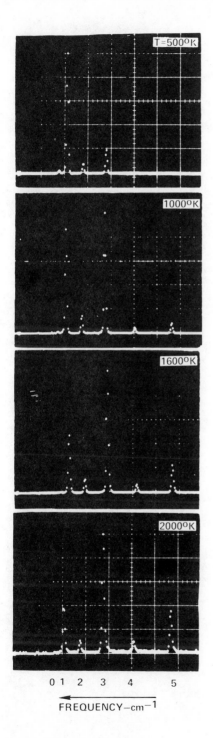

Figure 9.   CARS spectra of H_2 in a flat H_2–air diffusion flame at several temperatures determined from the identified Q-branch transitions

0 1 2   3   4   5

FREQUENCY—cm^{-1}

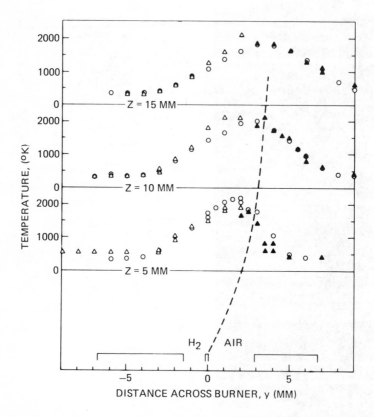

Figure 10.   Temperature measurements in flat H₂–air diffusion flame. The exit of the flat flame burner is shown schematically: (○), radiation-corrected thermocouple measurements; (△) H₂ CARS temperatures; (▲), O₂ CARS temperatures.

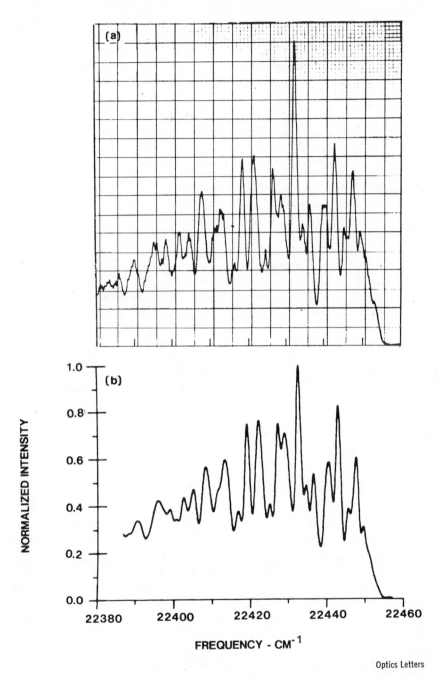

Optics Letters

Figure 11.   Comparison of experimental (a) and computed (b) CARS spectrum of H₂O in a premixed, methane–air flame at atmospheric pressure and 1675 K (21)

temperature sensitive (21) and should be very useful for thermometry.

    Concentration Measurements. Generally, species number
densities would be determined from the absolute intensity of the
spectrally integrated CARS spectrum (29). In certain concentration ranges however, species measurements can be performed from
the shape of the CARS spectrum. This is a unique feature of CARS
and arises from the interference of the desired resonant signal
with the background nonresonant electronic contribution (3). This
approach has been verified relative to microprobe sampling for CO
in flat premixed flames (23,25) and has been used to follow O_2
decay across a flat hydrogen/air diffusion flame (28). In Figure
12 are displayed computer calculations for O_2 at $2000^{\circ}K$. Various
oxygen concentrations illustrate the density dependence of the
spectral shape. In Figure 13, experimental O_2 CARS spectra
illustrative of this behavior are displayed. From such spectra
both O_2 concentration and temperature can be determined. When
the concentration becomes very low, the signal, i.e. modulation,
becomes imperceptible and concentration measurements are precluded
using this approach. In such instances, the nonresonant background
signal can be cancelled using polarization-sensitive CARS (30,31)
and the concentration obtained from the spectrally-integrated
CARS intensity. With polarization approaches, there is about a
factor of sixteen loss in signal strength. Coupled with the
density squared dependence of the signal at low concentrations,
such measurements can be performed only with considerable sacri-
fice in spatial and temporal resolution. Although CARS can be
electronically-resonantly enhanced (6), this approach is not
applicable to many molecules of combustion interest whose electronic
absorptions reside too far in the vacuum ultraviolet to be
spectrally accessible.

    CO_2 is the other dominant product of hydrocarbon-fueled
combustion and its CARS flame spectrum (23) is displayed in Figure
14. The spectrum is complicated by Fermi resonance and the fact
that the rotational transitions are closely overlapped. This
precludes treating them as independently broadened and so-called
collisional narrowing may need to be taken into account. Computer
modelling is currently in progress. Since all hydrocarbon fuels
are Raman-active, CARS should ultimately be capable of monitoring
total hydrocarbon concentrations during combustion as well.
    In summary, CARS is a powerful approach to the spatially-
and temporally-precise probing of combustion processes. It is
most suited to thermometry and major species concentration measurements and has been successfully demonstrated in a number of

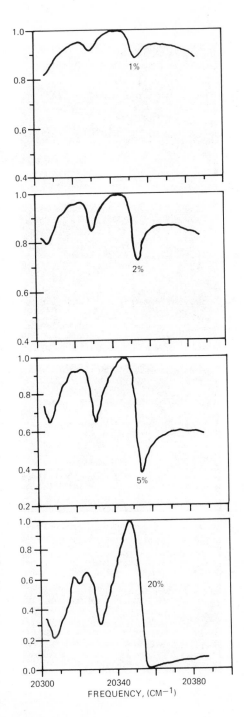

Figure 12. Computed CARS spectra of oxygen at 2000 K at various concentrations in a flame

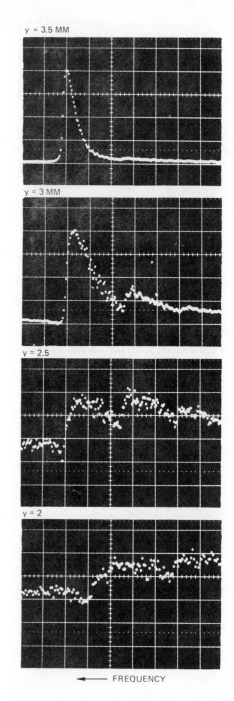

Figure 13.   CARS spectra of O_2 in flat H_2–air diffusion flame at various positions across the flame

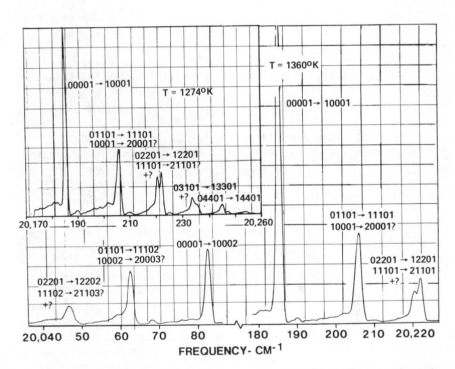

American Institute of Aeronautics and Astronautics

Figure 14.   Scanned CARS spectra of CO_2 in the postflame region of a $CO–O_2$ flame (23)

practical combustion environments. One may anticipate that CARS
will afford considerable advances in the combustion sciences in the
coming years.

Saturated Laser Fluorescence

With rare exceptions, CARS and spontaneous Raman scattering
are incapable of measuring species in very low concentrations,
i.e. ppm levels. Laser fluorescence has received considerable
attention recently in this regard, particularly for measurements
of flame radical concentrations as the papers in this volume
attest. Fluorescence is the spontaneous emission of radiation
from an upper electronic state excited in various ways; here
attention is restricted to excitation via absorption of laser
radiation only. Besides spontaneous radiative decay, the upper
state may also be deexcited due to collisions, a process termed
quenching, which reduces the level or efficiency of fluorescence.
Although analytic quenching corrections to fluorescence data are
possible, they are, most likely, feasible only in well characterized flames, where temperature and major quenching species concentrations are known. In less well characterized media, such
analytic corrections are probably quite inaccurate. Saturated
laser fluorescence (3, 32, 33) eliminates the need for quenching
corrections and, consequently, is undergoing detailed evaluation
at a number of laboratories.

Theory. In saturated laser fluorescence, the incident laser
intensity is made sufficiently large so that the absorption and
stimulated emission rates are much greater than the collisional
quenching rate which thus becomes unimportant. Another advantage
of working in the saturation regime is that the fluorescence
signal is maximized. For a given minimum signal detectability
level, saturation thus provides the highest species detection
sensitivity. Other approaches to avoid quenching corrections
(34,35) do not possess this advantage. Saturated fluorescence
has been observed in atomic species such as Tℓ and Zr,(36),and
Na (37,38) and in the molecular species MgO (38), C_2 (39), OH (40),
and CH and CN in our laboratory (41). Besides the molecules
listed, other candidate species for saturated fluorescence detection include CS, NH, NO, CH_2O, HCN, NH_2 and SO_2.
Saturated fluorescence is not without its complications and
difficulties. High laser intensities are required to achieve
saturation, but are difficult to obtain at certain wavelengths,
e.g. NO at ∿2265 Å . This is one reason OH has received much
study, since its absorptions reside at the frequency-doubled
wavelengths of powerful Rhodamine dye lasers. For many molecules

of interest, flashlamp-pumped dye lasers do not possess the
requisite spectral intensities to achieve saturation. Furthermore, the long pulse lengths, $O(10^{-6}\text{sec})$, permit more excited
state chemistry (42) to occur leading to measurement errors. Dye
lasers pumped by an appropriate harmonic of a Q-switched neodymium
laser generally possess the requisite intensities to saturate
and the short pulse lengths $O(10^{-8}\text{sec})$ minimize the potential of
laser induced chemistry. There has been some question whether
this is too short a pulse length to use. It is easy to show from
the time dependent rate equations in the saturation regime, that
the characteristic time for saturation to occur, i.e. the time
for absorption and stimulated emission to balance, is $\sim c/B_{12}I$
which is typically of $O(10^{-12}\text{sec})$. However, on the time scale
of the laser pulse, rotational equilibration does not necessarily
occur complicating the data analysis (43,44). Examples of rotationally nonequilibrated fluorescence spectra of CH and CN have been
obtained in our laboratory and will be displayed shortly. If
rotation is completely frozen or completely relaxed, the data
can be reduced in a straightforward manner. In between these
limits, the fractional population feeding the directly pumped
transition has to be estimated. A time dependent rate equation
model has been developed by Hall at UTRC to aid in estimating this
fractional population coupling. However,good rotational relaxation
rate data will be required to do this accurately, little of which
is available. Another problem concerns the focal intensity profile,
i.e. nonsaturation in the wings of the focussed laser beam. This
has been examined both analytically (45) and experimentally
(38,46). In 38, best agreement between fluorescence and absorption
data was obtained using the rectangular beam approximation.

CH, CN Investigations. In the previous studies of CH and CN
at our laboratory (41), the saturated fluorescence values were a
factor of two and five respectively below those determined by
absorption. If anything, the absorption measurements may have
been in error to the low side. The discrepancy, it was believed,
was due to the focussed laser beam diameter exceeding the radical
production region. This caused an overestimate of the fluorescence
sample volume, i.e. assumed to be the laser volume,and an under-
estimate of the species number density. Recently these experiments have been repeated using laser-pumped dye laserswhich have
a pulse width of 5-10 (10^{-9}) sec and a 10 Hz repetition rate.
The latter permits spectral scanning of the entire fluorescence
spectrum using boxcar averaging. In Figure 15 is shown a comparison of the normal CH flame emission from an oxy-acetylene slot
torch and the laser induced fluorescence spectrum. The CH was

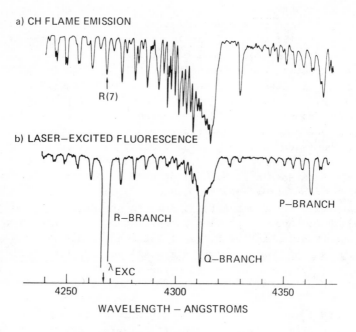

Figure 15. Comparison of CH flame emission and laser-excited fluorescence spectrum in an oxy-acetylene slot torch

excited by a Stilbene dye laser pumped by the third harmonic of
a Q-switched neodymium laser. As is apparent, the CH populations
do not have time to rotationally equilibrate during the laser
pulse. Saturation of the fluorescence was achieved and the data
reduced using the method developed in (39). In Figure 16 normal
CN flame emission and the laser induced fluorescence are compared
from an acetylene/nitrous oxide flame. To generate the appropriate radiation, a Rhodamine dye laser was pumped by the second
harmonic of Nd:YAG, then wavemixed in a crystal with the Nd:YAG
fundamental. As with CH, complete rotational equilibration does
not occur. This is apparent if one compares the 0-0 bandhead
width. In Figure 17, the saturation behavior of the fluorescence
is displayed. The fluorescence does not become independent of
laser intensity due presumably to an inability to achieve saturation in the wings. All of the measurements are summarized in
Table 1.

Table 1

Summary of Saturated Fluorescence Measurements

Species	ppm	$N(cm^{-3})$	$Q(sec^{-1})$
CH			
ABS*	57 $\pm$20	$1.6(10^{14})$	–
SF*(1)	23 $\pm$10	$7.1(10^{13})$	$3(10^{9})$
SF(2)	103 $\pm$50	$3.3(10^{14})$	$8(10^{11})$
CN			
ABS*	131 $\pm$34	$3.8(10^{14})$	–
SF*(1)	25 $\pm$11	$8(10^{13})$	$2(10^{9})$
SF (2)	380 $\pm$150	$1.1(10^{15})$	$2(10^{13})$

* Ref. (41).
SF(1) Flashlamp-pumped dye laser
SF(2) nxNd:YAG laser pumped dye laser

As can be seen the new fluorescence results are high compared to
absorption by a factor of two for CH and three for CN. The
fluorescence results are probably high since the initial Boltzmann
population fraction was used in the data analysis, i.e. frozen
rotation. However, in actuality, some rotational coupling from
adjacent rotational levels occurs during the pulse, and contributes
to the fluorescence. The fluorescence is thus larger than it
would be from a single level leading to an overestimate of the
total population. Dynamic modelling should allow estimates of the

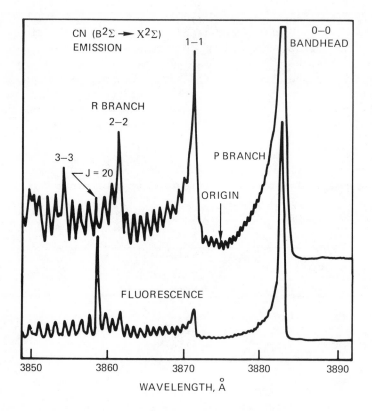

Figure 16.   Comparison of CN flame emission and laser-excited fluorescence spectrum in a nitrous oxide–acetylene slot torch

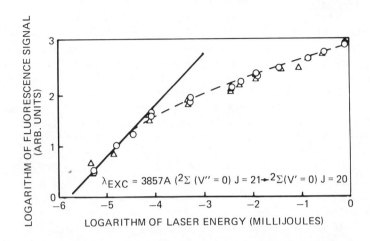

Figure 17. *Variation of laser-excited CN fluorescence intensity with laser spectral irradiance in a nitrous oxide–acetylene slot torch*

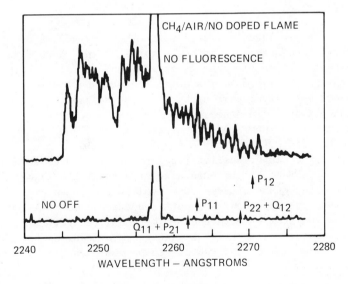

Figure 18. *Laser-excited NO fluorescence spectrum in an NO-doped methane–air premixed flat flame*

coupled population fraction for molecules where rotational relaxation data is available. In addition the absorption measurements are probably low since the absorbing beam waist was comparable to the radical gradient scale. In view of this and the other uncertainties mentioned earlier, saturated fluorescence at this stage of development appears to be accurate to within a factor of two.

NO Studies. To excite fluorescence in NO, radiation in the 2265 Å vicinity of the gamma bands is required. This is achieved by frequency-doubling a 2xNd:YAG pumped Rhodamine dye laser, and then sum frequency mixing with 1.06μ in a second crystal. Fluorescence was excited from NO doped in a premixed methane-air flame running near $2100^{\circ}K$. In Figure 18 is shown the laser excited NO fluorescence spectrum. Without the laser, no NO emission was observable. Up to the 0.1 mJ maximum pulse energy available corresponding to a spectral intensity of $6(10^6)$ $W/cm^2 cm^{-1}$, the variation of the NO fluorescence with laser energy was strictly linear. It is estimated (3) that laser spectral intensities of $0(10^8 W/cm^2 cm^{-1})$ will be required to saturate NO and efforts are currently directed toward improving the spectral intensity of the dye laser system.

Conclusions

CARS appears ideally suited to thermometry and major species concentration measurements in both practical and clean flame environments. It should see widespread application in both practical combustors and fundamental flame investigations, particularly where soot levels are high. Saturated laser fluorescence has great potential for the measurement of selected species in low concentration (ppm) in both practical and clean flames. Although the list of applicable species is limited, most are of extreme interest in combustion research. The fluorescence signals will be independent of gas quenching effects if the absorption resonances can be saturated. Two level models, when properly interpreted, are applicable to data reduction, but rotational relaxation/coupling effects need to be quantitatively evaluated. More fundamental research investigations are required to address these questions for this potential to be realized.

Acknowledgements

Separate portions of this research were supported by Project SQUID, the EPA and NASA (Langley). The author gratefully acknowledges the many contributions of his colleagues to the research described herein, notably Robert J. Hall, John A. Shirley, James F. Verdieck and Paul A. Bonczyk.

Literature Cited

1. Lapp, M.; Penney, C. M. "Laser Raman Gas Diagnostics";
   Plenum Press: New York, N. Y., 1974.

2. Lederman, S. Prog. Energy Combust. Sci., 1977, 3, 1.

3. Eckbreth, A. C.; Bonczyk, P. A.; Verdieck, J. F. Appl. Spect.
   Rev., 1978, 13,15.

4. Eckbreth, A. C. J. Appl. Phys., 1977, 48, 4473.

5. Aeschliman, D. P.; Setchell, R. E. Appl. Spect., 1975, 29,426.

6. Druet, S. A. J.; Attal, B.; Gustafson, T.K.; Taran, J.P.
   Phys. Rev. A, 1979, 18,1529.

7. Haraguchi, H.; Winefordner, J. D. Appl. Spect., 1977,31,330.

8. Bechtel, J. H. Appl. Opt., 1979, 18,2100.

9. Tolles, W. M.; Nibler, J. W.; McDonald, J. R.; Harvey, A. B.
   Appl. Spect.,1977, 31,253.

10. Durig, J. R., Ed. Vol.6; "Vibrational Spectra and Structure";
    Elsevier: Amsterdam, 1977.

11. Moore, C. B., Ed. "Chemical and Biological Applications of
    Lasers"; Academic Press: New York, N. Y., 1979.

12. Weber, A., Ed. "Raman Spectroscopy of Gases and Liquids";
    Springer-Verlag: Heidelberg, 1979.

13. Eckbreth, A. C. Combust. Flame, accepted for publication, 1980.

14. Switzer, G. L.; Roquemore, W. M.; Bradley, R. P.; Schreiber,
    P. W.; Roh, W. B. Appl. Opt., 1979, 18, 2343.

15. Switzer, G. L.; Roquemore, W. M.; Bradley, R. P.; Schreiber,
    P. W.; Roh, W. B., in this volume.

16. Stenhouse, I. A.; Williams, D. R.; Cole, D.R.; Swords, M. D.
    Appl. Opt., 1979, 18, 3819.

17. Eckbreth, A. C. Appl. Phys. Letts., 1978, 32, 421.

18. Laufer, G.; Miles, R. B. Opt. Comm., 1979, 28, 250.

19. Compaan, A.; Chandra, S. Opt. Lett., 1979, 4,170.

20. Hall, R. J. Combust. Flame, 1979, 35, 47.

21. Hall, R. J.; Shirley, J. A.; Eckbreth, A. C. Opt. Letts.,1979, 4, 87.

22. Roh, W. B.; Schreiber, P. W.; Taran, J. P. E. Appl. Phys. Letts., 1976, 29, 174.

23. Eckbreth, A. C.; Hall, R. J.; Shirley, J. A. AIAA Paper 79-0083, 1979.

24. Eckbreth, A. C.; Hall, R. J. Combust. Flame, 1979, 36, 87.

25. "Proceedings of the Seventeenth Symposium (International) on Combustion"; The Combustion Institute: Pittsburgh, Pa, 1979, p. 975.

26. Beattie, I. R.; Black, J. D.; Gilson, T. R. Combust. Flame, 1978, 33,101.

27. Eckbreth, A. C. Appl. Opt., 1979, 18,p. 3215.

28. Shirley, J. A.; Eckbreth, A. C.; Hall, R. J.,"Investigation of the Feasibility of CARS Measurements in Scramjet Combustion." presented at the 16th JANNAF Combustion Meeting, Monterey,Ca., 1979.

29. Roh, W. B.; Schreiber, P. W. Appl. Opt., 1978, 17, 1418.

30. Rahn, L. A.; Zych, L. J.; Mattern, P. L. Opt. Comm., 1979,30, 249.

31. Oudar, J. L.; Smith, R. W.; and Shen, Y. R. Appl. Phys. Letts., 1979, 34, 758.

32. Piepmeier, E. H. Spectrochim. Acta, 1972, 27B,431.

33. Daily, J. W., Appl. Opt., 1977, 16,569.

34. Stepowski, D.; Cottereau, M. J. Appl. Opt.,1979, 18,354.

35. "Proceedings of the Seventeenth Symposium (International) on Combustion"; The Combustion Institute: Pittsburgh, Pa., 1979, p.867.

36. Omenetto, N.; Benetti, P.; Hart, L. P.; Winefordner, J. D.; Alkemade, C. Th. J. Spectrochim. Acta, 1973, 28B,289.

37. Daily, J. W.; Chan, C. Combust. Flame, 1978, 33,47.

38. Pasternack, L.; Baronavski, A. P.; McDonald, J. R. J. Chem. Phys., (1978), 69,4830.

39. Baronavski, A. P.; McDonald, J. R. Appl. Opt., 1977, 16, 1897.

40. Lucht, R. P.; Laurendeau, N. M.; Sweeney, D. W. "Saturated Fluorescence Measurements on Diatomic Flame Radicals", in this volume.

41. Bonczyk, P. A.; Shirley, J. A. Combust. Flame, 1979, 34, 253.

42. Muller, C. K.; Schofield, K.; Steinberg, M. "Laser Induced Reactions of Lithium in Flames", 1978, Proceedings of the NBS 10th Materials Research Symposium, Gaithersburg, Md.

43. Lucht, R. P.; Laurendeau, N. M. Appl. Opt., 1979, 18,856.

44. Berg, J. O.; Shackleford, W. L. Appl. Opt., 1979, 18, 2093.

45. Daily, J. W. Appl. Opt., 1979, 17, 225.

46. Blackburn, M. B.; Mermet, J. M.; Boutilier, G. D.; Winefordner, J. D. Appl. Opt., 1979, 18, 1804.

RECEIVED February 1, 1980.

CARS Measurements in Simulated Practical Combustion Environments

GARY L. SWITZER—Systems Research Laboratories, Inc., Dayton, OH 45440

WILLIAM M. ROQUEMORE, ROYCE P. BRADLEY, and
PAUL W. SCHREIBER—Air Force Aero Propulsion Laboratory,
Wright–Patterson Air Force Base, OH 45433

WON B. ROH—Air Force Institute of Technology, Department of Physics,
Wright–Patterson Air Force Base, OH 45433

The measurement of temporally and spatially resolved temperature and species concentrations in the combustion zone of reacting gases presents a formidable problem. The conventional approach to solving this problem has been the use of mechanical probing techniques to obtain time-averaged data. Unfortunately, the perturbation of the reactive media due to the presence of probes remains an unknown factor. For this and other reasons, optical techniques that have the potential for real-time nonintrusive measurements are very desirable. By using the physical processes of radiation scattering or fluorescence, data can be collected from a temporally and spatially resolved point. However, either the data analysis required or the relatively low signal intensities involved may present serious problems. In overcoming these limitations, coherent anti-Stokes Raman spectroscopy (CARS) has been considered as a promising new method for combustion diagnostics. The signal levels generated by CARS may be orders of magnitude greater than those of spontaneous Raman scattering, and the data analysis does not significantly depend upon collisional quenching rates as it does with most fluorescence techniques. Although CARS has been used to study laboratory flames (1,2), it has only recently been applied to large-scale combustion environments (3,4). This paper reports the results of CARS measurements performed in a highly turbulent sooting flame produced in a large-scale practical combustion-type environment.

CARS measurements were made in a bluff-body stabilized flame with turbulent and recirculating flow characteristics similar to those found in many practical combustors. The combustor was operated at atmospheric pressure with inlet air temperatures between 280 and 300K, an air flow rate of 0.5 kg/sec, and an upstream Reynolds number 1.5×10^5. Gaseous propane was injected from a hollow-cone nozzle located at the center of the bluff-body combustor at a flow rate of 7.06 kg/hr. The flame consisted of a blue cone originating at the nozzle followed by a yellow-luminous tail.

0-8412-0570-1/80/47-134-303$05.00/0
© 1980 American Chemical Society

The CARS system used to measure temperature and species concentrations in the combustor zone is composed of a single-mode
ruby-laser oscillator-amplifier with a repetition rate of 1 Hz
and a ruby-pumped, near-infrared broad-band dye laser. The two
laser beams are combined collinearly and focused first into a
cell containing a nonresonant reference gas and then into the
sample volume (approximately 30-μ diam. × 2 cm) in the combustion
region. The anti-Stokes beams produced in the sample and reference volumes are directed to spatially separated foci on the
entrance slit of a spectrometer and detected by separate photomultiplier tubes. An optional means of detection is provided for
the sample signal in the form of an optical multichannel analyzer
(OMA), which makes it possible to obtain single-pulse CARS
spectra.

Q-branch spectra of N_2 and O_2 were obtained in the reaction
zone of the combustor during a single 15-ns pulse through the use
of the OMA and broad-band Stokes beam. An example of these
single-shot spectra is given for N_2 in Fig. 1. These spectra
suggest that it is feasible to obtain simultaneous single-pulse
measurements of temperature and species concentration in this
type of combustion environment. Although not presented here,
single-shot temperature determinations have indeed been made
during recent measurements.

Spatially resolved average temperatures were arrived at
through the use of time-averaged N_2 CARS spectra obtained by
stepping the spectrometer through the Q-branch spectra generated
in the combustion region. Temperature was determined by comparing the measured, normalized Q-branch spectrum, indicated by "+"
in Fig. 2, and the calculated spectra generated by adjusting the
temperature until the best fit was obtained. The measured temperature profiles are shown for the axial and the Y and X radial
dimensions in Figs. 3, 4, and 5, respectively. It should be
pointed out that the apparent discrepancy in the center-line
temperatures of Figs. 3 and 5 is the result of a slight modifica-
tion in the positioning of the fuel nozzle relative to the face
of the bluff-body combustor which was made between the measurement periods.

Single-shot integrated Q-branch intensity measurements were
performed to obtain the N_2 and O_2 molecular number densities
from the reaction products. These data were reduced with the
averaged CARS temperatures to obtain the species-concentration
profiles also shown in Figs. 3-5.

To obtain comparative temperature data, a Pt 13% - RhPt
thermocouple provided by NASA Lewis Research Center was used
to profile the propane flame at the Z = 50-cm downstream position. Comparison of the CARS and radiation-corrected
thermocouple-derived temperature profiles is shown in Fig. 6.
The agreement between CARS (solid curve) and thermocouple (dashed
line) temperatures appears to be quite reasonable for locations
within 3 cm of the combustor centerline. However, the CARS data

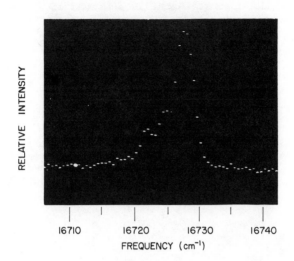

Figure 1. Single-pulse N_2 spectrum recorded on OMA during combustion

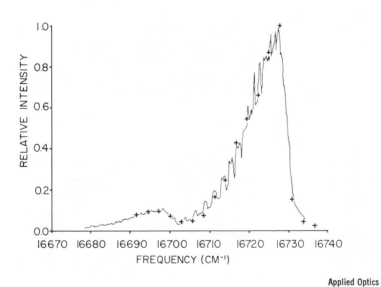

Applied Optics

Figure 2. Overlay of computer-generated spectrum of N_2 at 1700 K (————) onto measured spectra (+) (3)

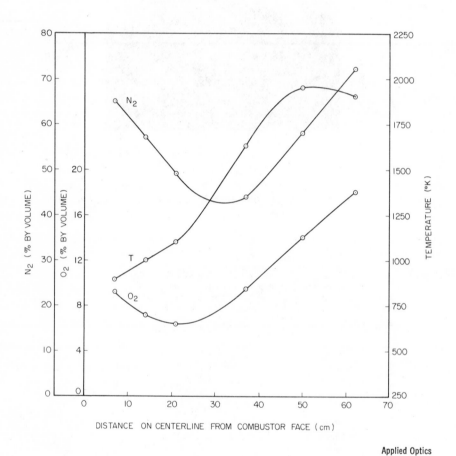

Applied Optics

Figure 3.   Axial profiles (3)

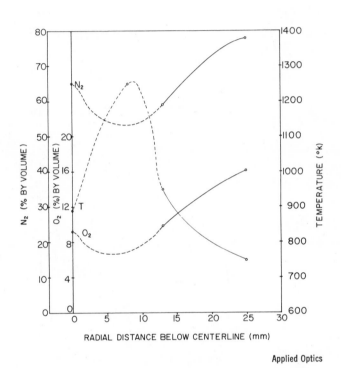

Applied Optics

Figure 4. Y *radial profiles at* X = 0 *and* Z = 7 cm (3)

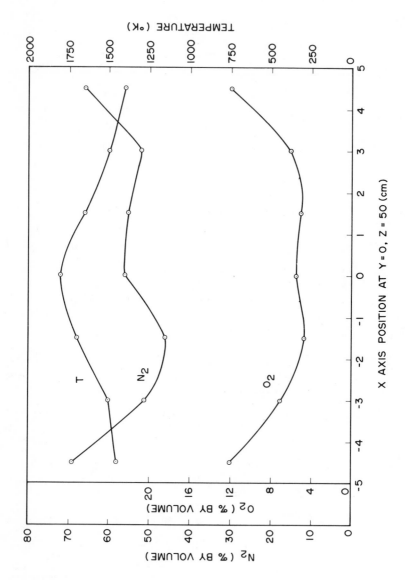

Figure 5.   X radial profiles at $Y = 0$ *and* $Z = 50 \, cm$

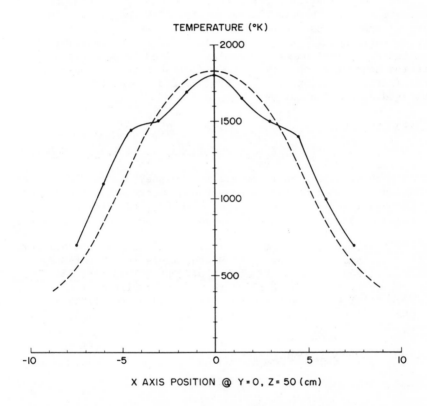

Figure 6. Comparison of CARS temperature (———) and average thermocouple temperature (– – –) at Y = 0 and Z = 50 cm

indicate that for distances of more than 3 cm from the centerline, some as yet unknown effects are contributing to the CARS
information and result, for example, in the knees shown in the
CARS temperature profile. Investigations to determine whether
these disturbances in the data are a consequence of optical mis-
alignment or improper averaging caused by flame turbulence or
whether they are of a more fundamental nature are presently
underway.

In conclusion, the data accumulated during these experiments
demonstrate that through the use of the CARS technique, combustion diagnostics can be performed in a large-scale practical combustor environment. Results also indicate the potential of CARS
for the determination of temporally resolved temperature and
species concentration. With the aid of such data, probability
distribution functions can be obtained from which true time-
averaged quantities may be determined.

Literature Cited

1. Moya, F.; Druet, S. A. J.; and Taran, J. P. E. Opt. Commun.,
    1975, 13, 169.
2. Eckbreth, A. C.; Hall, R. J.; and Shirley, J. A. "Investiga-
    tions of Coherent Anti-Stokes Raman Spectroscopy (CARS) for
    Combustion Diagnostics," Paper 79-0083, Presented at the
    17th AIAA Conference on Aerospace Sciences, New Orleans, LA,
    June 15-17, 1979.
3. Switzer, G. L.; Roquemore, W. M.; Bradley, R. P.; Schreiber,
    P. W.; and Roh, W. B. "CARS Measurements in a Bluff-Body
    Stabilized Diffusion Flame," Appl. Opt., 1979, 18, 2343.
4. Eckbreth, A. C. "Spatially Precise Laser Diagnostics for
    Practical Combustor Probing," Presented at the 178th ACS
    National Meeting, Washington, D.C., September 10-14, 1979.

RECEIVED February 1, 1980.

Update on CARS Diagnostics of Reactive Media at ONERA

M. PÉALAT, B. ATTAL, S. DRUET, and J. P. TARAN

Office National d'Etudes et de Recherches Aérospatiales, 92320 Châtillon, France

The CARS research effort at ONERA is divided into three parts :

continued development of the apparatus presently on hand;

in parallel, deployment of this equipment for measurement campaigns on practical burners;

fundamental and experimental work on resonance enhancement.

Development of CARS Spectrometer

The motivation of that effort is to improve spatial resolution, spectral resolution, measurement accuracy and sensitivity. Also of great interest is the implementation of single shot multiplex spectroscopy for turbulent combustion diagnostics. The CARS spectrometer consists in a portable, lightweight source assembly developed jointly with Quantel and in detection kits which can be adapted to various experimental problems. The source assembly comprises a frequency-doubled yag laser and a tunable dye laser ; space is provided on the optical table for various beam handling and combining optics (1).

Spatial resolution. Two modes of operation are possible. The conventional mode with superimposed beams gives medium spatial resolution of 5 to 10 mm with maximum signal generation. The crossed beam mode, or BOXCARS (2), gives a resolution adjustable between 1 and 3 mm, but with reduced signal strength (1/50 to 1/100). In both cases, the transverse resolution is better than 50 μm. A folded BOXCARS mode, with the ω_1 beams contained in a plane orthogonal to that of the ω_2 and ω_3 beams, has been tested and rejected for practical combustion work. In effect, although the anti-Stokes signal is spatially separated from the pump so that spectral filtering is greatly facilitated (for small Raman shifts especially), the net signal is reduced.

0-8412-0570-1/80/47-134-311$05.00/0
© 1980 American Chemical Society

Spectral resolution. The instrument readily gives a spectral
resolution less than 0.07 cm^{-1}, which betters that of the best
conventional Raman spectrographs, and is the best adapted for
sensitive detection in flames (one then takes full advantage of
the small Raman linewidths to maximize the signal). A resolution
of 0.7 cm^{-1} is also available for rapid scanning and for the
detection of major constituents. In addition, a broadband mode
can be used for multiplex CARS in conjunction with a spectrograph
and an OMA (3). In all cases, the single longitudinal mode character of the Yag laser oscillator and its frequency stability are
fundamental requirements, if one is to obtain reproducible
results. The Yag oscillator spectral properties are achieved by
inserting two temperature-controlled etalons in its cavity and by
using a passive Q-switch.

Measurement accuracy. Our r.m.s. shot to shot fluctuations are
± 5%, thanks to the use of reference. This accuracy does not
depend on the mode of operation (spatial resolution and choice of
field polarizations). It is not as good, however, in multiplex
CARS, or in dilute samples, or in the vicinity of narrow lines,
because other sources of noise such as photoelectron statistics
and laser frequency instabilities (however small) then play a
major role.

Detection sensitivity. Sensitivity depends on several parameters,
including background cancellation (4), measurement accuracy, stray
light emission from the flame, spectral interference from other
species, etc ... A very good indication is given by the room air
CO_2 spectra of Fig.1. Figure 1a gives a spectrum recorded with a
dye laser linewidth of 0.7 cm^{-1} in 0.1 cm^{-1} increments and averaging 10 consecutive measurements per increment. The strongest CO_2
line is shown. It appears as a 40% modulation on the background.
The CO_2 is easily detected because the measurement accuracy is
good (our detectivity is about 15 ppm here). Figure 1b is obtained
using background cancellation as described in (1) ; faint O_2 lines
now appear, giving an indication of the S/N improvement. Note that
the signal is reduced by a large factor (on the order of 16), and
that this technique therefore may not be applicable to very bright
flames. If we use a dye laser linewidth of 0.07 cm^{-1}, the lines in
Fig. 1a and 1b are approximately tripled, which shows the advantage of using the higher resolution available. Figure 2 was
obtained in broadband CARS. Ten spectra have been averaged for
easier comparison with Fig. 1b. The spectral resolution, which
depends both on spectrograph dispersion and OMA target resolution,
is slightly less than 1 cm^{-1}. The S/N ratio and the detectivity
are appreciably lower, but we have hopes of improving them.

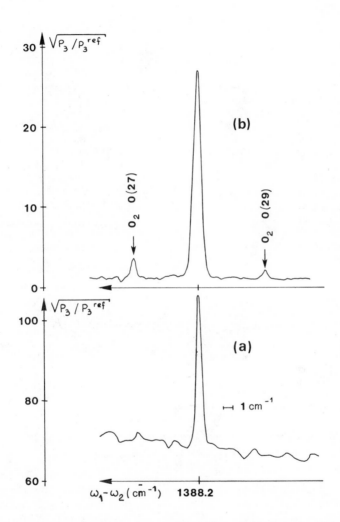

Figure 1.   BOXCARS spectra room air CO_2 with parallel polarizations (a). P_3 and $P_3{}^{ref}$ are the sample and reference anti-Stokes powers, respectively. Vertical scales are arbitrary with background cancellation (b).

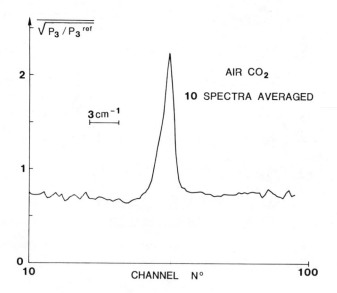

Figure 2.   Same spectrum as Figure 1b taken by OMA 2 with collinear pump beams

Practical Measurements

The feasibility of performing useful measurements in practical combustion system was demonstrated in December 1978 on a simulated jet engine combustor fueled by kerosene and having a nominal mass flow rate of 600 g/s with flow cross section of 10 cm x 40 cm (5). The spectrometer was set up close to the burner (Fig. 3). The noise level was 110 db, while the cell was flushed with air at temperatures ranging from 5 to 15°C. A plywood case was installed for the acoustic protection and an electric blanket was used to maintain the spectrometer's temperature above 13°C. Figure 4 presents an N_2 spectrum recorded in the exhaust. Computer data reduction yields a temperature T of 1150 ± 50 K and a mole fraction c of 78 ± 5%, in good agreement with thermocouple (T = 1050 K) and sampling probe (c = 77%) measurements. Data reduction was done using spectroscopic data from Gilson (6) and linewidth data from Owyoung (7). Careful matching of the non-resonant background is important for temperature retrieval, since a 20% error on the value used for the background may lead to an error in excess of 100 K on the temperature, while still permitting an acceptable fit to the data near the band center. We also monitored O_2 and CO_2 and showed that short bursts of trace CF_4 can be injected and used as a seed to measure residence times by CARS. CF_4 concentrations as low as 1% have been detected without background cancellation. We estimate that 300 ppm of this gas could be detected with background cancellation. Finally, anti-Stokes signal fluctuations of ± 20% are observed. We feel that they are caused mainly by local susceptibility fluctuations, and not by beam defocusing resulting from Schlieren effects. The optical path through the jet was 10 cm.

Explorations are being undertaken or planned on low pressure discharges in H_2, where a detection sensitivity of 10^{-7} atm. can be obtained, and on chemical lasers, piston engines and real jet engine combustors.

Resonance CARS

The detection sensitivity being limited at 100 to 1000 ppm for usual gases in flames using background cancellation, an effort has been undertaken in order to understand resonance enhancement mechanisms and in order to apply resonance CARS to trace species detection. The theory is now well understood (8) and an encouraging experimental verification has been reported with detectivity gains of 100 to 1000 (9). However, numerous experimental problems remain to be solved, among which are saturation and laser stability problems (10).

In conclusion, CARS is now a proven laboratory technique for chemical analysis and temperature measurements. These achievements have come through a very careful engineering of the laser sources and associated optics. The detection sensitivity has been improved

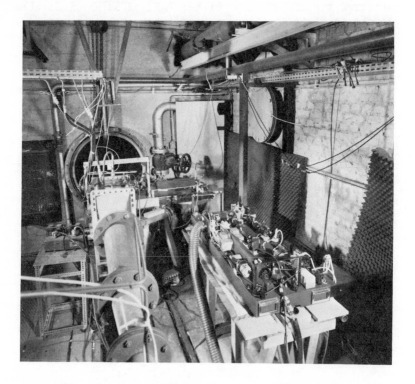

Figure 3. Quantel CARS spectrometer in combustor facility

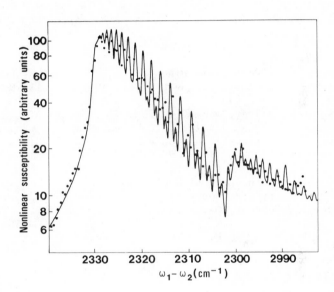

Figure 4.   Time-averaged spectrum of N_2 on exit plane of combustor recorded with collinear beams and without background rejection; 10 laser shots are averaged at each point.

thanks to the introduction of a background cancellation technique, but remains limited to 1000 ppm or thereabouts in practical situations. Hopefully, resonance CARS will help to overcome all these limitations and enable us to obtain ppm detectabilities.

Literature Cited

1. Péalat, M., Taran, J.P., Moya, F., Optics and Laser Technology, February 1980, to be published ; Attal, B., Péalat, M., Taran, J.P., to be presented at the 1980 AIAA Aerospace Sciences Meeting, Pasadena, Calif., January 14-16, 1980, AIAA Paper N° 80-282.
2. Eckbreth, A.C., Appl. Phys. Letters, 1978, 32, 421.
3. Roh, W.B., Schreiber, P.W., Taran, J.P., Appl. Phys. Letters, 1976, 29, 174.
4. Rahn, L.A., Zych, L.J., Mattern, P.L., Optics Comm., 1979, 30, 249.
5. Taran, J.P., "CARS Flame Diagnostics" presented at the CARS meeting of the Institute of Physics, AERE Harwell, March 1979.
6. Gilson, T., private communication.
7. Owyoung, A., "High Resolution Coherent Raman Spectroscopy of Gases", 4th International Conference on Laser Spectroscopy, Rottach-Egern, FRG, 11-15 June 1979.
8. Druet, S., Taran, J.P.E., "Coherent anti-Stokes Raman Spectroscopy" in Chemical and Biochemical Applications of Lasers, ed. by C.B. Moore, Academic Press, 1979.
9. Attal, B., Schnepp, O., Taran, J.P., Optics Comm., 1978, 24, 77.
10. Attal, B., Taran, J.P., to be published.

RECEIVED February 22, 1980.

The Application of Single-Pulse Nonlinear Raman Techniques to a Liquid Photolytic Reaction

WILLIAM G. VON HOLLE and ROY A. McWILLIAMS

University of California, Lawrence Livermore Laboratory, Livermore, CA 94550

Pulsed laser-Raman spectroscopy is an attractive candidate for chemical diagnostics of reactions of explosives which take place on a sub-microsecond time scale. Inverse Raman (IRS) or stimulated Raman loss (1,2) and Raman Induced Kerr Effect (3) Spectroscopies (RIKES) are particularly attractive for single-pulse work on such reactions in condensed phases for the following reasons: (1) simplicity of operation, only beam overlap is required; (2) no non-resonant interference with the spontaneous spectrum; (3) for IRS and some variations of RIKES, the intensity is linear in concentration, pump power, and cross-section.

This chapter describes the application of these techniques to a liquid photolytic reaction. The motivation was the assessment of the capabilities and limitations of single-pulse nonlinear Raman spectroscopy as a probe of fast reactions in energetic materials.

Theory

A complete discussion of the theory of the coherent Raman effects is not possible in the available space. There are many excellent introductions and reviews for a more detailed treatment (2,3,4). Let us simply outline some basic considerations pertinent to the following discussion. The electronic polarization of a medium can be expressed as a power series in the electric field as in Equation (1).

$$\vec{P} = \chi(1)\vec{E} + \chi(2)\vec{E}2 + \chi(3)\vec{E}3 \tag{1}$$

where P is the polarization per unit volume, the $\chi^{(i)}$ are dielectric susceptibility tensors of rank $(i + 1)$, and $\vec{E}$ is the applied electric field. At high fields in isotropic media, the second non-linear term becomes important. The third order susceptibility is a complex quantity which has resonant and non-resonant components according to Equation (2).

0-8412-0570-1/80/47-134-319$05.00/0
© 1980 American Chemical Society

$$\chi(3) = \chi' + i\chi'' + \chi^{N.R.} \tag{2}$$

$\chi^{N.R.}$, the non-resonant susceptibility, gives rise to the background interference in Coherent Anti-Stokes Raman Spectroscopy (CARS) (5). This interference which arises from solvents or closely spaced lines is responsible for the CARS band shape distortion observed under certain conditions.

For RIKES with circular pump polarization and for IRS, the background interference is surpressed. In the RIKES case, the non-resonant part of $\chi^{(3)}$ drops out according to Kleinman symmetry (3). In IRS only the imaginary part of $\chi^{(3)}$ contributes, but all of the probe intensity is admitted to the detector.

Experimental

The basic experimental arrangement is shown in Figure 1. A Q-switched ruby pump laser is frequency doubled to pump a broadband dye laser with a plano-spherical cavity. The remaining ruby power and the dye pulse were then used for the non-linear spectroscopy experiments. Spectra were recorded on film or plates by means of a Spex Model 1701 spectrograph equipped with a camera. The smoothness of the dye output intensity with wavelength, which determines the sensitivity of the single-pulse spectra, varied from shot to shot and depended on the dye. Also, the heterogeneity of the ruby power density cross section in the sample interaction volume caused a large shot to shot variation in the non-linear signal intensity. In the RIKES experiments, all optical elements were placed outside the crossed polarizers; only the sample cell windows remained. This arrangement prevented strain birefringence from interfering with the RIKES spectrum.

For the flash photolysis experiments, the ultraviolet pulse (20 μs FWHM) was delivered to the samples via four linear Xenon flashtubes surrounding the sample cell. The lamps were fired from the laser console through a variable delay to provide the desired time delay from the flash peak to the laser pulses.

Results

Schreiber (6) pointed out the usefulness of single-pulse CARS for combustion work. The apparatus described above can be easily adapted to perform a number of coherent Raman experiments with single 20 ns (FWHM) laser pulses. Following are examples of the application of single-pulse RIKES and IRS, first to static solutions, then to Xenon-lamp irradiated solutions.

Figure 2 shows two examples of RIKES spectra, demonstrating the attainable sensitivity. Smaller amounts can be detected with considerable loss in signal-to-noise ratio.

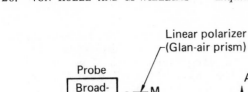

Figure 1. Diagram of experimental apparatus used to obtain IRS and RIKES spectra. The Q-switched ruby is frequency-doubled to pump the dye laser. RIKES and its variations require two polarizers, For IRS the analyzer can be removed and the quarter-wave plate removed or replaced by a half-wave plate.

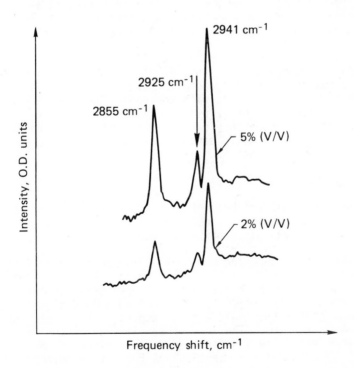

Figure 2. Microdensitometer traces of RIKES spectra of two solutions of cyclohexane in CCl₄ in the C–H stretching region. The peak ruby pump power was 150 MW cm⁻² and the pump beam was crossed with probe beam at about 10°.

Optical heterodyne detection of the RIKES signal (OHD-RIKES) is reported to greatly increase the signal-to-noise (S/N) ratio with photoelectric detection and single-frequency scanning (7). Figure 3 shows the results of the application of OHD-RIKES to multiplex experiments. In this case the local oscillator was introduced by a method suggested in reference (7). The analyzer was rotated to allow some of the dye probe pulse through and the ruby pump pulse was polarized linearly at 45⁰ to the dye. In the results shown in Figure 3, one sees a definite dependence of the line shape on the sense of rotation of the analyzer for the C-H stretching region, suggestive of polarization CARS (8). This effect could be useful for single pulse work in condensed media.

Let us now turn our attention to coherent Raman spectra of flashed solutions. Ten percent solutions of cyclohexane in $CC\ell_4$ were found to react to yield HCl when flashed as described in the experimental section. A typical time-resolved RIKES spectrum is shown in Figure 4. Superimposed on a large transient background signal there is evidence for new resonances, one of which may be due to cyclohexene. The background is sometimes coherent with the Raman signals, causing line shape distortion and even spectrum inversion as in Figure 5. Preliminary evidence indicates a definite peak in intensity of this background signal at about 19 microseconds after the Xe flash peak intensity.

Two inverse Raman spectra of the flashed cyclohexane solutions are compared in Figure 6. The 3026 cm⁻¹ absorption feature, suspected to be cyclohexene, agrees with the RIKES results. OHD-RIKES spectra with the ruby linearly polarized at 45⁰ to the linear dye probe polarization and the analyzer rotated are also shown in Figure 6. The position of the suspected cyclohexene line appears as well as other features present in the IRS spectra.

Discussion

Some additional evidence of the nature of the light-induced reaction in this system was obtained by mass spectrographic analysis of the gas given off as a result of Xe light flash. It was found to contain hydrogen chloride and possibly chloroform. Gas evolution is supported by the violence of the reaction when the flash intensity is high. All observations are consistent with a free radical mechanism in which cyclohexene is produced by cyclohexyl attack on $CC\ell_4$, regenerating an additional free radical. Cyclohexene could then be depleted by a number of plausible mechanisms. The lack of sensitivity of the present method does not allow a definitive determination of the mechanism of this interesting reaction. A repetitive method with photoelectric detection and signal averaging would perhaps provide the required sensitivity.

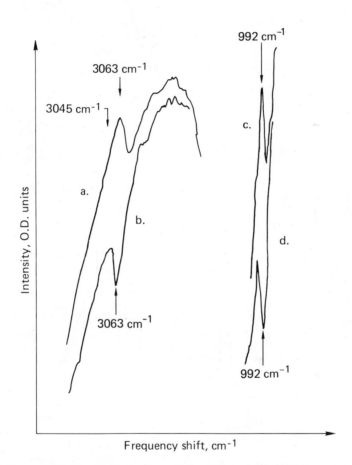

Figure 3. Microdensitometer traces of the OHD–RIKES single-pulse spectra of 10% (v/v) solutions of benzene. (a) C–H stretching region, analyzer rotated −4°; (b) conditions identical to (a) except analyzer rotated +4°; (c) 992 cm⁻¹ line with analyzer rotated −12°; (d) conditions identical to (c) except analyzer rotated +12°. A positive rotation is clockwise-observed from the spectrograph. Keep in mind that the apparent line shape is affected by the slope of the dye laser intensity.

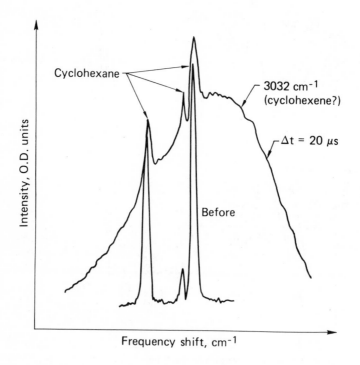

Figure 4. Microdensitometer traces of spectra from 10% (v/v) solutions of cyclohexane in CCl₄ taken before and at 20 μsec after the flash intensity peak

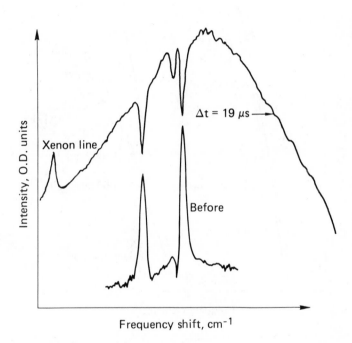

Figure 5.   Microdensitometer trace of spectra from 10% (v/v) solutions of cyclo-hexane in CCl₄ taken before and at 19 µsec after the flash intensity peak

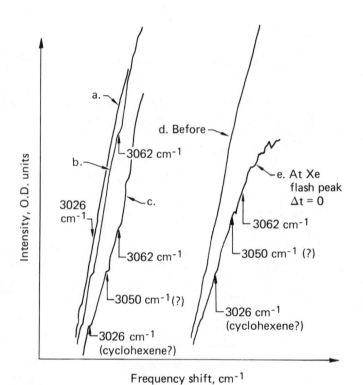

Figure 6.   Microdensitometer traces of IRS and OHD–RIKES spectra of 10% (v/v) cyclohexane solutions in CCl₄. (a) IRS spectra before flash; (b) 12 μsec after flash peak, peak ruby pump power, 150 MW/cm²; (c) 19 μsec after flash peak, peak ruby pump power, 700 MW/cm². In (c) the ruby pump is linearly polarized parallel to dye polarization but the analyzer is rotated +8° out of null; (d) and (e), microdensitometer traces of OHD–RIKES spectra of 10% (v/v) cyclohexane in CCl₄ before and at the peak of the flash, peak ruby pump power, 150 MW/cm². The ruby pump polarization was oriented 45° to the linear dye probe and the analyzer was rotated +8° from the null position.

The origin of the large flash-induced background produced in the RIKES experiments is not immediately evident. However, a rotation of the quarter-wave plate of about 10° yields nearly the same background as observed in the flash experiments. Perhaps turbulence produced by the reaction is the source of this induced ellipticity of the circularly polarized pump. An induced ellipticity combined with a slight rotation of the probe polarization could explain some of the flashed RIKES spectra.

IRS and OHD-RIKES are less effected by the background interference, but the signal levels tend to be low, and they are subject to noise in the broad-band dye laser probe. Attempts to eliminate the remaining noise by time dispersal of the Raman signal and local oscillator by means of a streak camera ($\underline{9}$) or a smoothing of the dye probe mode structure by intracavity phase shifting of the dye radiation are under consideration.

Planned experiments on shock-induced chemical reactions and detonations of explosives will be carried out with the nonlinear Raman techniques. Heterodyne detection of the transient products of such rapid reactions seems the most promising.

Literature Cited

1. Jones, W. J. and Stoicheff, B. P., *Phys. Rev. Lett.*, 1964, $\underline{13}$, 657.
2. Yeung, E.S., in "New Applications of Lasers to Chemistry," Hieftje, G. M., Ed., American Chemical Society, Washington, D.C., 1978, p 193.
3. Heiman, D., Hellworth, R. W., Levenson, M. D., and Martin, G., *Phys. Rev. Lett.*, 1976, $\underline{36}$, 189.
4. Levenson, M. D., *Physics Today*, May, 1977, 44.
5. Tolles, W. M. Nibler, J. W., McDonald, J. R., and Harvey, A. B., *Applied Spectroscopy*, 1977, $\underline{31}$, 253.
6. Roh, W. B., Schreiber, P. W., and J-P E. Taran, *Appl. Phys. Lett.*, 1976, $\underline{29}$, 174.
7. Eesley, G. L., Levenson, M. D., and Tolles, W. M., *I.E.E.E. J. of Quant. Elec.*, 1978, $\underline{QE-14}$, 45.
8. Koroteev, N. I., Endeman, M., and Byer, R. L., *Phys. Rev. Lett.*, 1979, $\underline{43}$, 398.
9. Levenson, M. D. and Eesley, G. L., *Appl. Phys.*, 1979, $\underline{19}$, 1.

Work performed under the auspices of the U.S. Department of Energy by Lawrence Livermore Laboratory under Contract No. W-7405-Eng-48.

RECEIVED February 1, 1980.

MODELLING AND KINETICS

Detailed Modelling of Combustion: A Noninterfering Diagnostic Tool

ELAINE S. ORAN, JAY P. BORIS, and M. J. FRITTS

Laboratory for Computational Physics, Naval Research Laboratory, Washington, D.C. 20375

Detailed modelling, or numerical simulation, provides a method we can use to study complex reactive flow processes (1). Predictions about the behavior of a physical system are obtained by solving numerically the multi-fluid conservation equations for mass, momentum, and energy. Since the success of detailed modelling is coupled to one's ability to handle an abundance of theoretical and numerical detail, this field has matured in parallel with the increase in size and speed of computers and sophistication of numerical techniques.

It is important to distinguish between empirical, phenomenological and detailed models. Empirical models are constructed from data obtained by experiments, summarized in analytical or numerical form, and subsequently tested against proven theoretical laws or other data. Phenomenological models are extrapolations from theory based on our physical intuition which must be tested against experimental data. The intuitive and experimental basis of empirical and phenomenological models have led to their widespread incorporation into simulations of combustor systems in spite of very serious shortcomings in the nature of the models themselves. Detailed modelling seeks to overcome these deficits by means of improved numerical techniques and the increased power of modern computers.

This paper will in no way constitute a review of current combustor models, but will instead attempt to elucidate the extensions and improvements made possible through detailed modelling. The purpose of this paper is to familiarize the reader with the goals, terminology and inherent problems in modelling fundamental combustion processes. The emphasis is not on presenting a full set of complicated multi-fluid equations or on explaining the numerical algorithms required to solve the governing equations. Instead we hope to impart a sense of the power and role of detailed modelling, an understanding of why physical insight must be built into numerical algorithms, and an indication of how to

This chapter not subject to U.S. copyright.
Published 1980 American Chemical Society

test these models at every stage of construction against both theory and experiment.

The shortcomings of the empirical models lie in their limited range of validity, while phenomenological models become more tenuous as they approach the complexities of real physical systems. Detailed models usually contain parts which may be empirical or phenomenological in origin. However, detailed modelling attempts to overcome these shortcomings by incorporating theoretical detail rich enough to approximate reality; detail far richer than could be summarized in any succinct analytical model, yet more theoretically sound than standard phenomenological models or empirical fits.

From experimental observation and approximate theoretical models we can postulate quantitative physical laws which we expect an effect to obey. These "laws" can be tested against reality by incorporating them in a detailed model which makes quantitative predictions for series of experimental measurements. Each calculation performed with a detailed model is like a unique experiment performed with one set from an infinity of possible sets of geometric, boundary, and initial conditions. Just as valid results can be extracted from an experiment only through an understanding of the effects and limitations of the instruments used in collecting the data, results obtained using detailed modelling must be examined in the light of the limitations inherent in its tools, both analytical and numerical. The first section of this paper will therefore deal with an exposition of the problems inherent in detailed modelling of combustion processes so that as we proceed we have a healthy respect both for the magnitude of the problems and the limitations of our methods.

The next section will concentrate on the choice of numerical algorithms used in the models. This process corresponds to the construction and design of an experimental apparatus which must reflect a good knowledge of the physics the experiment is to study. Modelling combustion systems has its own particular problems because of the strong interaction between the energy released from chemical reactions and the dynamics of the fluid motion. Release of chemical energy generates gradients in temperature, pressure, and density. These gradients, in turn, influence the transport of mass, momentum, and energy in the system. On a large scale, the gradients may generate vorticity or affect the diffusion of mass and energy. On a more microscopic scale, they may generate turbulence which drastically affects macroscopic mixing and burning velocities. In modelling shocks, detonations, or flame propagation, time and space scales of interest can span as many as ten orders of magnitude. Thus to obtain adequate resolution, the numerical methods must be computationally fast as well as accurate. Methods must be developed which rely on asymptotic solution techniques to follow short time and space scale phenomena on a macroscale. It is in this aspect that detailed modelling most closely approximates experiment. If our numerical apparatus cannot resolve the basic controlling physical processes, no meaningful calculations can be made of their effects.

Although detailed modelling does not directly provide the types of useful analytic relationships which guide our intuition and allow us to make quick estimates, it gives us the flexibility to evaluate the importance of a modelled physical effect by simply turning it off or on or changing its strength.   The model can also be used to test the sensitivity of the computed results to independent theoretical approximations.   Those analytic results which are available are valuable in benchmarking the model in various limits.   A series of tests which compare analytic results to numerical simulations may calibrate the simulation before it is compared to experiments or used for extrapolation. Conversely, a well-tested model serves as a very useful means of calibrating unknown parameters and form factors in approximate theories.

The last two sections of this paper will discuss this interplay between detailed modelling and both theory and experiment. The third section describes how a model must be tested in various limits for physical consistency to insure its accuracy.   The specific example chosen here is a comparison between an analytic solution and a detailed numerical simulation of a premixed laminar flame.   The last section shows how a comparison between model results and experiments can be used to calibrate the model and to guide further experiments.   The example chosen is a calculation of flow over an immersed object which is compared to both experimental and theoretical results.

Problems in Modelling Reactive Flows

Errors and confusion in modelling arise because the complex set of coupled, nonlinear, partial differential equations are not usually an exact representation of the physical system.   As examples, first consider the input parameters, such as chemical rate constants or diffusion coefficients.   These input quantities, used as submodels in the detailed model, must be derived from more fundamental theories, models or experiments.   They are usually not known to any appreciable accuracy and often their values are simply guesses.   Or consider the geometry used in a calculation.   It is often one or two dimensions less than needed to completely describe the real system.   Multidimensional effects which may be important are either crudely approximated or ignored. This lack of exact correspondence between the model adopted and the actual physical system constitutes the basic problem of detailed modelling.   This problem, which must be overcome in order to accurately model transient combustion systems, can be analyzed in terms of the multiple time scales, multiple space scales, geometric complexity,   and   physical complexity of the systems to be modelled.

Multiple Time Scales.   The first class of problems arises as

the result of trying to represent phenomena characterized by very
different time scales. In ordinary flame and detonation problems
these scales range over many orders of magnitude. When phenomena
are modelled that have characteristic times of variation shorter
than the timestep one can afford, the equations describing these
phenomena are usually called "stiff." Equations describing sound
waves are stiff with respect to the timestep one wishes to employ
when modelling a subsonic flame speed. Many chemical reaction
rate equations are stiff with respect to convection, diffusion,
or even sound wave timestep criteria. Two rather distinct modelling approaches, global implicit and timestep-split asymptotic,
have been developed to treat these temporally stiff equations.
These two approaches are briefly described later in this paper.

Multiple Space Scales. The second class of problems involves the huge disparity in space scales occurring in combustion
problems. To model the steep gradients at a flame front, a cell
spacing of 10^{-3} cm or smaller might be required. To model convection, grid spacings of 1 to 10 cm might be adequate. Complex
phenomena such as turbulence which occur on intermediate spatial
scales present a particular modelling problem. It would be a
pipedream to expect a numerical calculation to faithfully reproduce physical phenomena with scale lengths shorter than a cell
size. Therefore, to calculate realistic profiles of physical
variables, a certain cell spacing is required to obtain a given
accuracy. Choosing a method which maximizes accuracy with a
minimum number of grid points is a major concern in detailed
modelling.

Geometric Complexity. The third set of obstacles arises
because of the geometric complexity associated with real systems.
Most of the detailed models developed to date have been one-
dimensional, but this gives a very limited picture of how the
energy release affects the hydrodynamics. Even though many processes in a combustion system can be modelled in one-dimension,
there are others, such as boundary layer growth, or the formation
of vortices and separating flows, which clearly require at least
two-dimensional hydrodynamics. Real combustion systems are at
least two-dimensional , with unusual boundary conditions and internal sources and sinks. However, even with sixth generation
parallel processing computers available, what can be achieved
with two-dimensional detailed models is still limited by computer
time and storage requirements.
        In the current state-of-the-art, one-dimensional models can
best be used to look in detail at the coupling of a very large
number of species interactions in a geometry that is an approximation to reality. Processes such as radiation transport, turbulence, or the effects of heterogeneity of materials can be included either as empirically or theoretically derived submodels.
Two- and three-dimensional models are best used to study either

gross flow properties or detailed radiation transport. In these latter models, the chemical reaction scheme is usually quite idealized or parameterized.

Physical Complexity. The final set of obstacles to detailed modelling concerns physical complexity. Combustion systems usually have many interacting species. This leads to sets of many coupled equations which must be solved simultaneously. Complicated ordinary differential equations describing the chemical reactions or large matrices describing the molecular diffusion process are costly and increase calculation time orders of magnitude over idealized or empirical models. Table I lists some of the major chemical and physical processes which have to be considered for an accurate description of a complicated combustion system. The dashed line in the table indicates that multiphase processes such as surface catalysis and soot formation can be important even when we are primarily interested in gas phase combustion. For most interesting systems, one finds that the basic chemical reaction scheme, the individual chemical rates, the optical opacities, or the effects of surface reactions are not well known. Before a model of a whole combustion system can be assembled, each individual process must be separately understood and modelled. These submodels are either incorporated into the larger detailed model directly or, if the time and space scales are too disparate, they must be fit in phenomenologically. For example, diffusion and thermal conductivity between a wall and the reacting gas can be studied separately and then incorporated directly into a detailed combustion model. Turbulence, however, can be modelled on its own space scales only in idealized cases. These idealized, fundamental, "ab initio" turbulence calculations must be used to develop phenomenological models for use in the macroscopic detailed model. Resolution and computational cost prevent incorporating the detailed turbulence model directly.

Table I
Fundamental Processes in Combustion

	gas phase	multi-phase
Chemical kinetics		
Hydrodynamics-laminar		
Thermal conductivity, viscosity		
Molecular diffusion		
Thermochemistry		
Hydrodynamics-turbulent		
Radiation		
Nucleation		
Surface Effects		
Phase Transitions		
(Evaporation, condensation...)		

Often there are cases where the submodels are poorly known
or misunderstood, such as for chemical rate equations, thermochemical data, or transport coefficients. A typical example is
shown in Figure 1 which was provided by David Garvin at the U. S.
National Bureau of Standards. The figure shows the rate constant
at 300°K for the reaction HO + O_3 → HO_2 + O_2 as a function of the
year of the measurement. We note with amusement and chagrin that
if we were modelling a kinetics scheme which incorporated this
reaction before 1970, the rate would be uncertain by five orders
of magnitude! As shown most clearly by the pair of rate constant
values which have an equal upper bound and lower bound, a sensitivity analysis using such poorly defined rate constants would be
useless. Yet this case is not atypical of the uncertainty in
rate constants for many major reactions in combustion processes.

Gedanken Flame Experiment. In order to illustrate how the
problems caused by the requirements of temporal and spatial resolution and geometric and physical complexity are translated into
computational cost, we have chosen to analyze a gedanken flame
experiment. Consider a closed tube one meter long which contains
a combustible gas mixture. We wish to calculate how the physical
properties such as temperature, species densities, and position
of the flame front change after the mixture is ignited at one
end. The burning gas can be described, we assume, by a chemical
kinetics reaction rate scheme which involves some tens of species
and hundreds of chemical reactions, some of which are "stiff."
We will assume one-dimensional propagation along the tube. Boundary layer formation and turbulence will be ignored. We further
assume that the flame front moves at an average velocity of 100
cm/sec.

Table II summarizes the pertinent time and space scales in
this problem. Assuming the speed of sound is 10^5cm/sec, a timestep of about 10^{-9} sec would be required to resolve the motion of
sound waves bouncing across the chamber. Chemical timescales, as
mentioned above, are about 10^{-6} sec. This number may be reduced
drastically if the reaction rates or density changes are very
fast. It takes a sound wave about 10^{-3} seconds to cross the 1
meter system and it takes the flame front about one second to
cross. We further assume that the flame zone is about 10^{-2} cm
wide and that it takes grid spacings of 10^{-3} cm to resolve the
steep gradients in density and temperature in this flame zone.
In those portions of the tube on either side of the flame front,
we assume that 1 cm spacings are adequate.

To estimate the computational expense of this calculation,
we use 10^{-3} seconds of computer time as a reasonable estimate of
the time it takes to integrate each grid point for one timestep
(a single pointstep). This estimate includes a solution of the
chemical and hydrodynamic equations and is based on a detailed
model of a hydrogen-oxygen flame problem optimized for a parallel
processing computer. Figure 2 shows the information in Table II

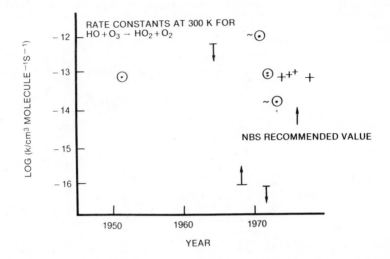

Figure 1. Measured values of the rate constant for $HO + O_3 \rightarrow HO_2 + O_2$ as a function of the year of measurement. The arrows with overbars and underbars indicate measured upper and lower bounds, respectively. The NBS-recommended value is the value with the smallest error bars (12).

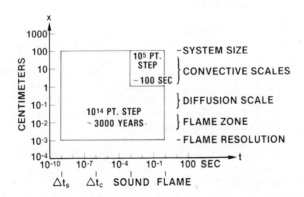

Figure 2. Space and time scales in the gedanken flame calculation. A naive direct solution of the problem could take 3000 years of computer time. The calculation should be possible in 100 sec.

cast into a graph of space versus time. Since the scales are logarithmic, a calculation of the number of pointsteps and then of the needed computer time requires exponentiation. Thus it appears that 3000 years of computer time is required to calculate the 10^{14} pointsteps needed to resolve the finest space and time scales everywhere!

Table II
Important Scales in Gedanken Flame Experiment

Timescales		Spacescales	
Δt	sec	Δx	cm
Sound Speed	10^{-9}	Flame Resolution	10^{-3}
Chemistry	10^{-6}	Flame Zone	10^{-2}
Sound Transit Time	10^{-3}	Diffusion Scale	10^{-1}
Flame Transit Time	1	Convective Scales	10
		System Size	100

$$V_f = 100 \text{ cm/sec}$$

Of course, this is unacceptable. Ideally such a simple calculation should take about 100 seconds (See Figure 3). What are needed are numerical algorithms which have the resolution in time and space only where it is required. Furthermore, these algorithms should be optimized to take advantage of what is known about the physics and chemistry of the problem. This will be discussed further below where it is shown how the application of various numerical algorithms can be used to reduce this flame system to a tractable computational problem.

Turbulence. Turbulence is one of the outstanding problems of reactive flow modelling and is another excellent example of the difficulty we have in resolving highly disparate time and space scales. Our understanding and eventual ability to predict the complicated interactions occurring in turbulent reactive flow problems is imperative for many combustion modelling applications. The presence of turbulence alters mixing and reaction times and heat and mass transfer rates which in turn modify the local and global dynamic properties of the system. What we need to resolve these problems are accurate yet compact phenomenological turbulence models which can be used to describe realistic combustor systems, open flames, and other turbulent reactive flows confidently and efficiently. These computational models must decouple the subgrid turbulence and microscopic instability mechanisms from calculations of the macroscopic flow. Below we list the important properties of an ideal turbulence model (2).
    1. Chemistry-Hydrodynamics Coupling and Feedback. Explicit energy feedback mechanisms from mixing and reactions to the turbulent velocity field and the macroscopic flow must be formulated.

The "laminar" macroscopic flow equations contain phenomenological terms which represent averages over the macroscopic dynamics to include the effects of turbulence. Examples of these terms are eddy viscosity and diffusivity coefficients and average chemical heat release terms which appear as sources in the macroscopic flow equations. Besides providing these phenomenological terms, the turbulence model must use the information provided by the large scale flow dynamics self-consistently to determine the energy which drives the turbulence. The model must be able to follow reactive interfaces on the macroscopic scale.

2. Modelling Onset and Other Transient Turbulence Phenomena. The model should be able to predict the onset of turbulence in initially laminar flow since bursts and other highly transient phenomena seem to be the rule in reactive flow turbulence. Gradients in density, temperature, and velocity fields in the reacting fluid drive the macroscopic fluid dynamic instabilities which initiate turbulence. Thus these gradients from the macroscopic calculation are bound to be key ingredients in determining the energy available to drive the turbulence.

3. Complicated Reactions and Flow. The ideal turbulence model must deal with multiscale effects within the subgrid model. If there is a delay as velocity cascades to the short wavelength end of the spectrum due to chemical kinetics or buoyancy, for example, the model must be capable of representing this. Otherwise bursts and intermittency phenomena cannot be calculated.

4. Lagrangian Framework. An ideal subgrid model should be constructed on a Lagrangian hydrodynamics framework moving with the macroscopic flow. This requirement reduces purely numerical diffusion to zero so that realistic turbulence and molecular mixing phenomena will not be masked by non-physical numerical smoothing. This requirement also removes the possibility of masking purely local fluctuations by truncation errors from the numerical representation of macroscopic convective derivatives. The time-dependent (hyperbolic) Lagrangian framework should also generalize to three dimensions as well as resolve reactive interfaces dynamically.

5. Scaling. Breaking a calculation into macroscopic scales and subgrid scales is an artifice to allow us to model turbulence. The important physics occurs continuously over the whole spectrum from k_0, the wavenumber corresponding to the system size, to k_{diss}, the wave number corresponding to a mean free path of a molecule. Thus the macroscopic and subgrid scale spectra of any physical quantity must couple smoothly at k_{cell}, the cell boundary wave number. If this number were to be changed, as might happen if numerical resolution were halved or doubled, the predictions of the turbulence model must not change.

6. Efficiency. Of course, the model must be efficient. The number of degrees of freedom required to specify the status of turbulence in each separately resolved subgrid region has to be kept to a minimum for the model to be generally useable. The

real fluid has essentially an infinite number of degrees of freedom to represent the state of the gas in each small element. We
would like to be able to do the job with a minimal number of degrees of freedom.

Choosing an Algorithm Based on the Physics of the Problem

In reactive flow calculations we are concerned with two flow
regimes which depend on the rate of energy release. When energy
is released quickly, shocks and detonations are formed. When
energy is released slowly, flames are formed. The former requires that the numerical algorithm used follow the changes of
the system on time scales determined by the speed of sound in the
material (Courant condition). If we follow this same acoustic
wave transit time scale in the flame case where the physical timescales of interest are much larger, the cost is exhorbitant. The
gedanken flame calculation described above cost so much partly
because we postulated the use of an explicit algorithm based on
timesteps determined by the Courant condition. For flame calculations, then, the answer is to use techniques in which the energy conservation equation is converted to a pressure equation
which is solved implicitly.

A major area of modelling concern is that of coupling into
one calculation all of the pertinent physical and chemical processes characteristic of a combustion system. Two distinct
approaches have evolved. In the first of these, often called
"global implicit" differencing, the complete set of nonlinear
coupled equations describing the physical system of interest is
cast into a simple finite-difference form. The spatial and temporal derivatives are discretized and the nonlinear terms are
linearized locally about the solutions obtained numerically at
the previous timestep. This process is valid only when the values of the physical variables change slowly over a timestep. A
rigorously correct treatment of the nonlinear terms requires
iteration and large matrices must be inverted at each timestep to
guarantee stability. In one spatial dimension, say x, the problem usually appears as a block tridiagonal matrix with M independent physical variables to be specified at N_x grid points.
Then an MN_x by MN_x matrix must be inverted at each iteration of
each timestep. The blocks on or adjacent to the matrix diagonal
are M × M in size so the overall matrix is quite sparse. Nevertheless, an enormous amount of computational work goes into advan-
cing the solution even a single timestep. Multidimensional problems, in this approach, lead to matrices which are MN_xN_y by
MN_xN_y in two dimensions and $MN_xN_yN_z$ by $MN_xN_yN_z$ in three dimensions. In complex kinetics problems with no spatial variation,
the M independent variables are the species number densities and
temperature in the homogeneous volume of interest. The Gear
method (3) is an example of this global implicit approach for
pure kinetics problems.

The second approach is a fractional-step method we call asymptotic timestep-splitting. It is developed by consideration of the specific physics of the problem being solved. Stiffness in the governing equations can be handled "asymptotically" as well as implicitly. The individual terms, including those which lead to the stiff behavior, are solved as independently and accurately as possible. Examples of such methods include the Selected Asymptotic Integration Method (4,5) for kinetics problems and the asymptotic slow flow algorithm for hydrodynamic problems where the sound speed is so fast that the pressure is essentially constant (6,7).

The tradeoffs between these two approaches are clear. The implicit approach puts maximum strain on the computer and minimal strain on the modeller. For this method, convergence of the computed solutions is easy to test with improved temporal and spatial resolution. Non-convergence of any particular calculation may be hard to spot since severe numerical damping has been introduced to maintain numerical stability and positivity. This damping changes the desired profiles quantitatively, although quickly detected qualitative errors are often smoothed out. Solutions may be wrong yet stable.

In contrast, the asymptotic approach puts minimal strain on the computer but demands more of the modeller. The convergence of the computed solutions is usually easy to test with respect to spatial and temporal resolution, but situations exist where reducing the timestep can make an asymptotic treatment of a "stiff" phenomenon less accurate rather than more accurate. This follows because the disparity of time scales between fast and slow phenomena is often exploited in the asymptotic approach rather than tolerated. Furthermore, the non-convergence of any particular solution is often easier to spot in timestep splitting with asymptotics because the manner of degradation is usually catastrophic. In kinetics calculations, lack of conservation of mass or atoms signals inaccuracy rather clearly.

The asymptotic approach usually leads to more modular simulation models than the global implicit approach. Hydrodynamics, transport, equation of state calculations, and chemical kinetics are tied neatly into individual packages. What is even more important, specialized techniques for enhancing accuracy can be incorporated at each stage and for each physical phenomenon being modelled separately. There is no need to use simpler methods which are suitable for inclusion into a single giant finite difference formula. Since each phenomenon is treated as an independent package, the full spectrum of numerical tricks is applicable.

These packages are relatively easy to test individually and can be very sophisticated. They can also be used directly in a number of totally different physical problems with little or no change and are hence more flexible than equivalent portions of a global implicit algorithm. The price for this flexibility is

the need to treat carefully all the couplings between the individual physical terms and effects. Using the asymptotic approach
one cannot sit back and turn a massive mathematical crank to get
an answer.

At this point the pros and cons of the two approaches seem
to roughly counterbalance. This apparent equity extends to most
accuracy criteria as well.   If a timescale is not resolved,
neither solution method can give detailed profiles of phenomena
occurring on that scale. Similarily, to compute spatial gradients
accurately they must be resolved with enough spatial grid points
in either type of calculation.

The fact that the asymptotic approach demands more work of
the modeller is counterbalanced by the work that must be done to
reduce the computational expense of using the global implicit
method. This calculational expense, above all else, is the factor which has caused us to employ asymptotic rather than global
implicit formulations. For example, solving a chemical kinetics
scheme for M species requires inverting a general matrix of size
M X M. This involves approximately M^3 operations. In contrast,
the selected asymptotic approach to solving the kinetics equations generally scales as M. It is one goal of detailed modelling to be able to include the full details of extremely complex
kinetics systems coupled to time-dependent fluid dynamics. Since
more complex problems can be solved for the same cost using asymptotics, we are willing to invest the effort in the physics mod-
ules and their coupling in order to be able to expand our computational abilities.

Using the information discussed so far, we can now return to
the gedanken flame experiment with the idea of considering modi-
fied numerical methods in order to reduce the computational cost.
The goal is to calculate the propagation of a flame front across
a one-meter tube using a one-dimensional geometry with a fixed
detailed chemical reaction rate scheme.

First, we recognize immediately that we are interested in
calculating a flame front moving at less than the local sound
speed. Thus either a slow flow approximation or any method which
treats pressure implicitly would eliminate the sound speed criterion on the timestep. By using the asymptotic slow flow technique described below and still assuming a uniform grid spacing,
the number of pointsteps is reduced from 10^{14} to 10^{11}. Thus
Figure 3 shows that the time required for the calculation is
reduced from 3000 to 3 years!

But this is still atrocious, and we must now face the problem of eliminating unnecessary grid points. Adaptive gridding
is currently a frontier in reactive flow modelling. As yet there
are no general, excellent techniques. The block on the graph in
Figure 4 shows the region spanned in the gedanken flame problem
by an adaptively gridded calculation. Here 100 cells of 1 cm
length are used and the region surrounding the flame front is
finely gridded with 100 additional cells of 10^{-3} cm length. The

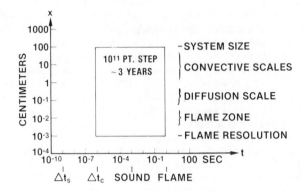

Figure 3.   Using the slow flow technique, which allows us to follow time scales larger than those required by an explicit solution of the energy equation, reduces the required computational time to three years

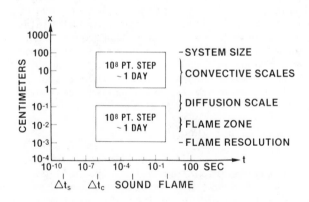

Figure 4.   An adaptive gridding method, in which the fine resolution is clustered around the flame front, reduces computational time to two days

timestep is still governed by the smallest cells, but by now only 200 cells are needed rather than 10^5. The saving, about a factor of 500, reduces the computational time to 2×10^8 pointsteps, or about two days.

Finally, Figure 5 summarizes the computational expense of performing the flame propagation problem using the possible, but as yet unexploited, technique of adaptive intermittent gridding. The idea here is that a finely gridded region is injected into the calculation at intermittent timesteps. This is done often enough to update the properties of the finely-spaced region which are then used as interior boundary conditions for the coarsely-spaced region. This is truly an asymptotic outer-inner-outer matching procedure. Now assume that 100 cells are needed to resolve the flame zone. Further, 100 short timesteps are enough to resolve changes in the flame zone brought about by the relatively slowly changing outer boundary conditions. During the imbedded calculation, the flame front moves only 10 of the fine zones, which is sufficient to determine flame speed and boundary conditions to be used in the coarsely spaced calculation. The imbedded calculation need be performed at most once in each large cell. Thus a total of $100 + (100)(10) = 1100$ seconds of computational time is required for the large scale simulation, a cost at least approaching our original naive estimate of 100 seconds.

This example has illustrated the importance of using the appropriate algorithm motivated by considerations of the actual problem that must be solved. It has further illustrated how much may be accomplished by developing the methods of adaptive gridding. One point that has not been mentioned, however, is that much of the cost of a detailed reactive flow calculation is taken up by the integration of the ordinary differential equations describing the chemical kinetics. Using the latest asymptotic techniques improves the picture painted above by a factor of two to four. But further improvements in these integration times without sacrificing accuracy is certainly an area where development is needed.

Testing the Model Against Theoretical Results

Analytical solutions, while often approximate, are extremely useful in providing functional relationships and generalizing trends. Below we show that by comparing numerical and analytic results, we can gain new insights into the controlling physical processes.

Ignition of a fuel-oxidizer mixture occurs when an external source of energy initiates interactions among the controlling convective, transport and chemical processes. Whether the process results in deflagration, detonation, or is simply quenched depends on the intensity, duration, and volume affected by an external heat source. Ignition also will depend on the initial ambient properties of the mixture which determine the chemical induction

time and the heat release per gram. Thus ignition is a complicated phenomena and its prediction for a specific mixture of homogeneous, premixed gases depends strongly on input parameters
which are often very poorly known. A convenient, inexpensive way
to estimate whether a mixture will ignite given a heat source intensity, duration, and volume would be a valuable laboratory tool
and a useful learning device.

A closed form similarity solution for the nonlinear time-
dependent slow-flow equations has been used as the basis for a
simple, time-dependent, analytic model of localized ignition
which requires minimal chemical and physical input ($\underline{8}$). As a
fundamental part of the model, there are two constants which must
calibrated: the radii, or fraction of the time-dependent similarity solution radius, at which the thermal conductivity and induction parameters are evaluated. This calibration is achieved
by comparison with the results of a detailed time-dependent numerical flame simulation model which is a full solution of the multi-
fluid conservation equations. The detailed model itself has been
checked extensively with respect to its various chemical, diffusive transport, and hydrodynamic components.

The basic similarity solution for this ignition problem is
derived from the slow flow ($\underline{6},\underline{7}$) approximation, characterized by
(1) flow velocities which are small compared to the speed of
sound, and (2) an essentially constant pressure field. The energy and velocity equations may then be written as

$$\frac{dP}{dt} \approx 0 = - \gamma P \underline{\nabla} \cdot \underline{v} + \underline{\nabla} \cdot \gamma Nk_B \kappa \underline{\nabla} T + S(t)e^{-k^2(t)r^2}, \qquad (1)$$

from which we can derive an algebraic equation for $\underline{\nabla} \cdot \underline{v}$. Here
P is the total pressure, $\underline{v}$ is the fluid velocity, T is the temperature, γ is the ratio of heat capacities C_p/C_v (assumed here to
be a constant) and κ is a function of the mixture thermal conductivity, λ_m,

$$\kappa \equiv \frac{\gamma-1}{\gamma Nk_B} \lambda_m(T), \qquad (2)$$

N is the total particle density and k_B is Boltzmann's constant.
The last term on the right hand side of Eq. (1) is the source
term. Proper choice of S(t) ensures that a given amount of ener-
gy, E_0, is deposited in a certain volume, $\frac{4\pi}{3} R_o^3$, in a time τ_o.
The choice of this Gaussian profile allows us to obtain a "closed"
form similaritv solution. If the fluid velocitv v is then expanded such that

$$v(r,t) \approx v_1(t)r \qquad (3)$$

and spherical symmetry is assumed, Eq. (1) may be solved anyti-
cally to obtain

$$T(r,t) = T_\infty \, e^{A(t)e^{-k^2(t)r^2}} \tag{4}$$

and

$$\rho(r,t) = \rho_\infty \, e^{-A(t)e^{-k^2(t)r^2}}, \tag{5}$$

where T_∞ and ρ_∞ are the background temperature and density far from the heat source. Solutions for $A(t)$ and $k(t)$ may then be obtained by solving two coupled ordinary differential equations

$$\frac{dk}{dt} = -kv_1 - 2\kappa k^3 \tag{6}$$

$$\frac{dA}{dt} = \frac{S(t)}{\gamma p} - 6\kappa k^2 A. \tag{7}$$

Invoking energy conservation yields an expression for the velocity coefficient v_1, and is effectively the first calibration in the model:

$$v_1 = \frac{S}{3\gamma P_\infty} \frac{F'(0)-F'(A)}{F(A)} + 2\kappa k^2 \frac{AF'(A)-F(A)}{F(A)} \tag{8}$$

where the function $F(A)$ is defined as

$$F(A) \equiv \int_0^\infty 4\pi x^2 [1-e^{-Ae^{-x^2}}]dx. \tag{9}$$

The model requires one further definition in order to predict ignition. A curve of chemical induction time as a function of temperature must be included in order to define the induction parameter,

$$I(t) = \int_0^t \frac{dt'}{\tau_c(T(r,t'))} \tag{10}$$

Ignition "occurs" when $I(t) = 1$ in this model, which is an exact result in the limit of large heat source and constant temperature near the center of the heated region. A simple analytic expression for $\tau_c(T)$ depending on three constants has been derived and can be calibrated using as few as three distinct values of τ_c at different temperatures.

The chemical reaction scheme used in the detailed model was used to generate a curve for $\tau_c(T)$. The values of thermal conductivity used in the detailed model were used to generate a function κ. Then a series of comparisons were made, in which the detailed model was configured in spherical symmetry with a Gaussian energy deposition.

We show results for several test cases. In one, $R_0 = 0.1$ cm and $\tau_0 = 1 \times 10^{-4}$ sec. The simple model predicts that 3.3×10^4 ergs is the minimum ignition energy and these results agree well with the simulation (Figures 6 and 7). Both models predict

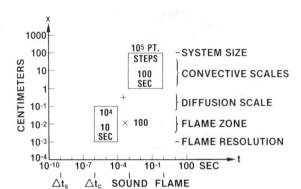

Figure 5. *The as yet undeveloped injected adaptive gridding method will reduce the computational time to 1100 sec.*

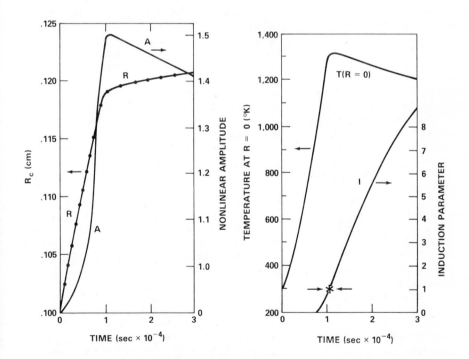

TIME (sec × 10⁻⁴)   TIME (sec × 10⁻⁴)

Figure 6. *Calculated results from the similarity solution plotted as a function of time. Here R_c is the characteristic radius for energy deposition, A is the nonlinear amplitude of the temperature and density functions, $T (R = 0)$ is the central temperature, and I is the induction parameter. The "*" indicates the predicted induction time; $\tau_o = 1.0 \times 10^{-4}$ sec; $E_o = 4.0 \times 10^4$ ergs; $R_o = 0.1$ cm.*

ignition at essentially the same time for a range of input ener-
gies. In the second example, R_o = 0.025 cm and τ_o = 1 $\times$ 10^{-4} sec.
The simplified model predicts a minimum ignition energy of $\sim$
8 $\times$ 10^2 ergs. The full simulation does not show ignition, but
predicts that some burning does occur and the flame is eventually
quenched (Figure 8). Thus in the regime for which both models
agree, we have in fact tested them both. In the regime where
they do not agree, we must then figure out what physics is miss-
ing from the similarity model. When this is done, we can, in
effect, use the detailed model to build accurate phenomenology
into the similarity solution. The similarity solution has tested
the full detailed model, and the full detailed model has shown
and helped extend the limits of the similarity solution.

Testing the Model with Experiment

    Although comparisons between analytic theory and model re-
sults can be used to extend our understanding of the controlling
processes in a system with limited physical complexity, many systems may preclude any analytic formulation. Then experimental
data provide the only means of checking the accuracy of the model.
Below we show a non-reacting case in which the results from an
experiment were used to test a numerical model. The model results then suggested new directions for the experiments.
    The determination of the effects of surface waves on sub-
merged structures has many practical applications, particularly
in an ocean environment. Due to the complexity of the problem,
analytic results are limited to idealized flows and geometries.
A major part of the complexity arises from the existence of the
free surface itself. Not only does the free surface dominate the
flow, but it may become multiply connected when sprays, wave-
breaking or cavitation occur. There is usually no steady state,
and a model for transient flow of the fluid over the obstacle must
be used.
    Here we describe the results of a model calculation of the
wave-induced pressure forces on a submerged half-cylinder, and
compare the results with experimental data. The implications of
the comparisons for both the validity of the model and the experimental procedure will be examined. Finally, the application of
the model to other fluid flows and to combustion problems will be
discussed.
    Figure 10 illustrates the initial conditions for the numeri-
cal model. A half-cylinder of radius "a" is submerged in a fluid
whose undisturbed free surface stands at a height h = 2a over the
bottom surface. A progressive wave with wavelength λ = 5a is
incident on the cylinder from the left. The sides of the computational region are periodic; that is, the physical system being
simulated is actually that of progressive waves over a series of
half-cylinders. Periodic boundary conditions were chosen to
avoid numerical damping and reflection at an outflow boundary.

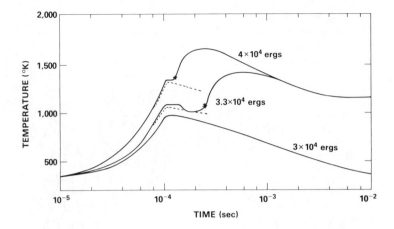

Figure 7. *Comparisons of the similarity (– – –) and detailed model (———) solutions for the central temperature as a function of time. The "*" marks the induction time predicted by the similarity solution;* $\tau_o = 1.0 \times 10^{-4}$ *sec;* $R_o = 0.1$ *cm.*

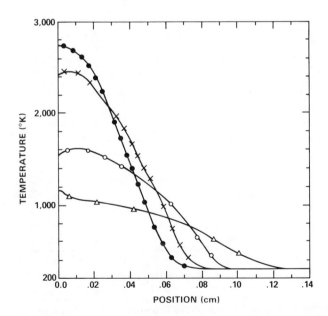

Figure 8. *Temperature as a function of position at four times in the detailed flame model. The flame dies out even though the similarity solution indicates it should not.* $R_o = 0.025$ *cm;* $E_o = 4 \times 10^2$ *ergs;* $T_o = 1 \times 10^{-4}$ *sec; (●), 7.127 \times 10^{-5} sec; (×), 1.113 \times 10^{-4}; (○), 2.256 \times 10^{-4}; (△), 7.123 \times 10^{-4}.*

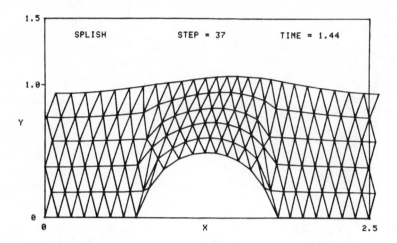

Figure 9.   The triangular grid early in a calculation of wave flow over a half-cylinder. The first wave of a wave train has passed over the cylinder from the left.

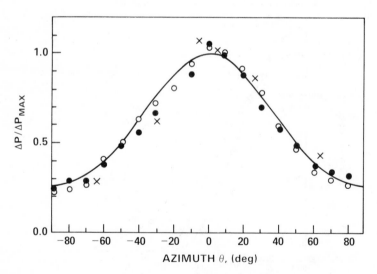

Figure 10.   A comparison of numerical results with experiment and with linear theory for the magnitude of pressure variations over a half-cylinder owing to a progressive wave with ka = 2.5 and kh = 5: (———), linear theory (3); (×), SPLISH results (NRL); (○, ●, △, ▲), experiments (NRL).

The calculation seeks to find the pressures at every point in the fluid as a function of time, and in particular the pressures and pressure gradient forces at points on the cylinder surface.

The numerical model is based on finite difference techniques for solving the equations for inviscid, incompressible fluid flow using a triangular grid which extends throughout the interior of the fluid (9). The free surface and rigid boundary shapes are approximated by straight lines which extend between points on those surfaces and which define the edges of the computational grid. The governing equations are cast in a Lagrangian formalism so that points originally lying on a surface will remain there at all times during the calculation. Points interior to the fluid will follow Lagrangian pathlines as if they were experimental marker particles in a real fluid. The equations are differenced such that vorticity is conserved identically at all times. Vertex pressures are chosen to keep the local fluid volumes divergence-free. These new pressures are in turn used to advance the velocities and update the grid positions.

The physical behavior of the governing equations can be preserved in the approximate difference equations being solved numerically by using a triangular grid. A Lagrangian grid will distort in any non-trivial flow field, and as grid distortion becomes severe the calculation quickly loses accuracy. However, a triangular grid can be manipulated locally in several ways to extend realistic calculations of transient flows (9,10). Each grid line represents a quadrilateral diagonal, and the opposite diagonal can be chosen whenever vertices move in the flow to positions which favor that connection. Such a reconnection involves just the four vertices describing the quadrilateral. No fluid moves relative to the quadrilateral, eliminating one form of numerical dissipation. Vertices may also be added or deleted to preserve the desired resolution by local algorithms which involve only those vertices in the vicinity of the grid anomaly. Major advantages of this technique are that the algorithms can be conservative, they permit a minimum of numerical dissipation and yet they require very little computer time since most of the grid remains unaltered.

Data for the experimental comparison was obtained through wave-tank experiments performed with a bottom-mounted half-cylinder so that pressure measurements could be compared directly to the numerical results (11). The obstacle was placed one-third of the tank length from a mechanical wavemaker and at the other end of the channel a sloping porous beach absorbed 95% or more of the incident wave energy.

Results of the experiment and of the numerical simulation are shown in Figure 10, together with the results of linear theory. The magnitude of the pressure fluctuations as measured by the experiment at different points on the cylinder ($\theta = 0°$ at the top of the cylinder) are compared with the predictions of the model. As shown in Figure 10, the comparison is quite good. Figure 11 compares the calculated and measured instantaneous pressure

distribution around the cylinder for the situation in which the
crest of the progressive wave is near the left side of the half-
cylinder. Again the comparisons look good, but now some differences become evident.

It was found that to within experimental error, all of the
observed discrepancies could be explained by two factors. The
first factor is that the model did not exactly describe the
physical situation in the experiment: the wave tank had a single
cylinder, whereas the calculation is for a series of cylinders.
The second factor was the surprising result that the roughly 5%
reflected wave from the wave tank significantly affected the experimental results due to modifications in the dynamic pressure
fluctuations. In this instance a detailed examination of the
model and experimental results has indicated that an experimental
effect thought to be small could in fact cause noticeable deviations in the data measured.

The application of this numerical technique to reactive flow
is relatively straightforward. Although the example presented
above is for homogeneous flows, the extension to include interfaces involves no basic changes to the underlying gridding scheme,
but only the provision that no interface sides are allowed to be
reconnected. Instead local gridding anomalies at the interfaces
are resolved by adding or subtracting vertices at the interface
and reconnecting grid lines leading to those vertices. That this
solution is viable is most easily shown by Figure 12 which shows
stages in the collapse of a Rayleigh-Taylor unstable fluid layer
calculated with the same model. Here the calculation can continue even though the originally simply-connected lighter fluid
performs a transition to a multiply connected fluid which includes "bubbles" which have been entrained by the heavier fluid.
Of course, for reactive flow calculations a new model would have
to be constructed based on these techniques which used instead
the equations governing compressible fluids and which contained
the added chemical reactions and diffusive transport effects.

Conclusion

Detailed modelling of laminar reactive flows, even in fairly
complicated geometries, is certainly well within our current capabilities. In this paper we have shown several ways in which
these techniques may be used. As the physical complexity we wish
to model increases, our footing becomes less sure and more phenomenology must be added. For example, we might have to add eva-
poration laws at liquid-gas interfaces or less well-known chemical reaction rates in complex hydrocarbon fuels.

Perhaps the biggest problem facing combustion modelling now
is turbulence: there are no excellent or even good methods of
including such effects in our calculations. At best we have a
number of phenomenological models with limited ranges of validity
and which imply a steady state.

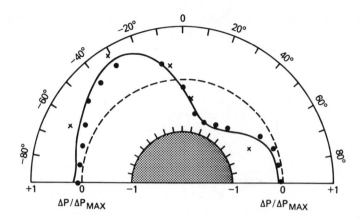

Figure 11. A comparison of numerical results with theory and experiment for instantaneous pressure distributions on a half-cylinder (ka = 2.5 and kh = 5) with the wave crest just to the left of the cylinder: (———), linear theory (3); (×), SPLISH results; (●), experiments (NRL)

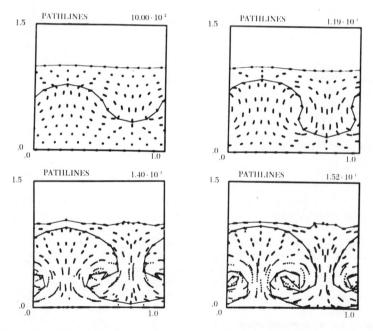

Figure 12. Lagrangian path lines at various stages of a Rayleigh–Taylor collapse for the case of two inviscid, incompressible fluids having a density ratio of 2:1. A free surface is present above the dense fluid and the interface between the fluids is indicated for each stage. The simulation shows how later evolution of the fluid flow is dominated by the strength and dynamics of the vortex pair created during the early stages of collapse.

We believe that devising a way to handle this difficult problem
of strongly coupled multiple time and space scales is the challenge we currently face.

Acknowledgments

This work has been funded by the Naval Research Laboratory
through the Office of Naval Research.

Abstract

A major goal of detailed numerical modelling of combustion
is to develop computational tools which can be used interactively
with laboratory diagnostic techniques to interpret and understand
experiments. Only through this quantitative interaction of prediction and measurement can we calibrate and extend our understanding of the fundamental physical and chemical processes involved. Modelling of flames and detonations is particularly
challenging because of the strong feedback between the energy released due to chemical reactions and the evolution of the hydrodynamic flow. Since it is generally not possible to decouple
these processes completely, special numerical techniques are required to solve the problems. In this paper, the basic steps and
decisions which must be made when constructing and using a numerical model will be discussed. These involve 1) choosing the
numerical algorithms based on the physics and chemistry of the
problem, 2) testing the model based on these algorithms in different limits against theoretical and analytical results, and 3)
using the model interactively with experiments. With the third
step specifically in mind, it will be shown how the species concentration, velocity, and temperature profiles obtained from
laser probe diagnostics can provide information for use in and
comparison with detailed calculations.

Literature Cited

1. Oran, E. S and Boris, J. P., Detailed Modelling of Combustion
    Processes, to appear in Prog. in Energy and Comb. Sci, 1980.

2. Boris, J. P. and Oran, E. S., Modelling Turbulence: Physics
    or Curve-Fitting, to appear in the Proc. of the International
    Symposium on Gas Dynamics of Explosions and Reactive Systems,
    Gottingen, 1979, AIAA; also NRL Memorandum Report,
    Naval Research Laboratory, Washington, D. C. 20375, 1979.

3. Gear, C. W.,"Numerical Initial Value Problems in Ordinary
    Differential Equations," Prentice-Hall, Englewood Cliffs,N.J.,
    1971.

4. Young, T. R., and Boris, J. P., A Numerical Technique for Solving Stiff Ordinary Differential Equations Associated with the Chemical Kinetics of Reactive Flow Problems, J. Phys. Chem., 1977, 81, 2424.

5. Young, CHEMEQ - A Subroutine for Solving Stiff Ordinary Differential Equations, NRL Memorandum Report 4091, Naval Research Laboratory, Washington, D. C. 20375, 1979.

6. Jones, W. W. and Boris, J. P., FLAME--A Slow-Flow Combustion Model, NRL Memorandum Report 3970, Naval Research Laboratory, Washington, D. C. 20375, 1979.

7. Boris, J. P. and Oran, E. S., Detailed Modelling of Reactive Flows, Proc. of the GAMNI International Conference on Numerical Methods for Engineering, Paris, France, 1978.

8. Oran, E. S. and Boris, J. P., Theoretical and Computational Approach to Modelling Flame Ignition, Proc. of the International Symposium on Gas Dynamics of Explosions and Reactive Systems, Gottingen, 1979; also NRL Memorandum Report, Naval Research Laboratory, Washington, D. C. 20375, 1979.

9. Fritts, M. J. and Boris, J. P., The Lagrangian Solution of Transient Problems in Hydrodynamics Using a Triangular Mesh, J. of Comp. Phys., 1979, 31, 173.

10. Crowley, Proceedings of the Second International Conference on Numerical Methods in Fluid Dynamics, Springer-Verlag, New York/Berlin, 1971.

11. Miner, E. W., Griffin, O. M., Ramberg, S. E., and Fritts,M.J., Numerical Calculation of Wave Effects on Structures, Proceedings of the Conference on Civil Engineering in the Oceans, IV, San Francisco, CA, 10-12 September 1979.

12. Garvin, David, personal communication.

RECEIVED February 1, 1980.

Rate of Methane Oxidation Controlled by Free Radicals

JOHN R. CREIGHTON

Lawrence Livermore Laboratory, Livermore, CA 94550

A simple model of the chemical processes governing the rate
of heat release during methane oxidation will be presented below.
There are simple models for the induction period of methane
oxidation (1,2,3); and the partial equilibrium hypothesis (4) is
applicable as the reaction approaches thermodynamic equilibrium.
However, there are apparently no previous successful models for
the portion of the reaction where fuel is consumed rapidly and
heat is released. There are empirical rate constants which, due
to experimental limitations, are generally determined in a range
of pressures or concentrations which are far removed from those
of practical combustion devices. To calculate a practical device
these must be recalibrated to experiments at the appropriate
conditions, so they have little predictive value and give little
insight into the controlling physical and chemical processes.
The model presented here is based on extending Semenov's model (2)
of the induction period to cover the period of heat release.
    Semenov's model considers any branching chain reaction. It
assumes that some initial dissociation of fuel leads to an
intermediate species. This species, or some of its products,
reacts with the fuel to create more of the intermediate species,
implying branching reactions. If recombination, or other chain
breaking reactions, are allowed one gets a rate equation for the
concentration of the intermediate species $[R]$.

$$d[R]/dt = A + B[R] - C[R]^2 \qquad (1)$$

The first term on the right represents the initial dissociation,
the second the branching chain reactions, and the third
recombination. Coefficients A, B and C are functions of the rate
constants and the concentration of fuel and oxidizer, but are
independent of the intermediate concentration.
    Creighton (3,5) has shown that the induction period of
methane oxidation is described by Semenov's model. Analysis of
the results of numerical calculations using a detailed chemical
kinetics reaction scheme showed that about eight reactions were
dominant, and that the rate of creation and consumption of

0-8412-0570-1/80/47-134-357$05.00/0
© 1980 American Chemical Society

species H, OH and O were balanced. This justified using a steady
state approximation on the corresponding rate equations yielding
algebraic equations which couple the concentration of these
species to one another and to $[CH_3]$. The rate limiting step was
found to be

$$CH_3 + O_2 = CH_3O + O \tag{2}$$

with rate constant $k_2 = 2 \times 10^{13} \exp(-14,500/T)$ $cm^3 mol^{-1} sec^{-1}$ (1).
The CH_3O immediately decomposes to CH_2O and H. It can be shown
(3,5) that branching reactions of H immediately result in three
new CH_3 radicals and branching reactions of O yield two more.
Thus branching reactions yield five new CH_3 for every one consumed
by reaction 2, a net increase of 4. This establishes a value for
Semenov's coefficient $B = 4k_2[O_2]$. (There is some controversy
concerning reaction 2 and the value of k_2, but it gives calculated
induction times in agreement with a wide variety of experiments
(3). The reaction may proceed via an intermediate complex,
CH_3O_2; but this gives the same calculated induction time provided
it decomposes in steps which give H and O atoms rather than OH,
and that decomposition is faster than reaction 2.)

Coefficient A depends on the initial dissociation step,

$$CH_4 + M = CH_3 + H + M \tag{3}$$

with rate constant $k_3 = 2 \times 10^{17} \exp(-44,200/T)$ $cm^3 mol^{-1} sec^{-1}$ (9).
Branching reactions of the H atom give three more CH_3 molecules
so $A = 4 k_4 [CH_4][M]$.

The rate of all recombination reactions will be proportional
to $[CH_3]^2$, because all radical concentrations are proportional to
$[CH_3]$. If we take the reverse of reaction 3 as a prototype, C
equals $k_{-3}[M]([H]/[CH_3])$, where the ratio $[H]/[CH_3]$ is determined
by the algebraic relations mentioned above and is a function only
of the rate constants and the concentration of fuel and oxidizer.
The magnitude of this ratio is about 10^{-3}. A value of
$C = 1.7 \times 10^{14}[M]$ has been found to give calculated induction times
in agreement with experiment at pressures above atmospheric, and
is unimportant at low pressures.

A rate equation for fuel consumption can also be written
(3,5).

$$d[CH_4]/dt = -A - 5/4 B[R] + C[R]^2 \tag{4}$$

The numerical factor 5/4 results from consumption of five
molecules of CH_4 for a net increase of four CH_3. (There should
also be a small numerical correction to the last term because an
H atom, as well as a CH_3, is consumed.)

Eqs. 1 and 4 constitute a model for the induction period of
methane oxidation, and can be integrated in closed form provided
the temperature is held constant. If they are integrated
numerically, along with appropriate thermochemistry to account for
the temperature change, the solutions are a semi-quantitatively
correct description of the fuel consumption, as well as induction.

Fig. 1 shows the results of such a calculation. (The model gives
a rate of fuel consumption which can be as much as an order of
magnitude too small. This occurs because the model neglects
additional reactions which increase the ratio of [OH] to [CH$_3$]
during fuel consumption.) Initially [R] is zero and A is the
dominant term in eq. 1, but B[R] rapidly becomes larger and [R]
increases exponentially as seen at the left of Fig. 1. Eventually
as [R] increases, C[R]^2 becomes comparable to B[R] and the time
derivative becomes very small, as in the middle of Fig. 1.
    The time derivative becomes small because [R] approaches a
quasi-equilibrium value [R]$_e$ = B/C. This might also be called
the steady state or stationary state, but all terminology seems
to lead to possible confusion. There are two important properties
of [R]$_e$. First, it is a stable solution of eq. 1 so [R] will tend
to stay near [R]$_e$. Second, it depends only on the rate constants
and the concentration of fuel, oxidizer and diluent, but not on
the radical concentration. It does have a strong temperature
dependence due to the large activation energy of B. A upper limit
to [R]$_e$ is shown as a dashed line in Fig. 2. This was calculated
holding [O$_2$] and [M] constant at their initial values of 5x10^{-5}
and 2x10^{-4} moles/cc. The actual value of [R]$_e$ will be lower than
this due to consumption of fuel and oxygen.
    To compare reactions with different time constants it is
useful to plot them as trajectories in a multi-dimensional phase
space whose coordinates are the species concentrations and the
temperature. Fig. 2 shows trajectories projected onto the
temperature vs. [R] plane for reactions with identical initial
fuel and air concentrations but different initial radical
concentrations and temperature. Trajectories beginning at the
left had no initial radicals, and the trajectory starting at
1200 K is represented in Fig. 1. The exponential increase of [R]
to [R]$_e$ is isothermal so it appears horizontal in Fig. 2. The
knee of the curve represents the relatively flat portion of Fig. 1
where [R] is approximately [R]$_e$. As the temperature increases [R]
remains approximately equal to [R]$_e$, which lies to the left of
the dashed line due to consumption of fuel and oxygen.
Trajectories beginning on the right had an initial radical
concentration equal to half the initial fuel concentration. The
radical concentration fell rapidly to [R]$_e$, releasing heat, and
then remained at [R]$_e$. A heat loss term was included in the
model with the result that trajectories which reach [R]$_e$ at
temperatures below 1050 K do not go to complete combustion because
the chemical heat release is less than the heat loss, and the
mixture cools.
    Fig. 2 shows clearly that the quasi-equilibrium radical
concentration sets the rate of fuel consumption and chemical heat
release. It also shows the stability. Whatever the initial value
of [R] it moves towards [R]$_e$ and remains there. It can only
increase as [R]$_e$ increases with temperature. Thus, though the
oxidation of methane is a branching chain reaction, fuel

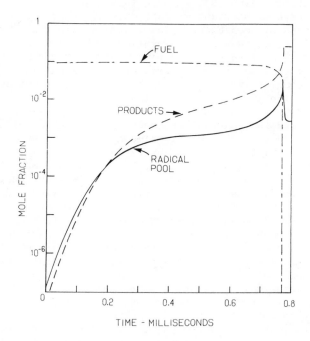

Figure 1. Calculated mole fractions of fuel, intermediate species, and products using Semenov's model for a stoichiometric methane–air mixture initially at 1200 K and atmospheric pressure

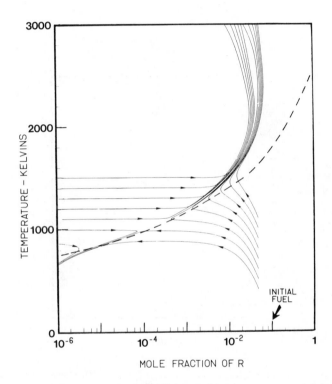

Figure 2. Reaction trajectories calculated with Semenov's model for a stoichiometric mixture at atmospheric pressure and various initial temperatures and radical concentration: (———), B:C calculated using constant initial fuel and oxygen concentrations but varying temperature

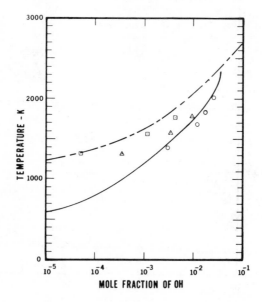

*Figure 3.   Reaction trajectories calculated using a detailed kinetics model. Symbols
are described in the text.*

consumption proceeds as a thermal reaction with the rate
determined by $[R]_e$.
The quasi-equilibrium radical concentration does not depend
on the assumptions of the Semenov model, although the model gives
the simplest explanation. Numerical calculations using the full
detailed kinetic reaction mechanism for methane oxidation (8) show
that each radical species concentration is in quasi-equilibrium.
During induction CH_3 has the largest concentration, but as other
reactions become important $[H]$, $[OH]$ and $[O]$ become larger. Fig. 3
shows reaction trajectories on a temperature versus $[OH]$ plane for
some detailed kinetics calculations. The solid line is a
simulation of a low pressure, laminar flame (6) with circles
representing experimental data (7). The dashed line is a constant
volume, adiabatic reaction for the same mixture, an approximate
simulation of fuel consumption in a shock tube or turbulent flow
reactor. The value of $[OH]_e$ for the flame differs from the
constant volume case because fuel is consumed at a lower
temperature in the flame and $[OH]_e$ depends strongly on the fuel
consumption. This is demonstrated by isothermal (constant volume)
calculations of $[OH]_e$ shown in Fig. 3 as triangles for fuel
concentrations equal to those in the flame as squares for the
adiabatic case. The value of $[OH]$ in the flame is somewhat
greater than $[OH]_e$ because radicals diffuse ahead of the flame.
   We conclude that free radical concentrations control both
the induction time and the rate of fuel consumption, and depend
only on a few critical rate constants and the concentration of
fuel and oxidizer. A more detailed report is being written and
Ref. 8 discusses the implications for ignition. This work was
performed at the Lawrence Livermore Laboratory for the U. S.
Department of Energy under contract No. W-4705-Eng-48.

Literature Cited

1.   Brabbs, T. A.; Brokaw, R. S., "15th Symp. (Int'nl) on Comb.",
Combust. Inst., Pittsburgh, PA, 1974, p. 893.
2.   Semenov, N. N., Compt. Rend. (Doklady) Acad. Sci. URSS, 1944,
XLIII, 342 and XLIV, 62.
3.   Creighton, J. R., J. Chem. Phys. 1977, 81, 2520.
4.   Kaskan, W. E.; Schott, G. L., Comb. Flame, 1962 6, 73.
5.   Creighton, J. R., Soc. Auto. Eng'rs. Paper 790249 (1979).
6.   Creighton, J. R.; Lund, C. M., "Proc. 10th Matl's Res. Symp."
U. S. Govt. NBS Spec. Publ. 561/1, 1979, p. 1223.
7.   Peeters, J.; Mahnen, G., "14th Symp. (Int'nl.) on Comb.";
Combust. Inst., Pittsburgh, PA, 1973, p. 133.
8.   Guirguis, R. H.; Karasalo, I.; Creighton, J. R.; Oppenheim,
A. K., "7th Int'nl. Symp. on Gasdynamics Explosions and Reactive
Systems", Gottingen, Ger., 1979 (in press).
9.   Hartig, R.; Troe, J.; Wagner, H. G., "13th Symp. (Int'nl) on
Comb."; Combust. Inst., Pittsburgh, PA, 1971, p. 147.

RECEIVED February 14, 1980.

The Detailed Modelling of Premixed, Laminar, Steady-State Flames. Results for Ozone

JOSEPH M. HEIMERL and T. P. COFFEE

Ballistic Research Laboratory, ARRADCOM, Aberdeen Proving Ground, MD 21005

The overall objective of these studies is to deli-
neate and validate the elementary gas phase kinetic mechanisms
involved in the combustion of the cyclic nitramines, HMX and RDX.
The modeling of premixed, laminar, steady state flames is the
approach taken. First, because the governing equations are simple
relative to other combustion processes one can focus upon the
kinetics. Second, because laser-based diagnostics enable species
and temperature profiles to be probed experimentally and one can
validate the model. Detailed comparisons of predicted and measured profiles of temperature and of species (particularly radicals) serve either to validate the model or to indicate refine-
ments. Given the pausity of reliable temperature dependent rate
coefficient data, the latter situation is anticipated. Such discrepancies can be exploited by using them to direct experiments on
or theoretical calculations of elementary rate coefficients. This
sequence of comparison and direction can be iterated. The ideal
end product is a validated network of elementary reactions that
can then be used with some confidence in more complex simulations.
A sequence of flame studies is planned; from the test case, ozone,
through the recognized intermediates, formaldehyde/oxides of
nitrogen, to the gas phase elementary networks that describe the
HMX and RDX flames. This paper discusses some of the ozone modeling results. A more complete description of the background, motivation and other details is available (1).

Equations and Solutions. The governing equations that describe a
one dimensional, premixed, laminar, unbounded flame for a multicomponent ideal gas mixture are (2, 3, 4):

$$(\rho)_t + (\rho u)_x = 0$$

$$\rho(Y_k)_t + \rho u(Y_k)_x = - (\rho Y_k V_k)_x + R_k M_k, \quad (k = 1, \ldots, N), \text{ and}$$

$$\rho(T)_t + \rho u(T)_x = c_p^{-1} \left[(\lambda T_x)_x - \sum_{k=1}^{N} (R_k M_k h_k + c_{pk} \rho Y_k V_k T_x) \right].$$

This chapter not subject to U.S. copyright.
Published 1980 American Chemical Society

For the kth species Y_k and X_k are the mass and mole fractions, respectively, R_k is the net rate of the production due to chemistry and M_k is the molecular weight. In addition the diffusion velocity V_k is given by the Stefan-Maxwell relation

$$(X_k)_x = \sum_{j=1}^{N} X_k X_j (V_j - V_k) D_{kj}^{-1}$$

and the other symbols have their usual meaning. The pressure through the flame is one atmosphere and constant (4, 5). We neglect effects of viscosity, thermal diffusion, body forces and radiation. To obtain a solution, we employ a relaxation technique and use the PDECOL package (6). PDECOL is based on a finite element collocation method employing B-splines. For computing efficiency we have developed a method of concentrating our break-points in the steep flame front where accuracy is necessary (7). Kinetic, transport and thermodynamic coefficients are required as input to the model. The kinetic mechanism is:

$$O_3 + M \leftrightarrow O + O_2 + M \tag{1}$$

$$O + O_3 \leftrightarrow O_2 + O_2 \text{ and} \tag{2}$$

$$O + O + M \leftrightarrow O_2 + M. \tag{3}$$

Expressions for the rate coefficients are taken from the literature (8, 9, 10) and are shown in Table I. Expressions for the transport coefficients (1, 11, 12, 13) are shown in Table II and the specific enthalpy, h_k, and specific heat capacity, c_{pk}, are obtained from Gordon and McBride (14). Each expression for the input coefficients is based on separate, independent measurements and the methodology for obtaining them has been discussed (1). Results and Discussion. Figure 1 shows the O, O_2, O_3 and temperature profiles computed for an initial ozone mole fraction of unity. No experimental profiles are known for comparison and so by our own definition the ozone flame remains unvalidated. We can however compare burning velocities. As can be seen in Figure 2 our computed burning velocities compare favorably with both the experimental results of Streng and Grosse (15) and the modeling results of Warnatz (16). (Warnatz has developed a finite difference model that also requires species dependent input coefficients.) The solid line in the figure is Streng and Grosse's fit to their data. Over the range of 0.25 to 1.0 initial ozone mole fractions our results are no more than 30% greater than Streng and Grosse's. Over the entire range shown our results agree with Warnatz's within ± 12%. This agreement does not imply that the sets of input coefficient used in the respective models are equivalent.

Figure 3 shows the ratio of the values of Warnatz' input coefficients to our corresponding values. This figure shows that at the higher temperatures (i.e., at the larger initial ozone mole fractions), the values for k_1 and k_2 differ markedly. The fact

TABLE I. KINETIC COEFFICIENTS

COEFFICIENT	EXPRESSION*	REFERENCES	REMARKS
k_1	$4.31 \times 10^{14} \exp(-11,161/T)$	Heimerl & Coffee, 8	$300K \leq T \leq 3000K$, $M=O_3$
k_{-1}	$1.2 \times 10^{13} \exp(+976/T)$		Derived from equilibrium const.
k_2	$1.14 \times 10^{13} \exp(-2300/T)$	Hampson, 9	$200K \leq T \leq 1000K$
k_{-2}	$1.19 \times 10^{13} \exp(-50600/T)$		Derived from equilibrium const.
k_3	$1.38 \times 10^{18} T^{-1} \exp(-171/T)$	Johnston, 10	$1000K \leq T \leq 8000K$, $M=O_2$
k_{-3}	$2.75 \times 10^{19} T^{-1} \exp(-59732/T)$	Johnston, 10	$1000K \leq T \leq 800K$, $M=O_2$

*centimeter-mole-second units.

TABLE II.   TRANSPORT COEFFICIENTS

TRANSPORT COEFFICIENT	EXPRESSION	REFERENCES
λ_1*	$1.60 \times 10^{-6} \, T^{0.71}$	Dalgarno and Smith, 11
λ_2	$5.74 \times 10^{-7} \, T^{0.827}$	Hanley and Ely, 12
λ_3	$3.90 \times 10^{-7} \, T^{0.842}$	Heimerl and Coffee, 1
pD_{12}**	$1.32 \times 10^{-5} \, T^{1.774}$	Marrero and Mason, 13
pD_{13}	$1.66 \times 10^{-5} \, T^{1.665}$	Heimerl and Coffee, 1
pD_{23}	$1.18 \times 10^{-5} \, T^{1.665}$	Heimerl and Coffee, 1

*Heat conductivity in units of cal-cm^{-1}-s^{-1}-K^{-1}.

**Pressure multiplied by binary diffusion coefficient in units of $atmos$-cm^2-s^{-1}.

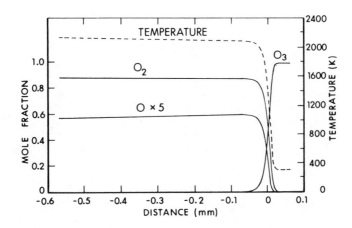

Figure 1. *Computed profiles for unity initial ozone mole fraction*

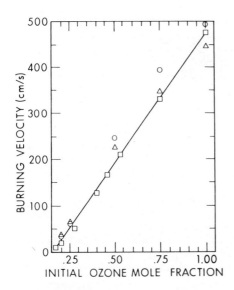

Figure 2. *Comparison of experimental (Streng and Grosse) and computed burning velocities over a wide range of initial ozone mole fractions: (○), this work; (△), Warnatz; (□), Streng and Grosse.*

that the ratios for k_1 and k_2 change in opposite directions (see Figure 3) suggests that the individual effects of k_1 and k_2 on the burning velocity offset each other. This hypothesis is checked by substituting Warnatz' expressions for k_1 and k_2 into our code. (This is done in such a way that the equilibrium constants remain unchanged.) For the case of the initial ozone fraction of unity we obtain a burning velocity of 459 cm/s. This value is to be compared to Warnatz' computed value of 445 cm/s. As an added bonus the difference in these two values provides a measure of the collective effects of differing transport coefficients.

In comparing profiles there are sensible differences in some model results. For an initial ozone mole fraction of unity Figures 4 and 5 show a comparison of the atomic oxygen and temperature profile respectively. (For ease in viewing the curves have been arbitrarily displaced from each other along the distance axis.) Following the method of comparison discussed above we substitute Warnatz' expressions for k_1 and k_2 into our code and find the dashed-line profiles. Thus, we find that differences in the model profiles are due mainly to the different expressions for k_1 and k_2.

We have recently and critically evaluated the available high temperature experimental data for the ozone decomposition reaction (8). The expression used here and shown in Table I is consistent with all the direct experimental data known to us and is valid over a decade range in temperature.

The expression for k_2 is another story. Figure 6 shows plots of the values of k_2 against reciprocal temperature. Warnatz developed and used his own expression and we have employed Hampson's (9). To use them in our codes we both have assumed that the respective expressions are valid for temperatures greater than 1000K, the upper limit of applicability of each.

In order to distinguish which expression for k_2, if either, is correct, high temperature measurements and/or *ab initio* calculations of the rate coefficient for reaction (2) are required. Alternately, the computed differences in the values for atomic oxygen and for the temperature in the burned region at an initial ozone mole fraction of unity appear to be large enough that profile measurements above such a flame may be sufficient to distinguish between the two expressions.

SUMMARY. We have shown that this model and its input parameters predict burning velocities that are in reasonable agreement both with the measurements of Streng and Grosse and with the computations of Warnatz. We have also demonstrated that agreement with burning velocities, even over a wide range of initial ozone mole fractions is a necessary but not sufficient condition to ensure that the input coefficients are realistic; for this reason profile measurements are vital to test a model's input coefficients. Finally, by a comparison of computed profiles we have indicated the need to measure or to calculate high temperature values for k_2.

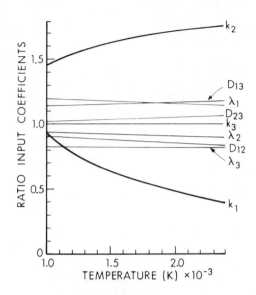

Figure 3.   Ratio of the values of Warnatz's and our input coefficients. Subscripts on the rate coefficients refer to reactions. Subscripts 1, 2, and 3 on the transport coefficients refer to O, O_2, and O_3, respectively.

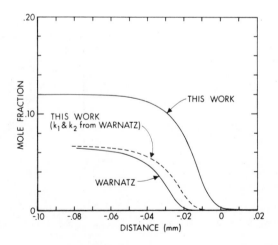

Figure 4.   Calculated atomic oxygen profiles for unity initial ozone mole fraction: (– – –), the result of substituting Warnatz's expression for k_1 and k_2 into our model.

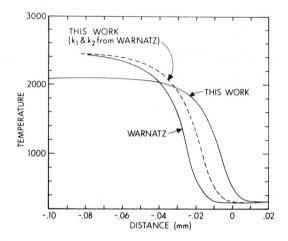

Figure 5. Calculated temperature profiles for unity initial ozone mole fraction:
(— — —), the result of substituting Warnatz's expression for k_1 and k_2 into our model.

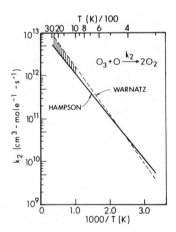

Figure 6. Expressions for the rate coefficient for Reaction 2: (———), the region of temperature over which both expressions have been assumed to be valid.

Literature Cited

1. Heimerl, J. M. and Coffee, T. P., to be published in Combust. Flame.

2. Hirschfelder, J. O., Curtis, C. F. and Bird, R. B., "Molecular Theory of Gases and Liquids", 2nd printing, corrected, with notes, John Wiley and Sons, NY, 1964.

3. Bird, R. B., Stewart, W. S. and Lightfoot, E. N., "Transport Phenomena", John Wiley and Sons, NY, 1960.

4. Williams, F. A., "Combustion Theory", Addision-Wesley, Reading, MA, 1965.

5. Fristrom, R. M. and Westenberg, A. A., "Flame Structure", McGraw-Hill, NY, 1965, p. 319.

6. Madsen, B. K. and Sincovec, R. F., Preprint UCRL-78263 (Rev 1), Lawrence Livermore Laboratory, 1977.

7. Coffee, T. P. and Heimerl, J. M., BRL Technical Report in press.

8. Heimerl, J. M. and Coffee, T. P., Combust. Flame, 1979, 35, 117-123.

9. Hampson, R. F. (ed), J. Phys. Chem. Ref. Data, 1973, 2, 267-312.

10. Johnston, H. S., NSRDS-NBS-20, September 1968.

11. Dalgarno, A. and Smith, F. J., Planet Spac. Sci., 1962, 9, 1-2.

12. Hanley, H. J. M. and Ely, J. F., J. Phys. Chem. Ref. Data, 1973, 2, 735-755.

13. Marrero, T. R. and Mason, E. A., J. Phys. Chem. Ref. Data, 1972, 1, 3-118.

14. Gordon, G. S. and McBride, B. J., NASA-SP-273, 1971, (1976 program version).

15. Streng, A. G. and Grosse, A. V., Sixth Symposium (International) on Combustion, Reinhold Pub. Co., 1957, pp. 264-273.

16. Warnatz, J., Ber. Bunsenges Phys. Chem., 1978, 82, 193-200.

RECEIVED February 1, 1980.

On the Rate of the O + N$_2$ Reaction

DANIEL J. SEERY and M. F. ZABIELSKI

United Technologies Research Center, East Hartford, CT 06108

The recent shock tube measurement of the O + N$_2$ reaction rate by Monat, Kruger and Hanson (1) provided a rate constant, k_1 = 1.84 x 10^{14} exp(-76, 250/RT) cm^3/mol sec. This rate is more than twice the most widely used value of k_1 = 7.5 x 10^{13} exp(-76, 250/RT) recommended by Baulch et. al (2). The latter value was based on an evaluation of two indirect measurements plus theoretical calculations and had an estimated uncertainty of about a factor of two. Further support for the lower value came from the flame study by Blauwens et. al (3) whose data indicated almost exact agreement with the recommendation of Baulch et. al and an uncertainty of only ± 30%. Because there are so many current analytical studies of NO formation, in which reaction (1) is of major importance, this confusion in the literature takes on added significance.

The present work involves measurement of k_1 in a 0.1 atmosphere, stoichiometric CH$_4$–Air flame. All experiments were conducted using 3 inch diameter water-cooled sintered copper burners. Data obtained in our study include (a) temperature profiles obtained by coated miniature thermocouples calibrated by sodium line reversal, (b) NO and N$_2$ composition profiles obtained using molecular beam sampling mass spectrometry and microprobe sampling with chemiluminescent analysis and (c) OH profiles obtained by absorption spectroscopy using an OH resonance lamp. Several flame studies (4) have demonstrated the applicability of partial equilibrium in the post reaction zone of low pressure flames and therefore the (OH) profile can be used to obtain the (O) profile with high accuracy.

The NO concentration profiles are shown in Fig. 1 for both cooled and uncooled quartz microprobes and include measurements of NO$_x$ obtained using a Mo catalytic converter.

0-8412-0570-1/80/47-134-375$05.00/0
© 1980 American Chemical Society

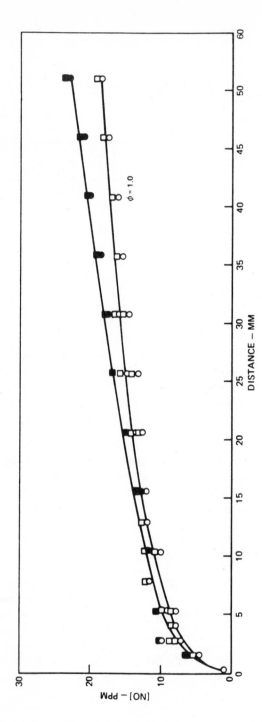

Figure 1. NO concentration profiles from chemiluminescent analyzer. CH₄–air flame: P = 0.1 atm; T = 2200 K. (○), NO; (□), NOₓ; (●, ■), water-cooled probe.

In addition to the data listed above the concentration profiles of all major species must be obtained in order to calculate
the average molecular weight throughout the flame. These data are
needed for the flame equations and were obtained using molecular
beam sampling-mass spectrometry. A sample of these concentration
profiles is shown in Fig. 2. The (H_2O) data are obtained by
oxygen balance and are accordingly less certain. The dashed line
through the H_2O data represents the formaldehyde which is
unmeasured and ties up large amounts of oxygen at this location in
the flame.

   If it is assumed that NO is formed exclusively by the reactions

$$O + N_2 \rightarrow NO + N \tag{1}$$

and

$$N + O_2 \rightarrow NO + O \tag{2}$$

and knowing that at the flame temperature ($T_f \sim 2200$ K) the rate
constant ratio k_2/k_1 is about 600, then neglecting back reactions
the rate of formation of NO is given by

$$\frac{d(NO)}{dt} = k_1 (O)(N_2) + k_2 (N)(O_2) \doteq 2 k_1 (O)(N_2) \tag{3}$$

Neglecting diffusion for downstream locations where the gradients
are small and converting to flame conditions the rate expression
becomes

$$\frac{dX_{NO}}{dx} = A \frac{n_o}{v_o} \left(\frac{T_o}{T}\right)^2 \frac{\bar{M}}{\bar{M}_o} 2 k_1 X_o X_{N_2} \tag{4}$$

where X_i is the mole fraction of species i at location x in the
flame of initial molar concentration, n_o, velocity, v_o, average
molecular weight, $\bar{M}_o$ and temperature, T_o. The area ratio A,
temperature, T, and molecular weight, $\bar{M}$, must also be known for
each location where the rate constant is to be evaluated. All the
terms in equation (4) are measured except k_1 and, therefore, we
can use the flame data to calculate the rate constant for several
locations in the flame. The results are presented in Figure 3
along with lines showing the recommended rate constants from Refs.
1 and 2.

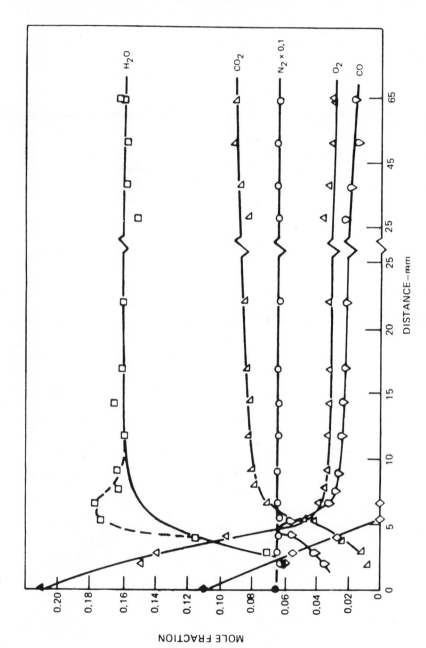

Figure 2.   Concentration profiles.   CH₄– air flame: φ = 1.0; P = 0.1 atm; T = 2200 K.

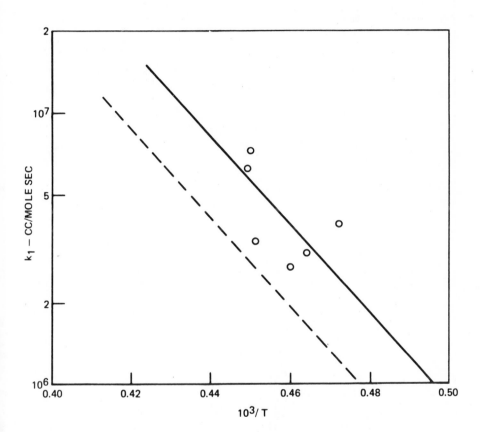

Figure 3.   Rate constants for O + N₂ → NO + N: (———), Monat et al. (Ref. 1); (– – –), Leeds (Ref. 2); (○), UTRC.

If the activation energy is fixed at 76.25 kcals, then the resulting pre-exponential factor for the flame data is calculated to be 1.76×10^{14}, almost identical with Ref. 1. The scatter in the flame data for the narrow temperature range covered has an average deviation of ± 28% from the value recommended by Ref. 1. While the cause(s) of the scatter are not known, the largest uncertainties are attributable to errors in temperature, O atom concentration and small errors in the concentration profiles which are magnified by taking numerical derivatives.

While the results from the present work do show scatter, they clearly support the higher rate of Monat et. al (1). The data of Monat et. al are the most direct measurements of the $O + N_2$ reaction and appear to be the most carefully obtained. Accordingly, these data should be used by those calculating NO formation from the Zeldovitch mechanism.

This work was sponsored by EPA under Contract 68-02-2188.

Literature Cited

1. Monat, J. P., Hanson, R. K. and Kruger, C. H., Seventeenth Symposium (International) on Combustion, p. 543, The Combustion Institute.

2. Baulch, D. L., Drysdale, D. D., Horne, D. G. and Lloyd, A. C., Evaluated Kinetic Data for High Temperature Reactions, Vol. 2. CRC Press, 1973.

3. Blauwens, J., Smets, B., and Peters, J., Sixteenth Symposium (International) on Combustion p. 1055, The Combustion Institute (1977).

4. Biordi, J. C., Lazzara, C. P. and Papp, J. F., Sixteenth Symposium (International) on Combustion, p. 1097, The Combustion Institute (1977).

RECEIVED February 1, 1980.

Reactions of C_2 ($X^1 \Sigma_g^+$) and ($a^3 \Pi_u$) Produced by Multiphoton UV Excimer Laser Photolysis

LOUISE R. PASTERNACK and J. R. McDONALD

Naval Research Laboratory, Washington, D.C. 20375

V. M. DONNELLY

Bell Laboratories, Murray Hill, NJ 07974

The kinetics and mechanisms of radical reactions important in combustion chemistry are best studied under conditions in which single reactions can be isolated rather than in flames where there are multiple pathways for formation and disappearance of the radicals. Reactions of C_2 are of particular importance since recent laser saturation measurements in our laboratory (1) have shown that C_2 $a^3\Pi_u$ is present in oxyacetylene flames at concentrations on the order of 10^{16} molecules/cm^3 (approximately 0.1 torr). Although concentrations of ground state C_2 in flames are unknown and cannot be measured by the same technique due to spectroscopic constraints, we expect that C_2 $X^1\Sigma_g^+$ populations are at least comparable. Because of these relatively large concentrations the reactions of both species are of considerable importance in combustion chemistry. However, until recently very little was known about these reactions due to the difficulty of producing a clean source of C_2 radicals.

In these experiments, we use multiphoton dissociation at 193 nm to generate C_2 radicals. C_2 concentrations are subsequently monitored using laser induced fluorescence. Disappearance rates of both C_1 $X^1\Sigma_g^+$ and $a^3\Pi_u$ are reported at ambient temperature with hydrocarbons (CH_4, C_2H_2, C_2H_4, and C_2H_6), hydrogen, oxygen, and carbon dioxide.

Experiment

Apparatus. The apparatus used in these experiments is shown schematically in Figure 1. C_2 radicals are generated by a two photon dissociation process using a mildly focussed pulsed ArF excimer laser (Tachisto Model XR 150). Hexafluorobutyne-2 ($F_3C-C\equiv C-CF_3$) is used as the precursor for C_2 $X^1\Sigma_g^+$ and acetylene is used as the precursor for $a^3\Pi_u$ radicals. Experiments are carried out in a 30 cm diameter cell with long side arms with scattered light baffling. Radical concentrations are monitored using laser induced fluorescence generated by a flashlamp pumped tunable dye

0-8412-0570-1/80/47-134-381$05.00/0

© 1980 American Chemical Society

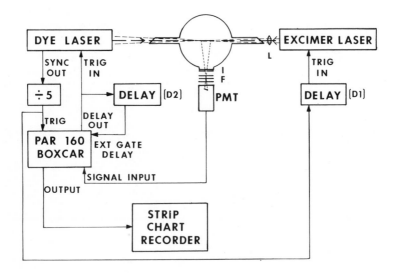

Figure 1.   Multiphoton UV-photolysis-laser fluorescence detection system

laser (Chromatux CMX-4). The two lasers propagate colinearly in opposite directions. Fluorescence is monitored at 90° using a photomultiplier tube and signal averaged with a box car integrator (PAR 160).

C_2 $X^1\Sigma_g^+$ populations are probed by exciting the $A^1\Pi_u$, $v' = 4 \leftarrow$ $X^1\Sigma_g^+$, $v'' = 0$ Phillips band transit on at 691 nm and detecting $v' = 4 \rightarrow v'' = 1$ fluorescence at 792 nm. C_2 $a^3\Pi_u$ populations are probed by exciting the $d^3\Pi_g \leftarrow a^3\Pi_u$ 0-0 Swan band transition at 516.5 nm and detecting dye laser induced fluorescence on the 0-1 vibronic band at 563 nm.

The rate of depletion of C_2 is monitored by varying the time delay between the photolysis and probe laser pulses. The process is automated by using the boxcar integrator gate to trigger the probe laser so that both the boxcar gate and the probe laser are synchronously scanned in time following the photolysis laser pulse. Further details of the experimental apparatus and technique are given in references (2) and (3).

Preparation of C_2 Fragments. The photodissociation of hexafluorobutyne-2 and acetylene yield a variety of products, many of which are in electronically excited states resulting in prompt fluorescence which is quenched by the presence of a buffer gas. The C_2 $X^1\Sigma_g^+$ and $a^3\Pi_u$ which are initially produced are also vibrationally and rotationally hot. Since we monitor only the $v'' = 0$ level of the fragments, it is necessary that the buffer gas either (a) completely thermalize the vibrational population in less than 20 μs or (b) not significantly relax $v'' > 0$ levels to the ground state on a time scale $\leq$ 300 μs. For C_2 $a^3\Pi_u$, condition (a) was easily met by a few torr of methane. However methane reacts with C_2 $X^1\Sigma_g^+$ and the other buffer gases we tried (Ar, N_2 and SF_6) vibrationally relaxed the C_2 $X^1\Sigma_g^+$ on the time scale of the reaction experiments. However over a range of He pressures up to 30 torr, no measurable relaxation of C_2 $X^1\Sigma_g^+$ $v'' > 0$ takes place during the time scale of the experiments (< 300 μs).

Results

The reactions are measured under the pseudo-first-order conditions of a large excess of reactant gases. A typical first order decay for C_2 $X^1\Sigma_g^+$ fragments using a large excess of H_2 reactive gas is shown in Figure 2. The disappearance of C_2 $X^1\Sigma_g^+$ is exponential over nearly two orders of magnitude in concentration. The slope of the data in Figure 2 gives the first order disappearance rate, k^I, for C_2 $X^1\Sigma_g^+$ reacting with 105.4 torr of hydrogen.

Bimolecular rate constants, k^{II}, are obtained by plotting, k^I vs reactive gas pressure, as shown in Figure 3 for the reaction of C_2 $(X^1\Sigma_g^+)$ + H_2. The slope of the least squares fit to the data gives k^{II} = (1.38 ± 0.06) x 10^{-12} cm^3 s^{-1} $molecule^{-1}$, the bimolecular disappearance rate for C_2 $X^1\Sigma_g^+$ in the presence of H_2, with the uncertainty representing ± 1σ. The nonzero in-

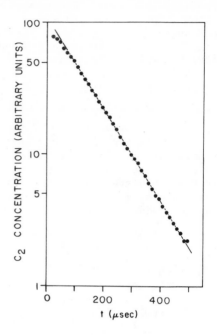

Figure 2.   *A plot of the C_2 $X^1\Sigma_g^+$ pseudo-first-order decay in the presence of excess H_2*

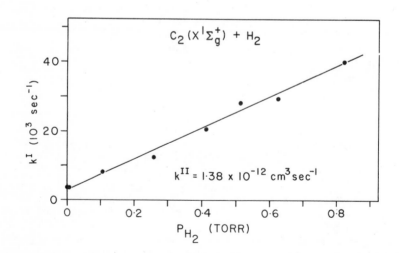

Figure 3.   *A plot of k^I vs. P for the reaction of $C_2(X^1\Sigma_g^+) + H_2$. The bimolecular rate constant, k^{II}, is obtained from the slope of the plot.*

tercept of the plot gives a value of the reaction rate of C_2 $X^1\Sigma_g^+$ with the precursor $CF_3C_2CF_3$ at 2.2 mtorr.

In Table I, the bimolecular rate constants for C_2 $X^1\Sigma_g^+$ and $a^3\Pi_u$ which we measured are tabulated. The experiments were carried out over a wide range of laser powers, buffer gas pressures, and precursor molecule pressures to assure that the experimental data does not contain artifacts due to three body reactions, vibrational quenching, fragment diffusion, or other fragment reaction.

Discussion

$\underline{C_2\ a^3\Pi_u\ reaction\ vs\ quenching.}$ The disappearance of C_2 $a^3\Pi_u$ may be due to either reaction or quenching to the ground state. However, quenching is spin forbidden; since the C_2 $a^3\Pi_u$ state lies only 1.7 kcal/mole above the ground state triplet-singlet resonance spin exchange processes are not possible with any of the reaction gases studied. Although there are instances where facile spin forbidden quenching is observed - notably the $O(^1D)$ + $N_2 \rightarrow O(^3p)$ + N_2 reaction (4) - such processes have been rationalized by postulating a long-lived RRKM complex (5). The long encounter time of the complex results in a curve crossing to the energetically favored ground state products despite the weak spin-orbit coupling. However, for C_2 $a^3\Pi_u$, if a long-lived complex is formed, the most energetically favored outcome is reaction with H_2 and hydrocarbons to form C_2H or C_2H_2 and not quenching to form C_2 $X^1\Sigma_g^+$. (3)

$\underline{C_2\ X^1\Sigma_g^+\ vs\ a^3\Pi_u\ reactions\ with\ hydrocarbons\ and\ hydrogen.}$ C_2 $X^1\Sigma_g^+$ reacts considerably faster than C_2 $a^3\Pi_u$ with hydrogen and hydrocarbons (see Table I) despite the presence of spin allowed exothermic pathways in both cases. We can use electronic orbital correlation arguments to explain these results. (See reference (2) for further details of these arguments.) Ground state C_2 has the electronic configuration π^4. The addition of a hydrogen 1s electron to a σ-bonding orbital correlates with the ground state of C_2H which has the electronic configuration $\pi^4 5\sigma$. In contrast, for the C_2 $a^3\Pi_u$ (electronic configuration $\pi^3 5\sigma$) to react with H, there would have to occur the unlikely transfer of a H 1s electron to a π-bonding orbital in order to form ground state C_2H. The reaction of $C_2(a^3\Pi_u)$ + H is more likely to form C_2H in the $\tilde{A}$ $^2\Pi$ state with electronic configuration $\pi^3 5\sigma^2$. This latter reaction is approximately 10 kcal/mole higher in energy than the ground state reaction. Either this difference in energy or the presence of a barrier for C_2 $a^3\Pi_u$ reactions may account for the difference in C_2 singlet and triplet reactivities with hydrogen and hydrocarbons. The difference in reactivities is less for reactions with unsaturated hydrocarbons where multiple pathways exist involving interaction with the C-C multiple bond as well as the direct reaction to strip an H atom.

Table I. Rate constants (cm^3 $s^{-1}molecule^{-1}$) for the disappearance of $C_2(X^1\Sigma_g^+)$ compared to $C_2(a^3\Pi_u)$ at 298 K.

Reactant	$k^{II}(C_2\ X^1\Sigma_g^+)^{a)}$	$k^{II}(C_2\ a^3\Pi_u)^{a)}$
H_2	$(1.38\pm0.06) \times 10^{-12}$	$<5 \times 10^{-15\ b)}$
CH_4	$(1.87\pm0.05) \times 10^{-11}$	$<1 \times 10^{-16\ b)}$
C_2H_6	$(1.59\pm0.05) \times 10^{-10}$	$(1.30\pm0.06) \times 10^{-12}$
C_2H_4	$(3.26\pm0.05) \times 10^{-10}$	$(1.44\pm0.06) \times 10^{-10}$
C_2F_4	$(5.99\pm0.14) \times 10^{-11}$	---
C_2H_2	---	$(9.6\pm0.3) \times 10^{-11}$
O_2	$(2.82\pm0.09) \times 10^{-12}$	$(2.96\pm0.07) \times 10^{-12}$
CO_2	--- no apparent reaction ---	

a) 1σ uncertainty.
b) 3σ uncertainty.

Table II. Product channels for $C_2 + O_2$

REACTION	ΔH_r^o (kcal/mole)
$C_2(X^1\Sigma_g^+) + O_2$	
$\rightarrow CO(X^1\Sigma^+) + CO(a^3\Pi)$	-113
$\rightarrow CO_2(\tilde{X}^1\Sigma^+) + C(^3P_o)$	-123
$\rightarrow C_2O(\tilde{X}^3\ \Sigma^-) + O(^1D)$	-26
$C_2(a^3\Pi_u) + O_2 \rightarrow$ in addition to products above	
$\rightarrow 2CO(X^1\Sigma_g^+)$	-254
$\rightarrow CO(X^1\ \Sigma^+) + CO(A^1\Pi)$	-68
$\rightarrow C_2O(\tilde{X}^3\ \Sigma^-) + O(^3P)$	-73
$\rightarrow CO_2(\tilde{X}^1\ \Sigma_g) + C(^1D)$	-95

$\underline{C_2 \ X^1\Sigma_g^+ \ vs \ a^3\Pi_u}$ reactions with oxygen. We observed the same rate constant (within experimental uncertainty) for the reaction of $C_2 \ X^1\Sigma_g^+$ and $a^3\Pi_u$ with O_2, (See Table I). Many possible channels exist for these reactions, (See Table II). Some of these products have been observed including $CO(A^1\Pi)$ by Filseth et al. (6), CO excited triplets by Wittig (7) and $C_2O(\tilde{X}^3\Sigma^-)$ by Donnelly and Pasternack (3). Because of the different reaction pathways that are accessible for $C_2 \ X^1\Sigma_g^+$ and $a^3\Pi_u$, the observation of identical disappearance rate constants seems to be a coincidence.

$\underline{C_2 \ X^1\Sigma_g^+ \ vs \ a^3\Pi_u}$ reactions with CO_2

We observed no room temperature reaction between either state of C_2 and CO_2. This result is somewhat surprising since both spin and symmetry allowed reactive channels exist for both C_2 fragments:

$$C_2(a^3\Pi_u) + CO_2(\tilde{X}^1\Sigma_g^+) \rightarrow C_2O(\tilde{X}^3\Sigma^-) + CO(X^1\Sigma^+)$$

$$\Delta H_r^o = -66 \pm 16 \ \text{kcal/mole}$$

$$C_2(X^1\Sigma_g^+) + CO_2(\tilde{X}^1\Sigma_g^+) \rightarrow C_2O(^1\Delta \ OR \ ^1\Sigma) + CO(X^1\Sigma^+)$$

$$\Delta H_r^o = -50 \ \text{kcal/mole}$$

These reactions will be studied at higher temperatures to assess their importance in combustion systems.

Literature Cited

1. Baronavski, A. P. and McDonald, J. R., J. Chem. Phys., 1977, 66, 3300; App. Optics 1977, 16, 1897.

2. Pasternack, Louise and McDonald, J. R., Chem. Phys. in press.

3. Donnelly, V. M. and Pasternack, Louise, Chem. Phys. 1979, 39, 427

4. Streit, G. E. and Johnston, H. S. J. Chem. Phys. 1976, 64, 95; Heidner, R. F., Husain, D., and Wiesenfeld, J. R., J. Chem. Soc. Faraday Trans. II 1973, 69, 927.

5. Tully, J. C., J. Chem. Phys. 1974, 61, 61.

6. Filseth, S. V., Hancock, G., Fournier, J., and Meier, K., Chem. Phys. Letters 1979, 61, 288.

7. Wittig, C., private communication.

RECEIVED March 31, 1980.

Pulsed-Laser Studies of the Kinetics of
$C_2O(\tilde{A}^3\Pi_i$ and $\tilde{X}^3\Sigma^-)$

V. M. DONNELLY,[1] WILLIAM M. PITTS, and A. P. BARONAVSKI

Chemistry Division, Naval Research Laboratory, Washington, D.C. 20375

Laser induced fluorescence is particularly well suited to combustion chemistry, as a sensitive "in-situ" probe for free radicals in flames; or under more controlled conditions in laboratory flash photolysis, discharge flow tube, or shock tube experiments. Using laser-saturation fluorescence previous studies from this laboratory ([1]) have shown that $C_2(a^3\Pi_u)$ is present in high concentrations in the hot region of an oxy-acetylene flame. $C_2(a^3\Pi_u$ and $X^1\Sigma_g^+)$ reacts with O_2.([2],[3],[4]) One of the products of this reaction (and/or the reaction of C_2H+O_2) is CCO.([2]) In the present study, we report $C_2O(\tilde{A}^3\Pi_i-\tilde{X}^3\Sigma^-)$ fluorescence excitation spectra, $\tilde{A}^3\Pi_i$ lifetimes and quenching rate constants, and $\tilde{X}^3\Sigma^-$ reaction rate constants.

PART 1: $\underline{C_2O(\tilde{A}^3\Pi_i)}$ Fluorescence Lifetimes and Quenching Rate Constants

$C_2O(\tilde{X}^3\Sigma^-)$ is formed by pulsed laser photolysis of carbon suboxide at 266 nm (4th harmonic Nd-YAG, 100 μJ/pulse, 30 Hz, 50 μsec). A second pulsed dye laser (5 mJ/pulse, 30 Hz, 1 μsec) excites the $\tilde{A}^3\Pi_i$ state. $\tilde{A}^3\Pi_i \rightarrow \tilde{X}^3\Sigma^-$ fluorescence is detected in the 750-900 nm region. Fluorescence excitation spectra are recorded using the apparatus shown in Figure 1. Fluorescence is detected as a function of dye laser wavelength at a fixed delay time between YAG and dye laser pulses (0-500 μsec). A sample spectrum is shown in Figure 2. The upper trace is recorded under near collision free conditions (50 mtorr total pressure, with no delay between laser pulses). The complex structure is indicative of the high degree of rotational and vibrational excitation in

[1] Present Address: Bell Laboratories, Murray Hill, NJ 07974.

0-8412-0570-1/80/47-134-389$05.00/0
© 1980 American Chemical Society

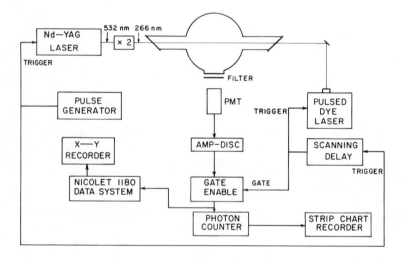

Figure 1. The experimental system used for measuring fluorescence excitation spectra and reaction rate constants

the initially formed fragment. In the lower trace, C_2O $(X^3\Sigma^-)$ is thermalized with Ar buffer gas prior to dye laser excitation. The sharp, strong features are identified using the absorption spectrum assignments of Devillers and Ramsay (5). New assignments not reported by Devillers and Ramsay are underlined.

Time resolved $\tilde{A}$ state fluorescence is measured with a transient recorder-signal averager, as described previously.(6) A typical fluorescence decay curve is shown in Figure 3. It is composed of at least three exponentials, independent of excitation wavelength. The short component lifetime is extracted by fitting approximately the first third of the decay to the expression.

$$I(t) = I_S^0 \exp(-t/\tau_S) + I_L^0 \exp(-t/\tau_L) \tag{1}$$

A plot of τ_s^{-1} vs P is linear. The intercept and slope yield a zero-pressure lifetime of 16.5($\pm$ 7.4, $-$2.9) μsec and a quenching rate constant (by C_3O_2) of 1.53×10^{-9} cm^3molecule^{-1}sec^{-1}.

The analysis of the long components can be broken into four pressure regions:

1) Between 0.5 and 2.0 mtorr C_3O_2, the long lived portion of the decay (e.g. 80-600 μsec in Fig. 3) is fit to equation (1). Varying amounts of Ar are added to fixed C_3O_2 pressures to slow diffusion of the recoiling C_2O fragment over these long times. Extrapolating τ_L^{-1} vs Ar pressure to zero gives the fluorescence lifetimes apart from diffusion effects (Δ in Figure 4).

(2) Between 2 and ~25 mtorr of C_3O_2, the C_2O fluorescence decay behaves as shown in Figure 3. Diffusion is sufficiently slow that Ar buffer gas is not required. The long component lifetimes of the long decay have been measured as a function of pressure and are included in Figure 4.

(3) Between ~25 and 100 mtorr the long-lived decay components collapse into a single exponential.

(4) At C_3O_2 pressures greater than 100 mtorr the decay becomes nonexponential.

The plot of τ_L^{-1} vs $P(C_3O_2)$ (Figure 4) gives a straight line. The data are extrapolated to obtain an estimate for the zero pressure fluorescence lifetime of 380 μsec. The slope of the data corresponds to a C_3O_2 quenching rate constant 2.88 x 10^{-11} cm^3molecule^{-1} sec^{-1}.

Long component quenching rate constants were measured for Ar, N_2, and O_2 over the pressure regions described above by (3). All C_2O lifetimes and quenching rate constants are given in Table 1.

Summary

1) C_2O $\tilde{X}^3\Sigma^-$ can be produced cleanly by a 266 nm laser photolysis of C_3O_2.

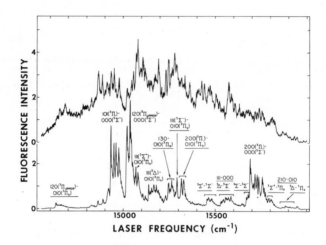

Figure 2. *The laser-induced fluorescence excitation spectrum of* $C_2O(A^3\Pi_i -$
$\overline{X}^3\Sigma^-)$ *recorded under nascent conditions* (top) *immediately following photolysis of*
C_3O_2 *and after cooling by a buffer gas*

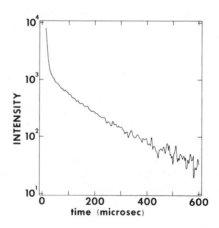

Figure 3. *Fluorescence decay for* C_2O
excited into the 101 level of the $A^3\Pi_i$.
Note the highly nonexponential behavior:
1.1 mtorr C_3O_2; *5.8 mtorr* Ar.

2) Laser-induced fluorescence is a very sensitive detection probe for C_2O ($\tilde{X}^3\Sigma^-$) ($\lesssim 10^9$ molecules/cm^3).

3) The extrapolated, $^3\Pi_i$ zero-pressure decay of $\tilde{A}$-state fluorescence contains at least 3 distinct lifetime components. This behavior is reminiscent of other triatomic molecules such as NO_2,(7) SO_2,(8) CS_2,(9) and NH_2,(6) and is likely due to variable coupling between vibrational levels of the $\tilde{A}^3\Pi_i$ state and levels of the $\tilde{X}^3\Sigma^-$, $\tilde{a}^1\Delta$(0.5 eV) and/or $\tilde{b}^1\Sigma^+$ ($\sim$0.8 eV) states.(10)

4) The extremely fast quenching of C_2O $\tilde{A}^3\Pi_i$ by C_3O_2 is probably rotational relaxation of the highly rotationally excited C_2O photofragment. The slower process for C_3O_2 is some unknown combination of vibrational and electronic quenching and reaction. Likewise, contributions of vibrational and electronic relaxation to the observed quenching by Ar, O_2, and N_2 are not determined.

Part 2: C_2O ($\tilde{X}^3\Sigma^-$) Reaction Kinetics

Pseudo-first-order decays of C_2O ($\tilde{X}^3\Sigma^-$) in the presence of large excess reactant gases are measured with the system shown in Figure l, by scanning the delay of the probe dye laser with respect to the YAG photolysis pulse. The first order rate constant, k^I is plotted vs reactant pressure to obtain the bimolecular rate constant k^{II} (Figure 5). Table 2 lists the absolute rate constants measured in this study. The C_2O+NO rate constant should be regarded as approximate due to an observed dark reaction between C_3O_2 and NO, which complicates the disappearance kinetics of C_2O $\tilde{X}^3\Sigma^-$ The ratio of reaction rates NO:O_2:isobutene of 386:2.95:1 is in excellent agreement with the ratio of 362:2.68:1 of Williamson and Bayes.(11) Since they measured relative rates, our values in Table 2 can be used to obtain absolute rate constants for all the reactants investigated by Williamson and Bayes.

Summary

C_2O has been shown to be formed when C_2, and/or C_2H react with O_2.(2) Reactions such as (11,12)

$$O+C_2H_2 \rightarrow C_2O+H_2$$

and (13,14)   (2)

$$O+C_2H_2 \rightarrow HC_2O+H$$ (3a)

$$HC_2O+R \rightarrow C_2O+RH \text{ (R = radical)}$$ (3b)

have also been shown to generate C_2O. The reactions of C_2 and C_2O with O_2 are fast enough at room temperature to be important pathways converting fuels to CO and CO_2 at combustion temperatures. The fast reaction with NO should also be considered as

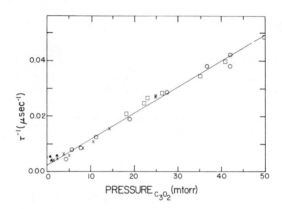

Figure 4.   The Stern–Volmer plot for the quenching of the $A^3\Pi_i$ state long decay component is shown

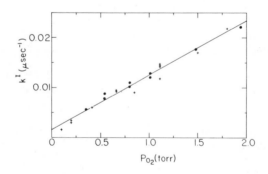

Figure 5.   The pseudo-first-order disappearance rate constants for ground-state C_2O are plotted as a function of O_2 pressure for two pressures of C_3O_2. The results fall on a straight line and the slope gives a second-order disappearance rate constant for C_2O reacting with O_2 of $3.30 \pm 0.12 \times 10^{-13}$ cm³ molecule⁻¹ sec⁻¹: (●), 10 mtorr C_3O_2; (+), 5 mtorr C_3O_2. $K^{II} = 3.30 \pm 0.12 \times 10^{-13}$ cm³ sec⁻¹.

TABLE 1: C_2O Lifetime and Quenching Data

TABLE 1A: RADIATIVE LIFETIMES FOR $C_2O(\tilde{A}^3\Pi_i)$

SHORT COMPONENT: $\tau=16.5(+7.4,-2.9)$ µsec

LONG COMPONENT: $\tau=\sim100$ to >200µsec
(non exponential)

TABLE 1B: QUENCHING RATE CONSTANTS FOR $C_2O(\tilde{A}^3\Pi_i)$

LONG COMPONENT QUENCHING RATE CONSTANT

QUENCHER GAS	$k_q(cm^3molecule^{-1}sec^{-1})$
C_3O_2	$2.88\pm0.10 \times 10^{-11}$
O_2	$5.67\pm0.21 \times 10^{-12}$
N_2	$4.00\pm0.07 \times 10^{-12}$
Ar	$1.95\pm0.04 \times 10^{-12}$

SHORT COMPONENT QUENCHING RATE CONSTANT

C_3O_2	$1.53\pm0.11 \times 10^{-9}$

TABLE 2: C_2O $(\tilde{X}^3\Sigma^-)$ REACTION RATE CONSTANTS AT 298 K

REACTANT GAS	$k^{II}(cm^3molecule^{-1}sec^{-1})$
NO	$(4.43\pm0.12) \times 10^{-11}$
O_2	$(3.30\pm0.12) \times 10^{-13}$
ISOBUTENE	$(1.12\pm0.05) \times 10^{-13}$
H_2	$<2\times10^{-14}$
CO_2	$<1\times10^{-14}$
C_2H_4	$<1\times10^{-14}$

potentially important in NO_x pollution chemistry in combustion. However, before the importance of C_2O can be established, temperature dependences of reaction rates must be measured, and its concentration levels in "in-situ" flames should be estimated.

Literature Cited

1. Baronavski, A.P.; McDonald, J.R. J. Chem. Phys., 1977, 66, 3330.
2. Donnelly, V.M.; Pasternack, Louise Chem. Phys., 1979, 39, 427
3. Pasternack, Louise; McDonald, J.R. Chem. Phys., 1979, 43, 173.
4. Pasternack, Louise; Donnelly, V.M.; McDonald, J.R., following paper.
5. Devillers, C.; Ramsay, D.A. Can. J. Phys., 1971, 49, 2839.
6. Donnelly, V.M.; Baronavski, A.P.; McDonald, J.R. Chem. Phys., 1979, 43, 283.
7. Donnelly, V.M.; Kaufman, F. J. Chem. Phys., 1977, 66, 4100; 1978, 69, 1456.
8. Brus, L.E.; McDonald, J.R. J. Chem. Phys., 1974, 61, 97.
9. Brus, L.E. Chem. Phys. Letters, 1977, 12, 116.
10. Bayes, K.D. J. Am. Chem. Soc., 1963, 85, 1730.
11. Williamson, D.G.; Bayes, K.D. J. Am. Chem. Soc., 1967, 89, 3390; 1968, 90, 1957.
12. Becker, K.H.; Bayes, K.D. J. Chem. Phys., 1968, 48, 653.
13. Bayes, K.D. J. Chem .Phys., 1970, 52, 1093.
14. Williamson, D.G. J. Phys. Chem., 1971, 75, 4053.

RECEIVED February 11, 1980.

Kinetics of CH Radical Reactions Important to Hydrocarbon Combustion Systems

J. E. BUTLER, J. W. FLEMING, L. P. GOSS, and M. C. LIN

Chemistry Division, Naval Research Laboratory, Washington, D.C. 20375

One of the important hydrocarbon combustion reaction intermediates is the CH radical. Although CH chemiluminescence ($A^2\Delta \rightarrow X^2\pi$) has been observed in many hydrocarbon flames, the mechanism of CH formation and its reaction kinetics have been difficult to unravel *in situ* due to the low steady-state concentrations and the complex nature of combustion reactions. This project was undertaken to investigate a means of CH radical production and to study its reactions with various important species so that an overall picture of the oxidation processes, particularly with regard to the mechanism of NO_x formation, may be better understood.

Production and Detection

One of the most effective methods of CH production is the multiphoton dissociation of $CHBr_3$(1). A high-power ArF excimer laser (193 nm) was used to dissociate a $CHBr_3$:Ar gas mixture ($\sim 1:10^5$) slowly flowing through the reaction cell at pressures of 30-100 torr. A high-power tunable dye laser pumped by a tripled Nd:YAG laser was employed to monitor the production and decay of the CH radical formed in the dissociation process via laser induced fluorescence of the CH ($A \rightarrow X$) transition near 430 nm. The ArF beam, dye laser probe beam and fluorescence collection optics were mutually perpendicular as shown in a schematic diagram given in Figure 1. Figure 2 shows a typical laser excitation spectrum and the rotational assignments.

Kinetics

Kinetic measurements were made by monitoring the laser-induced fluorescence of CH following the excitation in the (0-0) band of the $X \rightarrow A$ transition as a function of the time delay after the ArF laser dissociation. In the absence of any added reactants, CH had a decay time of 100 to 300 μsec at a total pressure of 30 to 100 torr ($CHBr_3$ pressures of 1 to 10 mtorr) which can be attributed mainly to the CH + CH reaction. The addition of the reactants listed in Table I shortened the CH radical decay times considerably, indicative of some removal process involving a bimolecular mechanism since the total pressure was always maintained constant. Least squares plots of the inverse lifetimes of CH radicals versus the partial pressure of the added reactant yielded

This chapter not subject to U.S. copyright.
Published 1980 American Chemical Society

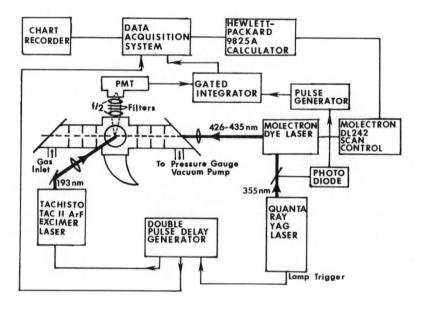

Figure 1. Apparatus for the productions and detection of CH radicals and the measurement of their reaction kinetics

a second order rate constant for each reactant. These results are summarized in Table I and compared with previously published values for selected molecules.

Table I. Rate Constants for CH Radical Reactions
at Room Temperature with Selected Molecules
Relevant to Hydrocarbon Combustion

Reactant	$k \times 10^{11}$ (cm^3/molecule · sec)^a		
	I	II	III
H_2	2.6 ± 0.5	0.10	1.74 ± 0.20
O_2	5.9 ± 0.8	–	4.0
NO	29 ± 7	–	–
CO	2.1 ± 0.3	–	0.48
N_2	0.093 ± 0.01	0.0071	0.10 ± 0.02
CO_2	0.19 ± 0.04	–	–
CH_4	10 ± 3^b	0.25	3.3 ± 0.08
C_2H_6	40 ± 1	–	–
C_4H_{10}	58 ± 5	–	13 ± 1

a. I— This work (100 Torr total system pressure); II — Braun et al. (2); III — Bosnali and Perner (3).
b. The value reported in Ref. (1) was too high by a factor of 3 due to errors in CH_4 concentration calculations.

Previous studies on the reactions of CH employed either the vacuum ultra-violet photodissociation (2) or the electron beam dissociation (3) of CH_4 to generate the radical. The formation and decay of the CH was monitored by UV absorption spectroscopy on the $C \leftarrow X$ transition at 314 nm. The results of the former study (2), which relied partly on final product analysis, are considerably smaller (by a factor of 10 to 40) than the values of Bosnali and Perner (3) and our present data for the reactions with H_2, N_2 and CH_4. The agreement between ours and those of Bosnali and Perner, although significantly better, is only fair and lies within a factor of 2 to 5. Further work is certainly needed in order to reconcile these two sets of data.

Comments on CH + N_2

The CH + N_2 reaction has now been generally considered as one of the most important precursor reactions for "prompt" NO formation in high temperature hydrocarbon combustion systems. The rate constant of this reaction at room temperature was found to be pressure-dependent (see Figure 3) and is considerably higher than the value extrapolated from the expression, 1.3×10^{-12} exp $(-11,000/RT)$ cm^3/molecule · sec, obtained from the rate of NO production in several hydrocarbon flame fronts (4). These findings could be understood by the thermochemistry of the CH + $N_2 \rightleftharpoons CHN_2 \rightleftharpoons HCN + N$ system (5). Since the production of HCN + N is endothermic by 3 kcal/mole, it probably occurs with a relatively high activation energy (such as the value, 11 kcal/mole, obtained from the high temperature flame studies mentioned above). The formation of the CHN_2 radical adduct, which is expected to be pressure-dependent as was found experimentally, probably can proceed with little or no activation energy. The following mechanism can at least qualitively account for the overall reaction:

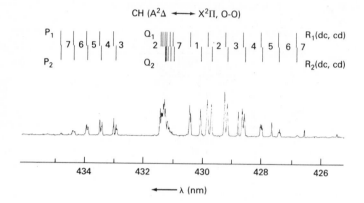

Figure 2.   Laser-induced fluorescence spectrum of the Ã²Δ ⟷ X̃²π transition of the ground-state CH radicals·in 100 torr of Ar buffer gas

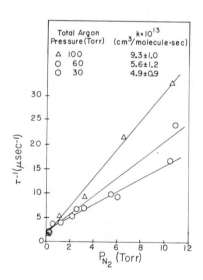

Figure 3.   Pressure effect on the rate of the CH + N₂ reaction at room temperature

$$CH + N_2 + M \rightleftharpoons HC = N = N + M \qquad \text{(low T)}$$
$$\qquad\qquad\qquad \longrightarrow HCN + N \qquad\qquad \text{(high T)}$$

Although the formation of HCN + N is not spin-conserved, the formation of the CHN_2 radical intermediate is expected to overcome and facilitate the doublet → quartet conversion. At room temperature, the CHN_2 adduct probably disappears by secondary reactions with active species present in the system. The effects of both temperature and pressure on the rate of this important reaction will be thoroughly investigated in the near future.

Conclusions

In this work, we have demonstrated that the CH radical can be generated with sufficiently high concentrations by means of the multiphoton dissociation of $CHBr_3$ at 193 nm for kinetic measurements. The formation and decay of the CH radical was monitored by the laser-induced fluorescence technique using the $(A^2\Delta \leftarrow X^2\pi)$ transition at 430 nm. Several rate constants for the reactions relevant to high temperature hydrocarbon combustion have been measured at room temperature. One of the key reactions, $CH + N_2$, has been shown to be pressure-dependent, presumably due to the production of the CHN_2 radical at room temperature.

Literature Cited

1. Butler, J.E., Goss, L.P., Lin, M.C. and Hudgens, J.W., Chem. Phys. Lett. (1979) **63**, 104.

2. Braun, W., McNesby, J.R. and Bass, A.M., J. Chem Phys. (1967) **46**, 2071.

3. Bosnali, M.W. and Perner, D., Z. Naturforsch. (1971) **26a**, 1768.

4. Blaurvens, J., Smets, B. and Peeters, J., Sixteenth Symposium (International) Combustion (The Combustion Institute, 1977) p. 1055.

5. Benson, S.W., Sixteenth Symposium (International) on Combustion (The Combustion Institute, 1977) p. 1062.

RECEIVED February 1, 1980.

Carbon Monoxide Laser Resonance Absorption Studies of O(³P) + 1-Alkynes and Methylene Radical Reactions

W. M. SHAUB and M. C. LIN

Chemistry Division, Naval Research Laboratory, Washington, D.C. 20375

The CO laser resonance absorption technique is a useful tool for studying the dynamics of chemical reactions that involve the initial production of vibrationally excited CO molecules. We have recently applied this technique to study various atomic and free radical reactions related to combustion and electronic-to-vibrational energy transfer processes (1–6). In this brief account, we discuss mainly the dynamics of O(^{3}P) + 1-alkynes and associated free radical reactions.

Experimental Technique

Figure 1 is representative of the general experimental configuration which has been used in this laboratory. Briefly, a cw, mode-stabilized CO laser, which is line-tuned to preselected vibrational-rotational lines, is concentrically directed along the axis of a flash photolysis tube (which may be of Pyrex, quartz or Vycor construction). The temperature of the flash tube is controlled ($\pm$ 1 K) by means of a regulator. O(^{3}P) atoms were produced from photodissociation of NO_2 in a Pyrex tube ($\lambda \geqslant 300$ nm). For the 1-alkyne reactions, for example, mixtures of NO_2, SF_6 (to ensure rotational relaxation) and an alkyne were flash-photolyzed at appropriate energies, typically 0.5-1.0 kJ. The temporal evolution of the vibrational population of the product CO was then monitored by recording the transient CO absorption utilizing appropriate IR detectors. Typically, data collection was made either by oscilloscope photography (in single shot experiments) or by signal averaging (in multiple shot experiments) in conjunction with a transient recorder. The CO product vibrational populations were determined by analyzing the initial portions of the absorption curves via a computer solution of the gain equation (1, 2). Initial population distributions were evaluated by careful extrapolation of the N_v/N_o ratios to the appearance time of the earliest absorption. This procedure eliminates effects due to vibrational relaxation and secondary reactions. Stable product analysis (3, 4) and isotope labeling experiments (2, 6) have also been used in conjunction with this technique to further clarify reaction pathways and the nature of reaction intermediates. For those reactions that occur via long-lived complexes, the experimentally observed CO product vibrational energy distributions have been compared to those expected on the basis of simple statistical models (3). These comparisons help elucidate the mechanisms of the reactions studied.

O(^{3}P) + 1-Alkynes

Figure 2 shows the results of two typical experiments. The observed CO vibrational energy distribution produced from the reactions: O(^{3}P) + C_2H_2 and

This chapter not subject to U.S. copyright.
Published 1980 American Chemical Society

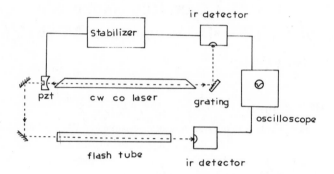

Figure 1. A schematic for the CO laser resonance absorption apparatus

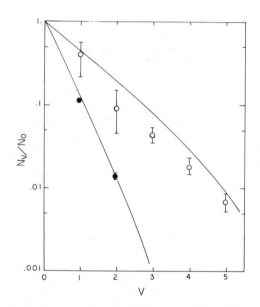

Figure 2. Vibrational energy distributions of the CO formed in $O(^3P)$ + C_2H_2 (O) and $O(^3P)$ + $C_4H_9C_2H$ reactions; (———), are statistically predicted distributions based on the model discussed in Ref. (3).

$O(^3P) + n - C_4H_9 C_2H$ are shown. The solid lines represent the statistically expected population distributions based on the assumption of complete randomization of internal energy based on the general mechanism (7):

$$O(^3P) + RC_2H \rightarrow RCHC = C = O\dagger \rightarrow CO(v) + R\ddot{C}H$$

$$\Delta H ° = -47 \text{ to } -50 \text{ kcal/mole.}$$

Generally, good agreement was found between the experimentally observed and the statistically predicted population distributions assuming diradicals (as shown above) are formed initially instead of their alkene isomers. Table I summarizes results which show the comparison between the experimentally determined average product CO vibrational energies and the statistically expected values.

The validity of the above mechanism for alkyne reactions is further strengthened by the observation that the reaction of $O(^3P)$ with allene, an isomer of propyne, generates nascent CO molecules that carry 3 times as much vibrational energy as that formed in the propyne reaction, although the overall exothermicities for the formation of the major products, C_2H_4 + CO, in both reactions are nearly the same. The observed CO vibrational energy distribution in $O(^3P)$ + allene can be quantitatively accounted for by our statistical model based on the mechanism that assumes direct formation of C_2H_4, rather than CH_3CH as in $O(^3P)$ + propyne, in which the energy released in the isomerization reaction, $CH_3CH \rightarrow C_2H_4$, 68 kcal/mole is not available for CO product excitation (3).

Since the rate constants for the reaction of $O(^3P)$ with C_2H_2 and CH_3C_2H are known (8), we have also used the intensity of CO absorption to evaluate the unkown rate constants, employing either $O + C_2H_2$ or $O + CH_3C_2$ H as a reference reaction. The results obtained from this study are also summarized in Table I. It is assuring that the rate constants for butyne, pentyne and hexyne (which were obtained from experiments using different reference reactions) agree very closely.

The RCH diradicals formed in this series of 1-alkyne reactions with $O(^3P)$ atoms are statistically expected to possess a large fraction of available energies. Accordingly, with the exception of the CH_2 formed in $O + C_2H_2$, they disappear rapidly via unimolecular isomerization processes: $RCH \rightarrow R'CH=CH_2$, producing excited 1-alkenes. There is no evidence that any highly excited CO has been produced by the secondary reactions involving RCH with $O(^3P)$, NO and NO_2 which are present in these flashphotolyzed systems in the early stages of reactions. In the $O(^3P)$ + propyne reaction, the results of gas product analysis indicated that the C_2H_2/C_2H_4 ratio is strongly pressure-dependent (3). Here, C_2H_2 was produced from the decomposition of vibrationally excited C_2H_4 derived from the isomerization of CH_3CH. Interestingly, the C_2H_4 formed in the $O(^3P)$ + allene reaction was found to be significantly less excited, as expected from the mechanism of the reaction mentioned above (3).

In the $O + C_2H_2$ reaction, the CH_2 radical formed initially cannot decay unimolecularly as indicated above for larger RCH diradicals. When NO_2 was used as the source of $O(^3P)$, CH_2 radicals appeared to be scavenged effectively by NO or undissociated NO_2, producing no CO in the early stages of photodssociation reaction (7). This is also indicated by the absence of highly excited CO (v $\geqslant$ 6) molecules, contrary to that observed in the photolysis of an SO_2-C_2H_2-SF_6 mixture in a quartz tube ($\lambda \geqslant$ 200 nm). Here SO_2 is used as an effective source of $O(^3P)$ atoms. In fact, CO laser emission (v $\leqslant$ 13) has been reported for the SO_2-C_2H_2 system (9). The highly excited CO (i.e., $v \geqslant 6$) produced in this system is believed to have resulted from the secondary reaction,

$$O(^3P) + CH_2 \rightarrow CO + 2H \text{ (or } H_2),$$

which is highly exothermic ($\Delta H° = -74$ and -178 kcal/mole for H and H_2 formation, respectively).

Reactions of CH_2 with $O(^3P)$, O_2 and CO_2

In separate studies ($\underline{10}$, $\underline{11}$), we have investigated the dynamics of CO production from reactions of CH_2 with $O(^3P)$, O_2 and CO_2 employing CH_2I_2 or CH_2Br_2 as the source of CH_2. For the $O(^3P) + CH_2$ reaction, mixtures of CH_2I_2, SO_2 and SF_6 were flash photolyzed in a quartz tube and the initial CO vibrational energy distribution was measured with the CO laser absorption method. The results of this experiment are shown in Figure 3, together with those of CH_2 reactions with O_2 and CO_2 carried out with a Vycor flash tube ($\underline{10}$). The Vycor tube was employed for these two reactions to avoid photodissociation of both O_2 and CO_2. CH_2I_2 was again used as the source of CH_2 for these two reactions. The average CO vibrational energies and possible channels for CO production in these three reactions are summarized in Table II.

The reaction of $O(^3P)$ with CH_2 probably occurs via a vibronically excited formaldehyde intermediate which possesses as much as 180 kcal/mole of internal energy due to formation of the C = O bond. Because of this excess amount of energy, the excited intermediate is expected to decompose into either CO + 2 H (channel a) or CO + H_2 (channel b) as shown in Table II. The CO formed in these processes carries about 17 kcal/mole of vibrational energy with excitation up to $v = 18$, having a vibrational temperature of $\sim 10^4$K. Since channel (a) can only excite CO up to the maximal level of $v = 13$, we can conclude that the molecular channel (b) also occurs simultaneously. A rough estimate based on the distribution shown in Figure 3 indicates that both channels occur concurrently to similar extents, if the CO's formed in these channels have near statistical vibrational energy distributions ($\underline{12}$).

The reaction of CH_2 with O_2 is believed to occur via a $\cdot CH_2OO \cdot$ diradical or a $\overline{CH_2O_2}$ ring intermediate which subsequently rearranges into excited HCOOH, possessing over 184 kcal/mole internal energy ($\underline{10}$). In this reaction, CO can be formed by two possible paths as indicated in Table II. A rough estimate based on the observed CO vibrational energy distribution shown in Figure 3 indicated that about 30% the reaction occurs by channel (a) and 70% by channel (b).

The reaction of CH_2 with CO_2 was first postulated by Kistiakowsky and Sauer ($\underline{13}$) as taking place via an α-lactone intermediate. The occurrence of this reaction was subsequently demonstrated by Milligan and Jacox ($\underline{14}$) in low temperature matrices. These low temperature matrix isolation experiments, however, could not determine definitely the structure of the CH_2 CO_2 intermediate. The result of our laser absorption experiment shows that the CO is vibrationally excited up to $v = 4$ with a distribution close to the one predicted by a statistical model assuming the existence of a long-lived CH_2 CO_2 complex. This calculation, however, is insensitive to the structure of the complex assumed. Since the ground state triplet CH_2 is known to be less reactive and kinetically behaves like CH_3 ($\underline{15}$, $\underline{16}$), which does not react readily with CO_2, the singlet $^1A_1$ CH_2 is assumed to be involved in the reaction.

In this brief report, we have discussed the utility of a simple laser absorption technique for the elucidation of the mechamisms of CO production from the reactions of

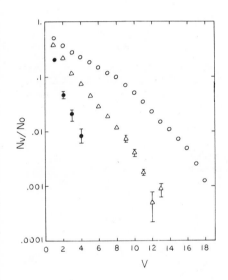

Figure 3. Vibrational energy distributions of the CO formed in O(^{3}P) + CH$_2$ (○), CH$_2$ + O$_2$ (△) and CH$_2$ + CO$_2$ (●). The data for the latter two reactions are those of Hsu and Lin (10).

Table I. Average CO Vibrational Energies and Absolute Rate Constants
for CO Production at 300K for $O(^3P)$ + 1-Alkynes.

Reaction	$< E_v >$* calc.	$< E_v >$* expt.	k $(cm^3/mole-s)$	Ref.
$O(^3P) + C_2H_2 \rightarrow CH_2 + CO$	4.51	3.23 ± 0.69	0.4×10^{10}	(7)
$O(^3P) + CH_3C \equiv CH \rightarrow CH_3CH + CO$	2.20	2.28 ± 0.28	4.2×10^{11}	(3,4)
$O(^3P) + C_2H_5C \equiv CH \rightarrow C_2H_5CH + CO$	1.18	0.97 ± 0.04	5.0×10^{11}	(4)
$O(^3P) + C_3H_7C \equiv CH \rightarrow C_3H_7CH + CO$	1.04	0.92 ± 0.04	4.9×10^{11}	(7)
$O(^3P) + C_4H_9C \equiv CH \rightarrow C_4H_9CH + CO$	0.80	0.73 ± 0.03	3.6×10^{11}	(7)

* The average energies (in kcal/mole) were calculated using $\Delta H° = -47$ kcal/mole for $O + C_2H_2$ and -50 kcal/mole for the reactions involving higher members of the homolog.

Table II. Average CO Product Vibrational Energies
Measured for the Reactions of CH_2
with $O(^3P)$, O_2 and CO_2

Reaction	$-\Delta H°$*	$<E_v>$	Reference
$O(^3P) + CH_2 \xrightarrow{a} CO + 2H$	74	16.9	This work
$\xrightarrow{b} CO + H_2$	178		
$CH_2 + O_2 \xrightarrow{a} CO + H_2O$	175	7.1	(10)
$\xrightarrow{b} CO + H + OH$	56		
$CH_2 + CO_2 \rightarrow CO + H_2CO$	61	1.9	(10)

*Exothermicities and average CO vibrational energies (in units of kcal/mole)
were computed by assuming that the reactions involve triplet CH_2, except with
CO_2 which is reactive only with excited singlet CH_2 (see the text).

$O(^3P)$ atoms with 1-alkynes and from the reactions of CH_2 with $O(^3P)$, O_2 and CO_2. These reactions can not be readily studied with other techniques such as chemiluminescence (due to low product concentrations as well as the difficulty in producing radical species in flow experiments). With the aid of simple statistical models, the dynamics and branching ratios of these reactions can be reasonably interpreted and crudely estimated. More detailed discussion of the application of this technique to many other examples including energy transfer reactions has recently been reviewed elsewhere (17, 18).

Literature Cited

1. Lin, M. C. and Shortridge, R. G., Chem. Phys. Lett. (1974), **24**, 42.
2. Shortridge, R. G. and Lin, M. C., J. Chem. Phys. (1976), **64**, 4076.
3. Lin, M. C., Shortridge, R. G. and Umstead, M. E., Chem. Phys. Lett. (1976), **37**, 279.
4. Umstead, M. E., Shortridge, R. G. and Lin, M. C., Chem. Phys. (1977), **20**, 271.
5. Hsu, D. S. Y. and Lin, M. C., Chem. Phys. Lett. (1976), **42**, 78.
6. Hsu, D. S. Y. and Lin, M. C., J. Chem. Phys. (1978), **68**, 4347.
7. Shaub, W. M., Burks, T. L. and Lin, M. C., Chem. Phys. (1980).
8. Herron, J. T. and Huie, R. E., J. Phys. Chem. Ref. Data (1974), **2**, 46.
9. Lin, M. C., in "Chemiluminescence and Bioluminescence," M. J. Cormier, D. M. Hercules and J. Lee, Eds., p.61, Plenum Press, N. Y., 1973.
10. Hsu, D. S. Y. and Lin, M. C., Int. J. Chem. Kinet. (1977), **9**, 507.
11. Shaub, W. M., Hsu, D. S. Y., Burks, T. L., and Lin, M. C., to be published.
12. Hsu, D. S. Y., Shortridge, R. G. and Lin, M. C., Chem Phys. (1979), **38**, 285.
13. Kistiakowsky, G. B. and Sauer, K., J. Am. Chem. Soc. (1958), **80**, 1066
14. Milligan, D. E. and Jacox, M. E., J. Chem. Phys. (1962), **36**, 2911.
15. Frey, H. M. and Kennedy, G. T., J. C. S. Chem. Comm. (1975), 233.
16. Meadows, J. H. and Schaefer, H. F. III., J. Am. Chem. Soc. (1976), **98**, 4383.
17. Lin, M. C., Adv. Chem. Phys. (1980), in press.
18. Baronavski, A. P., Umstead, M. E. and Lin, M. C., Adv. Chem. Phys. (1980), in press.

RECEIVED February 1, 1980.

OTHER DIAGNOSTIC TECHNIQUES

Absorption Spectroscopy of Combustion Gases Using a Tunable IR Diode Laser

R. K. HANSON, P. L. VARGHESE, S. M. SCHOENUNG, and P. K. FALCONE

High Temperature Gasdynamics Laboratory, Department of Mechanical Engineering, Stanford University, Stanford, CA 94305

Experimental studies of combustion chemistry require measurements of species concentrations, often under conditions where in situ spectroscopic techniques are desirable or necessary. Among other new methods, tunable laser absorption spectroscopy using infrared diode lasers offers prospects for improved accuracy and specificity in concentration measurements, when a line-of-sight technique is appropriate. The present paper discusses diode laser techniques as applied to a flat flame burner and to a room temperature absorption cell. The cell experiments are used to determine the absorption band strength which is needed to properly interpret high temperature experiments. Preliminary results are reported for CO concentration measurements in a flame, the fundamental band strength of CO at STP, collision halfwidths of CO under flame conditions, and the temperature dependence of CO and NO collision halfwidths in combustion gases.

Experimental Arrangement

Details of the experimental arrangement and procedures have been described in a series of previous papers dealing with flames (1,2,3,4) and shock tube flows (5,6). A schematic of the single-beam optical system usually employed is shown in Figure 1. In this arrangement, the laser beam passes through the flame, into a monochromator for laser mode selection and wavelength identification, and is then split into two beams: one passing through a room temperature absorption cell and the other through a Fabry-Perot etalon used for measuring changes in wavelength. This configuration is suitable for either flame measurements or absorption cell experiments. The laser is repetitively modulated using a sawtooth current waveform. In the present experiments with steady absorption conditions, the detector (D_1 and D_2) output signals are recorded with a signal averager (PAR 4202).

The problem of laser power variations with wavelength is overcome either by taking the difference between separately recorded absorbed and non-absorbed transmitted intensity records

0-8412-0570-1/80/47-134-413$05.00/0
© 1980 American Chemical Society

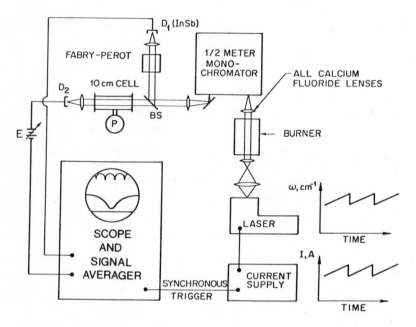

Figure 1.   Optical arrangement for tunable diode laser absorption spectroscopy

or by employing a double-beam optical system (4) which gives the
real-time difference between the absorbed beam and a reference
(non-absorbed) beam. Repetition rates for laser modulation can be
varied to over 5 kHz but are typically 200–500 Hz in the current
work. Since the recorded signal is actually a measure of the
absorbed intensity, $\Delta I = I° - I$ where I and $I°$ are respectively the transmitted intensities with and without absorption,
the quantity $I°$ must also be measured in order to obtain the
quantity of interest, the transmissivity $I/I°$ across the fully
resolved absorption line. $I°$ can be measured directly (with no
absorption present) either by chopping the laser beam or by bias-
ing the detector output (with the voltage source E) to give zero
signal with the laser beam blocked.

Two flat flame burners have been employed, a 4 cm × 10 cm
burner with a ceramic-lined chimney for NO measurements (4) and a
2.6 cm × 8.6 cm open-faced burner with a nitrogen shroud flow for
CO measurements. Both burners operate at atmospheric pressure
with laminar, premixed methane-air mixtures. These burners work
satisfactorily over a broad range of fuel-air equivalence ratios,
but both have cold boundary regions which cause non-uniform conditions along the optical axis that can be important in the data
analysis (4).

Absorption Theory

The theory required to interpret the experimental absorption
data is well established. The governing equation which links the
measured transmissivity, T_ν, at wavenumber ν, to the absorbing
species concentration and its absorption line parameters is the
Bouguer-Lambert law of absorption

$$T_\nu = (I/I°)_\nu = \exp\left[-\int_0^L \beta_\nu P_j \, dx\right]$$

where P_j is the partial pressure (atm) of the absorbing species
and L is the pathlength (cm) across the flow. The absorption
coefficient β_ν is the product of the line strength S ($cm^{-2} -$
atm^{-1}) for the transition of interest and the lineshape factor
$g(\nu-\nu_0)$ (cm), which is a function of the non-resonance, $\nu-\nu_0$;
i.e., $\beta_\nu = S \, g(\nu-\nu_0)$, where

$$\int_{line} g(\nu-\nu_0) \, d\nu \equiv 1$$

and

$$S = \int_{line} \beta_\nu \, d\nu .$$

Thus the line strength, sometimes called the line intensity, is
simply the area under a curve of the absorption coefficient. When

uniform conditions along the line of sight can be assumed, the
absorption law reduces simply to

$$T_\nu = \exp(-S \ g(\nu-\nu_0) \ P_j \ L) \ .$$

When absorption lines overlap, the absorption coefficient is a
summation: $\beta_\nu = \Sigma \ S_i \ g_i$.

It is clear from the above equation that both the line
strength and the lineshape factor must be known to convert a measurement of transmissivity to a species partial pressure. In our
work, the lineshape factor is determined directly in each experiment by recording the fully resolved absorption profile. The line
strength for a given vibration-rotation transition is a known function (7) of the flame temperature and the band strength evaluated
at a reference temperature, typically 273.2 K. Controlled laser
absorption experiments at room temperature with known levels of CO
are conducted to determine the band strength, which is used together with a measured temperature to specify the individual line
strengths at flame conditions.

In the case of a transition $(v'' + 1, J'' \pm 1 \leftarrow v'', J'')$ in the
fundamental band $(\Delta v = 1)$ of CO, the relation between the line
strength, S, the band strength, S°, and the temperature, T, is:

$$S_{v'' \rightarrow v''+1}^{J'' \rightarrow J'' \pm 1} = [S°(STP)(273.2/T)](\nu_0/\overline{\nu})(v''+1)\{\exp(-T_e(v'',J'')hc/kT]\}\cdot$$

$$\cdot[1 - \exp(-h\nu c/kT)][S^J/Q(T)]$$

where

$$S^J = J \ , \qquad P \ branch \ (J'' - 1 \leftarrow J'')$$

$$S^J = J + 1 \ , \qquad R \ branch \ (J'' + 1 \leftarrow J'')$$

and

$$Q(T) = \sum_{v,j} (2J + 1) \ \exp(-T_e(v,J) \ hc/kT) \ .$$

Here S°(STP) is the fundamental band strength at 273.2 K and Q
is the partition function for vibration and rotation. $T_e(v,J)$ is
the energy of the (v,J) state, in cm^{-1}; ν_0 is the wavenumber at
line center for the specific transition, and $\overline{\nu}$ is an average
wavenumber for the band, usually taken as the band-center value
although small corrections can be calculated (8). The quantities
h, c and k are Planck's constant, the speed of light and
Boltzmann's constant, respectively. For most purposes, it is
sufficiently accurate (an error of a few per cent or less) to use
rigid-rotor, harmonic oscillator relations for the partition function. Recent determinations of the CO and band strength by other
workers have been in the range S° = 260-280 cm^{-2} - atm^{-1} at
273.2 K (9).

The lineshape function $g(\nu-\nu_0)$ is defined assuming a Voigt
profile (7) which allows for a combination of Doppler and collision

line broadening. The Voigt function is computed (10) as a function of the parameter a, where

$$a = (\ln 2)^{1/2} \Delta\nu_C/\Delta\nu_D .$$

The Doppler-broadened linewidth (FWHM) is given by

$$\Delta\nu_D = 7.16 \times 10^{-7} (T/M_{CO})^{1/2} \nu_o$$

where T is the temperature (K) and M_{CO} is the molecular weight of the absorbing species (gm/mole). The collision-broadened linewidth is expressed in terms of the collision halfwidth 2γ, i.e., the collision-broadened linewidth (FWHM) per unit pressure of the broadening species, and the pressure P:

$$\Delta\nu_C = (2\gamma, cm^{-1} - atm^{-1})(P, atm) .$$

The collision halfwidth for a given transition is a function of temperature and the broadening species. In the present diode laser experiments, the temperature (and hence $\Delta\nu_D$) is known so that it is straightforward to infer values for the parameters a and $\Delta\nu_C$, and hence 2γ, from the observed absorption linewidths.

The temperature dependence of the collision halfwidth, $2\gamma(T)$, is of fundamental and practical interest and has not previously been investigated at elevated temperatures. In the past, most determinations of collision halfwidth have been made near room temperature, and high-temperature values have been obtained by extrapolation, usually assuming a $T^{-0.5}$ temperature dependence so that

$$2\gamma = 2\gamma°(300/T)^{0.5}$$

where $2\gamma°$ is the collision halfwidth at 300 K. This temperature dependence is based on hard-sphere collision theory arguments and is known to be incorrect. Tunable diode laser spectroscopy applied to a variety of gas conditions, including a room temperature static cell, shock tube flows (5,6) and flames, therefore provides a unique opportunity to study the temperature dependence of 2γ. The present paper provides initial results for selected CO and NO transitions in combustion gas mixtures.

Room Temperature Cell Experiments

An extensive series of room temperature experiments recently has been completed in our laboratory to determine the fundamental band strengths of CO and NO and to measure CO and NO collision halfwidths for N_2, Ar and combustion gas broadening as a function of rotational and vibrational quantum number. Some preliminary results for CO are reported here.

In these experiments, a gas mixture of known proportions was introduced into the room temperature cell, and the temperature and pressure were measured. The signal-averaged absorption and non-absorption records were differenced (by the PAR 4202) and the output was displayed on a chart recorder together with the output of the Fabry-Perot etalon. The non-absorbed signal, $I°$, was measured separately. A careful calibration of all system components was performed. Data were reduced by obtaining a computer-generated best Voigt fit to each absorption line profile.

A typical result for CO highly dilute in N_2 is shown in Figure 2. The cell conditions were: T = 294.2 K, P = 51.5 torr, CO mole fraction = 0.00350, cell length = 15.0 cm. The quantity plotted is: $S g P_{CO} L$ versus the normalized non-resonance, $\Delta = (\nu-\nu_o)/2(\ln 2)^{1/2} \Delta\nu_D$, for the $v = 1 \leftarrow 0$, R(1) transition at 2150.86 cm^{-1}. The best Voigt fit, shown as the solid curve, yields values of $S(294.2$ K$) = 4.86$ cm^{-2}-atm^{-1}, a = 1.73 and $2\gamma(294.2$ K$)_{CO-N_2} = 0.153$ cm^{-1}-atm^{-1}. Details of the experimental procedures and additional results will be published separately (11).

Each line strength determination was converted to a value for the band strength at 273.2 K; preliminary results for $S°(STP)$ as a function of the rotational quantum number are shown in Figure 3. Each of the points plotted actually represents the average of several determinations; experimental scatter in $S°$ for a given line was less than ± 3%. The average of the determinations at 10 values of m is $S°(STP) = 279.4$ cm^{-2}-atm^{-1}, which is in good agreement with recent work by Varanasi and Sarangi ($S°(STP) = 273 \pm 10$ and 277 ± 4 cm^{-2}-atm^{-1}) using a different experimental technique (9). It should be noted that our data are for specific absorption lines of $^{12}CO^{16}O$, and we have not yet applied a correction to account for the presence of other isotopes in the CO sample. This is equivalent to assuming negligible levels of other CO isotopes, which is in error by about 1%.

Flat Flame Burner Experiments

Experiments are currently in progress to measure CO and NO concentrations in a flat flame burner by diode laser spectroscopy. Comparative measurements are also being made using microprobe sampling with subsequent analysis by non-dispersive infrared and chemiluminescent techniques. Some preliminary laser absorption results for CO are reported here; initial results for NO have been published separately (4). Also reported are initial data for collision halfwidths in combustion gases.

The experimental and data reduction procedures are essentially the same as for the static cell experiments. The gas temperature is obtained using a fine wire, radiation-corrected thermocouple. The cold mixing layer at each flame boundary is accounted for by using an effective pathlength (8.0 - 8.2 cm, depending on the fuel-air equivalence ratio) which differs slightly from the actual burner length of 8.6 cm. Fuel-air equivalence ratios of

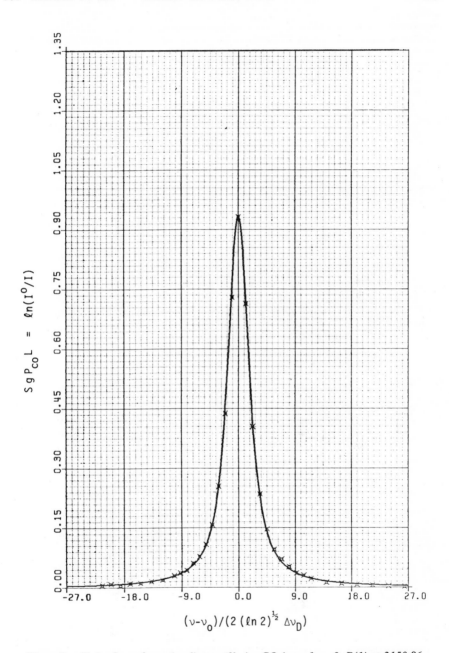

Figure 2. *Voigt fit to absorption line profile for CO ($v = 1 \leftarrow 0$, R(1) at 2150.86 cm^{-1}) in room-temperature cell experiment. Cell conditions are:* T $= 294.2$ K, P $= 51.5$ torr, CO–N_2 mixture with CO mole fraction $= 0.00350$. Inferred results are $S(294.2$ K$) = 4.86$ cm^{-2} atm^{-1}, a $= 1.73$ and 2γ $(294.2$ K$) = 0.153$ cm^{-1} atm^{-1}.

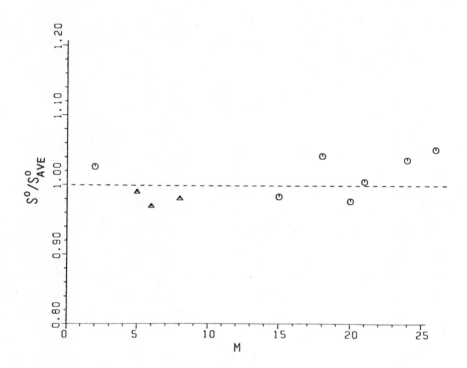

Figure 3.   CO fundamental band strength at 273.2 K determined in room-temperature cell experiments: ($\triangle$), M = J: P-branch; ($\bigcirc$), J + 1: R-branch. $S°_{ave}$ = 279.4 cm^{-2} atm^{-1}.

ϕ = 0.8 - 1.4 are being investigated. Measurements are made in the postflame gases 1 cm above the burner surface. The recorded signals represent the average of 8-32 laser modulation cycles.

Typical experimental results and the best Voigt fit for a CO absorption line in a flame with ϕ = 1.24, T = 1850 K are shown in Figure 4. The observed transition is the (v = 1 ← 0, P(7)) line of CO at 2115.63 cm^{-1}. The line strength S is calculated from the previously determined band strength, and the pathlength L is known, so the unknown parameters are P_{CO} and g which can both be inferred from the best Voigt fit analysis. The results are P_{CO} = 0.0421 atm, a ≐ 2.66 and $2\gamma(1850)_{CO-Comb gas}$ = 0.0393 cm^{-1} - atm^{-1}. As with the static cell experiments, the Voigt profile is seen to provide a good fit throughout the observed portion of the absorption line.

In fuel-rich flames, the CO should be in local chemical equilibrium, and hence the partial pressure of CO can be calculated from the local temperature and the measured fuel and air flowrates. Thus, a comparison between measured and calculated CO levels can serve as a validation of the diode laser technique for flame measurements. Such a comparison is shown in Figure 5 for equivalence ratios in the range ϕ = 1.04 - 1.37. The data points shown represent the average of several observations on separate lines including ground state (v″ = 0) and excited state (v″ = 1) transitions. The agreement is consistently within the experimental uncertainty of ± 5%.

Results showing the dependence of the CO collision halfwidth in combustion gases on the vibrational and rotational quantum numbers are shown in Figure 6. The data were obtained with a flame temperature of 1875 K and equivalence ratios in the range 1.2 - 1.4. Although too few data points are available for a detailed analysis, it is clear that 2γ decreases with increasing m and that values for 2γ are nearly equal (within 5%) for ground state and excited state transitions.

The temperature dependence of the collision halfwidth for combustion gas broadening is also of interest. Results for specific transitions in CO and NO are given in Table I. In the

TABLE I. TEMPERATURE DEPENDENCE OF COLLISION HALFWIDTHS

A. CO [v = 1←0, P(7))]

2γ (300 K)$_{CO-N_2}$ = 0.131 cm^{-1}/atm

2γ (1850 K)$_{CO-Comb Gas}$ = 0.039 cm^{-1}/atm   (CH$_4$/air, ϕ = 1.24)

n   = 0.67

B. NO [Ω = 3/2, v = 1←0, R(13/2)]

2γ (300 K)$_{NO-Comb Gas}$ = 0.137 cm^{-1}/atm ⎫
                                                        ⎬ CH$_4$/air, ϕ = 0.65
2γ (1700 K)$_{NO-Comb Gas}$ = 0.043   ⎭

n   = 0.67

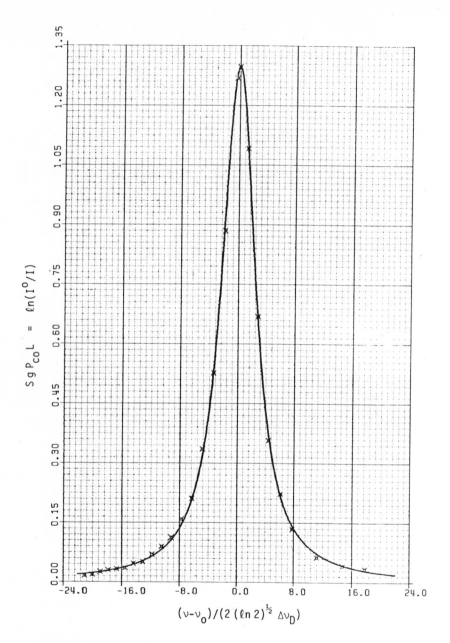

Figure 4.   Voigt fit to absorption line profile for CO ($v = 1 \leftarrow 0$, P(7) at 2115.63 cm^{-1}) in CH$_4$–air flat flame. Flame conditions are: T = 1850 K, P = 1 atm, and $\phi = 1.24$. Inferred results are P$_{CO}$ = 0.0421 atm, a = 2.66 and $2\gamma(1850)$ = 0.0393 cm^{-1} atm^{-1}.

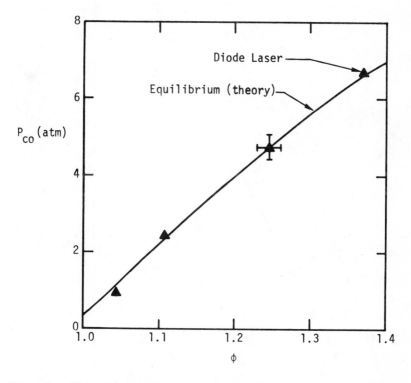

Figure 5. CO partial pressure in CH₄–air flat flame as a function of fuel–air equivalence ratio: T = 1875 ± 25 K; P = 1 atm.

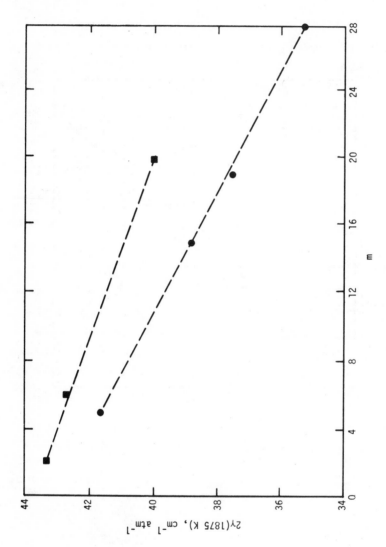

Figure 6. CO collision halfwidth in CH₄–air flat flame as a function of rotational quantum number: $T = 1875$ K; $\phi = 1.24 - 1.4$. $J + 1$: R-branch; $m = J$: P-branch; (■), $v'' = 0$; (●), $v'' = 1$.

case of the CO transition, the collision halfwidth obtained in a
flame is compared with the room temperature value for broadening
by N_2. This procedure should not introduce a large error since
the combustion products are primarily N_2. In the case of NO, the
collision halfwidth was actually measured in the same combustion
gas sample, first in the high temperature gases using the in situ
technique, and subsequently in low temperature gases which were
extracted by a sampling probe and sent to the absorption cell.
The temperature dependence can be expressed in terms of an expo-
nent n, where

$$2\gamma(T) \;\;=\;\; 2\gamma°(300/T)^n$$

with the result, for both the CO and NO transitions studied, n =
0.67. The exact agreement for these CO and NO lines must be considered fortuitous, but the similarity of the results with the
value found in a previous shock tube study of CO broadening by Ar
(6) suggests that n ~ 0.7 may provide a reasonable estimate for
temperature dependence. Work is presently underway to determine
n for a larger number of CO and NO transitions with N_2, Ar and
combustion gas broadening. Such data should be of use to
theoreticians interested in modelling collisional interactions of
molecules at elevated temperatures.

Concluding Remarks

These experiments demonstrate that tunable diode laser absorption spectroscopy is well suited for in situ measurements of
species concentrations in combustion flows, when a line-of-sight
technique is appropriate, and for accurate measurements of
spectroscopic parameters needed to characterize high-temperature
absorption lines. The technique is sensitive, species specific
and applicable to a large number of important combustion species
including reactive intermediates, and hence it should prove to be
a useful tool in future studies of combustion chemistry. The
potential of tunable laser absorption spectroscopy in particle-
laden flows should also be noted (12), in that modulation of the
laser wavelength on and off an absorption line allows simple discrimination against continuum extinction by particles.

Acknowledgements

This work has been supported by the Department of Energy
under grant EY-76-S-03-0328, PA58 and by the Air Force Office of
Scientific Research under Contract F49620-78C-0026.

Literature Cited

1. Hanson, R. K., Kuntz, P. A., and Kruger, C. H., "High-Resolution Spectroscopy of Combustion Gases Using a Tunable IR Diode Laser"; Applied Optics, 1977, 16, 2045.

2. Hanson, R. K. and Kuntz-Falcone, P. A., "Temperature Measurement Technique for High-Temperature Gases Using a Tunable Diode Laser"; Applied Optics, 1978, 17, 2477.

3. Hanson, R. K., "Combustion Gas Measurements Using Tunable Laser Absorption Spectroscopy"; AIAA Preparint No. 79-0086, 17th Aerospace Sciences Meeting, New Orleans, Jan. 1979.

4. Falcone, P. K., Hanson, R. K., and Kruger, C. H., "Measurement of Nitric Oxide in Combustion Gases Using a Tunable Diode Laser"; Paper No. 79-53, Western States Section/Combustion Institute Meeting, Berkeley, Ca., Oct. 1979.

5. Hanson, R. K., "Shock Tube Spectroscopy: Advanced Instrumentation with a Tunable Diode Laser"; Applied Optics, 1977, 16, 1479.

6. Hanson, R. K., "High-Resolution Spectroscopy of Shock-Heated Gases Using a Tunable Infrared Diode Laser"; in "Shock Tube and Wave Research"; University of Washington Press, 1978, p. 432.

7. Penner, S. S., "Quantitative Molecular Spectroscopy and Gas Emissivities"; Addison-Wesley: Reading, Mass., 1959.

8. See Ref. 7, Chapter 7, Section 5, and the references listed there.

9. Varanasi, P. and Sarangi, S., "Measurements of Intensities and Nitrogen-Broadened Linewidths in the CO Fundamental at Low Temperatures"; Jour. of Quant. Spectrosc. Radiat. Transfer, 1975, 15, 473.

10. Drayson, S. R., "Rapid Computation of the Voigt Profile"; Jour. of Quant. Spectrosc. Radiat. Transfer, 1976, 16, 611.

11. Varghese, P. L. and Hanson, R. K., "Diode Laser Measurements of CO Collision Halfwidths and Fundamental Band Strength at Room Temperature"; to be published.

12. Hanson, R. K., "Absorption Spectroscopy in Sooting Flames Using a Tunable Diode Laser"; submitted to Applied Optics.

RECEIVED February 1, 1980.

Multiangular Absorption Measurements in a Methane Diffusion Jet

ROBERT J. SANTORO and H. G. SEMERJIAN

Thermal Processes Division, National Bureau of Standards, Washington, D.C. 20234

P. J. EMMERMAN, R. GOULARD, and R. SHABAHANG

School of Engineering and Applied Science, The George Washington University, Washington, D.C. 20052

The mathematical reconstruction of a property field, $F(x,y)$, from its projection in the θ direction is the basis of "Computerized Tomography" ($\underline{1},\underline{2}$). An identical technique can be used to reconstruct a field of linear absorption coefficient functions in a combusting flow field from multiangular path integrated absorption measurements. The linear absorption coefficient is the familiar $N_i Q_i$ product, where N_i is the concentration of species i and Q_i is the absorption cross section of species i at the frequency ν. The Bouguer-Lambert-Beer law states that

$$I = I_o \exp - \int N_i Q_i ds \qquad (1)$$

where we assume a single monochromatic light source of frequency ν and that N_i is the only constituent in the measured domain with absorption at this frequency. Taking the natural logarithm of both sides yields

$$-\ln I/I_o = \int N_i Q_i ds \qquad (2)$$

If two-dimensional functions in the coordinate system defined in Figure 1 are considered and letting

$$F(x,y) = N_i(x,y) Q_i$$
$$P(r,\theta) = -\ln I/I_o = \int F(x,y) ds \qquad (3)$$

where the coordinate systems are related by

$$x = r \cos \theta - s \sin \theta$$
$$y = r \sin \theta + s \cos \theta \qquad (4)$$

it can be shown that the Fourier transforms of the function taken in the two coordinate systems are related by

$$\hat{F}(\hat{x},\hat{y}) = \hat{F}_\theta(\hat{r},\hat{s}) \qquad (5)$$

0-8412-0570-1/80/47-134-427$05.00/0
© 1980 American Chemical Society

where ^ represents a Fourier transform operation. Thus the two
Fourier transforms are the same if the transform axes $(\hat{r}, \hat{s})$ are
a rotation of $(\hat{x}, \hat{y})$ in the frequency domain by the angle θ

$$\hat{P}(\omega, \theta) = \hat{F}(\omega, \theta) \tag{6}$$

where ω is 2π times the spatial frequency. This is the central
slice theorem of Fourier transforms, which states that the transform of a projection taken at angle θ is equal to a line through
the center of the two-dimensional transform domain of the function
F which makes an angle θ with the $\hat{x}$ axis. Therefore knowledge of
all projections would define the transform of the function and by
taking the corresponding inverse transform, the function can be
evaluated at any point in its domain. Applying a convolution
theorem ($\underline{3},\underline{4}$) yields the following equation

$$F(x,y) = \frac{1}{4\pi^2} \int_0^\pi d\theta \int_{-\infty}^\infty \hat{P}(\omega, \theta) e^{i\omega(x \cos \theta + y \sin \theta)} |\omega| d\omega \tag{7}$$

The linear absorption coefficient functions can thus be reconstructed using only projection data.

Experiment

     A methane-argon asymmetrical diffusion jet has been analyzed
for the methane concentration mapping in a steady flow condition.
The jet apparatus consists of a 12.7mm i.d. brass tube located
19mm from the center line of a 15.25cm circular brass plate. The
jet is supplied from a 10cm diameter cylindrical chamber containing a glass bead mixing section. A combination of flow
straighteners and screens are contained in the jet section to
provide a uniform exit flow profile. The jet/plate combination
is mounted on a milling bed which allows for accurate three
dimensional positioning of the unit.
     Absorption by methane of the 3.39μ line of a HeNe laser is
used to determine the methane concentration across the jet. The
optical arrangement used is shown in Figure 2. The HeNe laser
output (∿1mw) is attenuated with a 1.0 neutral density filter so
as not to exceed the maximum irradiance specification of the
detector. The beam is interrupted using a mechanical light
chopper operating at 1015 Hz. Transmitted radiation is detected
by an uncooled PbSe detector. A preamplifier with a gain of ten
is used to impedance match the resulting signal to an oscilloscope and lock-in analyzer. The output from the lock-in analyzer
is displayed on a digital voltmeter.
     In the present experiments a 10% CH_4 - 90% Ar mixture was
used. All experiments were done under conditions of atmospheric pressure and room temperature. The flow rate was measured
to be 0.57 liters/s. A series of absorption measurements were
taken across the jet at intervals of .64 mm (.025 inches) at a
height of 12.7 mm above the jet. The jet was then rotated 15°
and the experiment was repeated with the result that data for

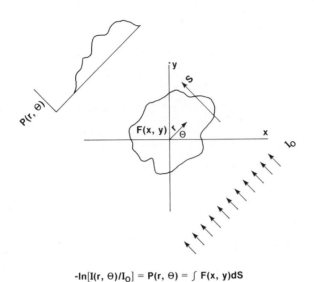

$$-\ln[I(r, \Theta)/I_0] = P(r, \Theta) = \int F(x, y)dS$$

Figure 1. Projection P(r,Θ) in polar coordinates

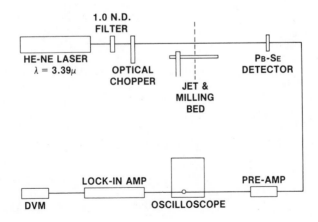

Figure 2. Schematic of apparatus for methane jet experiment

twelve separate angles was obtained to map out the jet region.
Spatial resolution was limited by the 2 mm laser beam
diameter. Average absorption measurements were obtained by
using a 4.0 s time constant on the lock-in detector.

The absorption data obtained was input to a minicomputer
for analysis, using a discrete form of equation 7.

Results

An objective of this series of experiments is to provide a
preliminary assessment of the tomographic reconstruction approach
for combustion diagnostics. In order to minimize experimental
difficulties introduced by combustion, a simple flow configuration has been chosen for this initial study. It is mathematical truism that any bandlimited function can be accurately
reconstructed from its projections if both the number and the
signal to noise ratio of these projections approach infinity. In
any real combustion situation, both of these conditions will be
severely limited. The present results provide insights into the
measurement capabilities of the tomographic reconstruction approach under such limitations.

The reconstructed linear absorption values for a cross
section of the jet (12.7mm above the exit plane) for the case of
twelve angles is shown in Figure 3. Qualitatively the reconstructed field shows the expected jet profile. Using an
absorption coefficient of 10 atm^{-1} cm^{-1} extrapolated from the
results of McMahon et.al. (5), the center line concentration of
methane is found to be 9.6%. Since the measurements were taken
close to the jet exit the potential core of the flow is observed
to survive to the measurement point. Thus the methane concentration would be expected to be the 10% initially introduced.
Therefore good agreement is observed with respect to concentration measurement. The position of the jet center line was
obtained from the location of the peak values of the linear
absorption coefficients. This approach yielded a position of
19±1.3 mm which compares quite well with the known position of
19 mm. This is further confirmation of the accuracy of the
reconstruction algorithm.

The analysis was repeated using projections for only six
angles and the results are shown in Figure 4. As can be seen
there is a marked increase in the absolute values of the linear
absorption coefficient in regions away from the jet as compared
to the twelve angle case. Since there is no methane present
in these regions, these values are a result of "ringing"
generated by the algorithm. These results are in qualitative
agreement with previous analytical studies using simulated
absorption functions (6).

Work is in progress to obtain results with improved spatial
resolution and to extend the range of measurements to much lower
concentrations. Further work will also consider more complicated
flow geometries with a longer range goal of studying combustion

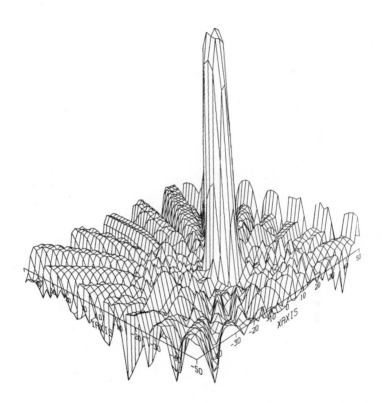

Figure 3. Reconstruction of methane jet cross section for 12-angle case

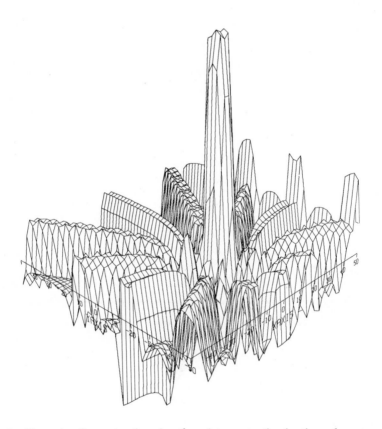

Figure 4. Reconstruction of methane jet cross section for six-angle case

experiments. In these experiments the goal is to achieve "instantaneous" time concentration profiles in two and three dimensions while using measurement aperature times as short as 50 μsec.

Literature Cited

1. Gordon, R., Herman, G.T., Johnson, S.A., Sci. Am. (1976), 233, 56.
2. Brooks, R.A., DiChiro, G., Phys. Med. Bio., (1976), 21, 689.
3. Ramachandran, R.N., Lakshmenarayanan, Proc. Natl. Acad. Sci. U.S., (1971), 68, 2236.
4. Shepp, L.A., Logan, B.F.. IEEE Trans. Nucl. Sci., (1974), 21, 21.
5. McMahon, J., Troup, G.J., Hubbert, G., Kyle, T.G., J. Quant. Spectrosc. Radiat. Transfer, (1972), 12, 797.
6. Goulard, R., Emmerman, P.J., 17th Aerospace Sciences Meeting, (1979), 79-0085.

RECEIVED February 1, 1980.

Temperature Measurement in Turbulent Flames Via Rayleigh Scattering

ROBERT W. DIBBLE, R. E. HOLLENBACH, and G. D. RAMBACH

Sandia Laboratories, Livermore, CA 94550

Using laser Rayleigh scattering, temperature measurements have been made, on a turbulent jet diffusion flame, with a frequency response, DC-5 kHz,unachievable by any other present day laser based technique. The flame reactants and products had nearly equal Rayleigh scattering cross section so that temperature could be inferred directly from the scattering intensity from a point on a probe laser beam. Probability densities, means, higher moments, and power spectrum are generated from the time series of temperature.

The intensity of Rayleigh scattered light can be written as ($\underline{1},\underline{2},\underline{3},\underline{4}$),

$$I_s = K' \ N \left(\frac{d\sigma}{d\Omega}\right)_{eff} = \frac{K}{T} \left(\frac{d\sigma}{d\Omega}\right)_{eff} \tag{1}$$

where the constants, K' and K, contain experimental details such as solid angle of optics, slit opening, phototube quantum efficiency, pressure, etc. N, T, and $(d\sigma/d\Omega)_{eff}$ are respectively number density, temperature, and effective Rayleigh scattering cross section, defined by:

$$\left(\frac{d\sigma}{d\Omega}\right)_{eff} = \sum_i x_i \left(\frac{d\sigma_i}{d\Omega}\right)$$

where x_i and $d\sigma_i/d\Omega$ are respectively the mole fraction and Rayleigh scattering cross section for species i.

Assuming the laser intensity to be constant, variations in the Rayleigh scattered intensity are a result of both temperature and species variations. Hence, an unambiguous interpretation of the Rayleigh scattered intensity requires that experiments be contrived into one of three cases. These are:

Case 1: Isothermal - In this case, variations in the Rayleigh scattering intensity are attributed to variations in the effective Rayleigh cross section, i.e., variations in mole fractions. The feasibility of making such time-resolved mea-

0-8412-0570-1/80/47-134-435$05.00/0
© 1980 American Chemical Society

surements of the dispersion of a turbulent, isothermal, nonreact-
ing methane jet into air was demonstrated by Graham and coworkers (5). Quantitative measurements of the dispersion of
a non reacting propane jet into air was made by Dyer (6) as a
preliminary test of a diagnostic that was used to map fuel-air
distribution in an I.C. engine combustion simulator.

Case 2: Constant Rayleigh Cross Section - In this case,
variations in the Rayleigh scattering intensity are attributed
to variations in temperature. A natural compliment to the
isothermal mixing investigations identified in Case 1 would be
to measure the time-resolved temperature in a submerged jet of
heated air. For premixed flames, the variation of Rayleigh
intensity is primarily due to variation in temperature, which
can vary by a factor of 7.

The variation in the effective cross section from reactants,
to intermediates, to products is often less than 10% for simple
hydrocarbon systems. This is largely due to the major species
being nitrogen in both products and reactants. As for the
remaining species, contributions of a given atom to the Rayleigh
cross section are roughly independent of which molecule that atom
is part of (1,3). In any event, the change in Rayleigh scattering
cross section from reactants to products can be incorporated
into data reduction to produce a refined temperature measurement.

Robben and co-workers have exploited these facts to measure
mean and rms temperature fluctuations in a turbulent flat flame
(7) and above a catalytic surface (8). By measuring the postflame temperature on a flat flame burner, as a function of
reactant flow rate, a precise measurement of laminar flame speed
was reported by Muller-Dethlefs and Weinberg (9).

Case 3: Both temperature and species variation - In this case,
additional information is required. This could be obtained from
another diagnostic or a mathematical model. Smith (10) used an
extensive mathematical model of a laminar hydrogen diffusion
flame to predict the species distribution throughout the flame;
having this, the temperature could be inferred from the Rayleigh
scattering intensity.

EXPERIMENT: A schematic diagram of the experimental
apparatus for temperature measurement is shown in Figure 1.
Scattered light collected by the F1.5 lens (f = 30 cm) is
relayed to the photomultiplier tube via 1 mm^2 slits, a 1 nm
bandwidth interference filter and a polarization filter, to
reduce background from flame luminescence. The PDP-11/34
computer instructs the A/D convertor to make a conversion every
100 μsec. The resulting digital data are stored sequentially in
core memory. The memory is saturated at 16,000 temperature
measurements, at which time the data are transferred to a hard
disk memory. The data in this transfer constitute one time

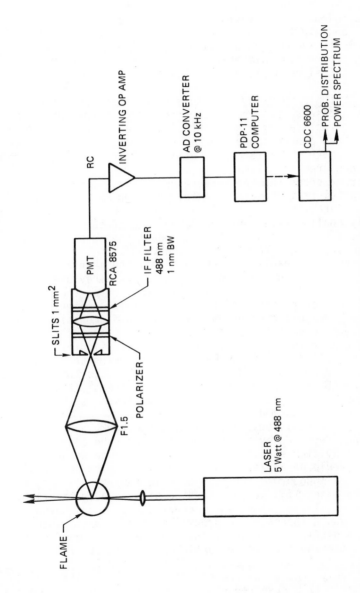

Figure 1.   Laser-scattering apparatus

series. Storage of the time series is a significant improvement
of this data collection apparatus. The power spectrum, auto-cor-
relation and probability histogram are calculated from each time
series. From the probability histogram, calculation of the
mean, rms, and higher moments is straighforward.
     A turbulent jet diffusion flame was investigated. The
apparatus and experimental procedure are described in detail in
the article by Rambach et al. (11). The fuel jet had the
following properties: diameter of 1.6 mm, Reynolds number of
4400, and a fuel composition of 37% methane and 63% hydrogen.

     DISCUSSION: By using a mixture of hydrogen and methane as
the fuel, a case 2 (see above) experiment has been contrived,
i.e., the effective Rayleigh scattering cross section of the
fuel, air, and products are nearly ($\leq$ 2%) equal. It was then
possible to make the first application of laser Rayleigh thermometry to non-premixed flames.
     Figure 2 shows the temperature probability distributions
along a radial traverse at the axial position L/D = 65. When
one contemplates these probability distributions convolved with
the highly nonlinear temperature dependence of the reaction
rates, the futility of attempting to model the mean reaction
rates with any single temperature is obvious.
     From these probability distributions, the mean temperature
and the rms temperature are easily generated. They are displayed
in Figures 3 and 4, respectively. The symmetry of the data is a
result of reflecting the data through the axial centerline,
i.e., data were collected on one side of the flame only. The
main point of this paper is to illustrate the feasibility of
obtaining temperature data in a turbulent diffusion flame.
     Wider application of the Rayleigh scattering technique for
temperature or concentration measurements will, to some extent,
rest on the ability to overcome two problem areas: flame
luminescence and Mie scattering from particles. Neither
problem appears insurmountable.
     At certain positions in the flame, the background flame
luminescence received by the photomultiplier tube can be
15% of the Rayleigh scattered intensity. A large reduction of
this noise would be achieved by replacing the 1 nm bandpass
filter with a monochrometer. Use of a multipass cell (12) or
intracavity laser (13) would raise the signal above the flame
luminescence. In addition, the increased scattered photon
count rate would increase the precision of each measurement.
     A more difficult problem is due to the presence of particles.
The Mie scattering from these particles in room air is about
equal to the Rayleigh scattering. Filtered air from a compressor
was effectively free of particles and was used in this work.
Pitz and Daily (14) have reported a promising method of suppress-
ing photon counts from Mie scattering. Basically, their method
increases the electronic dead time of the count system. This

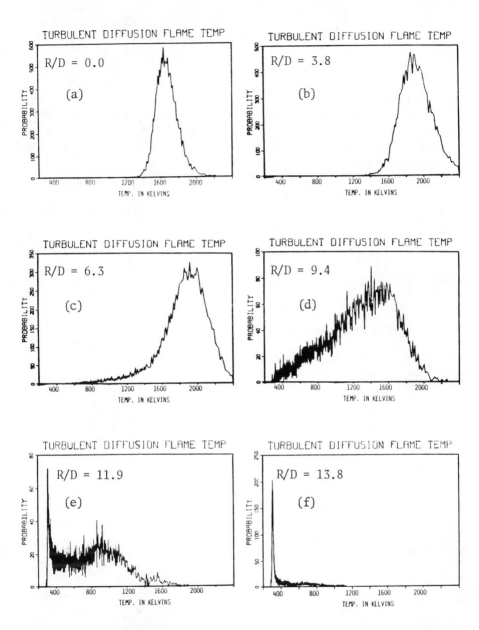

Figure 2. Probability of temperature at L/D = 65; R/D = 0.0, 3.8, 6.3, 9.4, 11.9, 13.8 for a, b, c, d, e, and f, respectively.

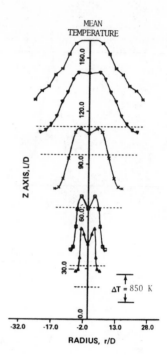

Figure 3. Temperature profile in turbulent jet diffusion flame: (– – –), position of radial traverse. Temperature at $L/D = 110$, $r/D = 0$ is 1980 K (n.b.: data collected on one side of flame only).

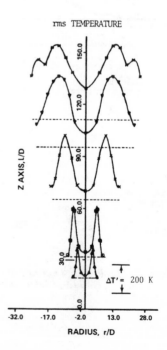

Figure 4. RMS temperature profile in turbulent jet diffusion flame: (– – –), position of radial traverse. RMS temperature at $L/D = 110$, $r/D = 12.5$ is 475 K (n.b.: data collected on one side of flame only).

dead time is tuned such that it has little effect on the Rayleigh count rate, but grossly undercounts the high photon arrival rate, i.e., the burst of photons, associated with particles.

This work supported by the Department of Energy, Division of Basic Energy Sciences.

1. Robben, F., " Combustion Measurements in Jet Propulsion Systems, p. 179, R. Goulard, Ed., sponsored by Project Squid, December 1975.
2. Kerker, M., The Scattering of Light, Academic Press, New York, 1969.
3. Born, M. and Wolf, E., Principles of Optics, p. 90, Pergamon Press, Oxford, 1975.
4. Jenkins, F. A. and White, H. E., Fundamentals of Optics, p. 457-463, McGraw-Hill, New York, 1957.
5. Graham, S. C., Grant, A. J., and Jones, J. M., AIAA Journal, 12, 1140-1142, 1974.
6. Dyer, T. M., AIAA Journal, 17, 912-914, 1978.
7. Namer, I., Agrawal, Y., Cheng, R. K., Robben, F., Schefer, R., and Talbot, L., "Interaction of a Plane Flame Front with the Wake of a Cylinder," presented at Fall Meeting, Western States Section of the Combustion Institute, Stanford, CA, October 1977.
8. Robben, F., Schefer, R., Agrawal, V., and Namer, I., "Catalyzed Combustion in a Flat Plate Boundary Layer I. Experimental Measurements and Comparison with Numerical Calculations," Presented at Fall Meeting, Western States Section, The Combustion Institute, Stanford, CA, October 1977.
9. Müller-Dethlefs, K. and Weinberg, F. J., "Burning Velocity Measurement Based on Laser Rayleigh Scattering," Seventeenth Symposium (International) on Combustion, p. 985, The Combustion Institute, Pittsburgh, PA, 1979.
10. Smith, J. R., "Rayleigh Temperature Profiles in a Hydrogen Diffusion Flame," Proceedings of SPIE Vol 158 Laser Spectroscopy (1978) p. 84-90.
11. Rambach, G. D., Dibble, R. W., Hollenbach, R. E., "Velocity and Temperature Measurements in Turbulent Diffusion Flames," paper no. 79-51 Fall Meeting of Western States Section of the Combusion Institute, Berkeley, CA, 1979.
12. Hill, R. A. and Hartley, D. L., Applied Optics, 13, p. 186, 1974.
13. Neely, G. O., Nelson, L. Y., and Harvey, A. B., Applied Spectroscopy, 26, p. 553, 1972.
14. Pitz, R. W. and Daily, J. W., "Experimental Studies of Combustion in a Free Shear Layer," Presented at 2nd International Symposium on Turbulent Shear Flows, Imperial College, London, Great Britain, July 1979.

RECEIVED February 22, 1980.

Droplet-Size Measurements in Reacting Flows by Laser Interferometry

UMBERTO GHEZZI

Politecnico, p. Leonardo da Vinci, 32-Milano, Italy

ALDO COGHE and FAUSTO GAMMA

C.N.P.M., v. F. Baracca, 69–Peschiera Borromeo, Milano

The current status of prediction and modelling in the area of fuel spray combustion requires, among other parameters, the measurement of droplet or solid particle size distribution and the relative velocity between the fuel and the surrounding gas. Many optical techniques, based on laser light scattering, have been investigated to this purpose (Refs.1,2,3,4,5,6 and 7), but the only system able to simultaneously determine the size and the velocity is the dual-beam laser Doppler velocimeter shown in Figure 1.

It has been demonstrated (Refs.8,9 and 10) that a particle crossing the intersection of two laser beams scatters a modulated light signal containing velocity and size information. The first is related to the modulation frequency, the second to the ratio of the A.C. to the mean amplitude. This ratio is called visibility

$$(1) \qquad V = \frac{I_{Max} - I_{min}}{I_{Max} + I_{min}} = \frac{D}{P}$$

where I_{Max}, I_{min}, D and P are defined in Figure 1.

It is possible to express the visibility in a functional form

$$(2) \qquad V = V(\frac{\pi d}{\lambda}, m, \frac{d}{\sigma}, \Omega, \gamma, \xi)$$

where d is the diameter of the scattering particle, m is the complex refractive index of the particle, β is the cross-angle of the two laser beams, λ is the laser wavelength, γ and ξ are angles defining the position of the axis of the collecting aperture, Ω is the collecting solid angle and $\sigma = \lambda /(2 \sin(\beta /2))$.

The other parameters being fixed, the relation V(d) can be numerically evaluated following the Mie scattering theory applied to individual spherical particles, supposed in the central region of the crossover volume of two coherent beams (Ref.10). Because of the large number of possible arrangements of the parameters appearing in eq.(2), it is impossible to give a complete representation of the visibility behaviour. A parametric analysis covering also

0-8412-0570-1/80/47-134-443$05.00/0
© 1980 American Chemical Society

many off-axis light collection directions and particle diameters up to 100 μm has been performed (Ref.11). In the range of interest of fuel sprays it has been found that in forward scatter, with small cross-beam and collecting angles, the refractive index of the particle has no significant influence and a one-to-one correspondence between visibility and particle diameter can be obtained, as shown in Figure 2.

The major limit to the practical use of these results is the maximum particle concentration, N, that can be allowed in order to preserve the single scattering condition. Roughly, $N_{max}(m^{-3}) \sim V_p^{-1}$ and the probe volume, V_p, is varying as $1/\sin\beta$. With $\beta \simeq 1°$, the maximum concentration can be about 10^8 m^{-3}. With higher β, the range of a monotonic relation between visibility and particle size is dramatically reduced. It has been found possible, however, to overcome these difficulties and to measure particle size distributions in fuel sprays.

Particle sizing measurements were performed in water sprays produced by a pressure-jet nozzle and in an industrial furnace. In the first case (Ref.12), a 5mW He-Ne laser and the LDV geometry relative to the visibility curve of Figure 2, with $\beta = .5°$ was used. Because of the high droplet concentration and the large probe volume dimensions, we used two glass probes coaxial to the LDV optical axis and facing each other (see Figure 3). The two probes were centered on the central region of the cross-beam region and kept 5 mm apart to reduce the effective cross-section and thus the measured particle rate. The allowed maximum concentration was about 10^9 m^{-3} and hence reliable measurements were obtained only in the outher part of the jet. The particle number concentration can be determined by the relation

$$(3) \qquad\qquad \dot{n} = N \, \overline{U} \, S$$

where $\dot{n}$ is the particle rate (s^{-1}), S the cross-sectional area of the effective probe volume, normal to the measured mean velocity component, $\overline{U}$.

The visibility of several hundreds individual scattered signals has been measured by a storage oscilloscope and severe validation criteria were defined to fulfill the hypothesis of the theoretical model. The actual size distribution function, f(d), was directly deduced by the measured visibility distribution, g(V), through the V(d) relation of Figure 2. Results of Table 1 demonstrated the capability of this technique to resolve different injection pressures.

Pressure (atm)	S.M.D. (μm)	$\overline{X}_{RR}$ (μm)
2	66.4	70.0
10	54.3	58.5
15	49.4	52.5

Table 1: Experimental results with a pressure-jet nozzle. S.M.D. is the Sauter mean diameter, $\overline{X}$ the Rosin-Rammler mean diameter.

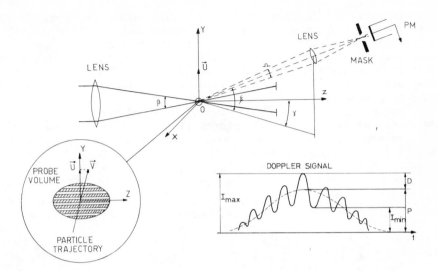

Figure 1. Geometry of the dual-beam LDV optics with the enlarged probe volume and the shape of a single-particle Doppler signal

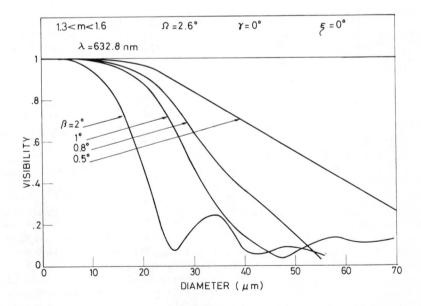

Figure 2. Typical V(d) curves for different β angles. Forward scatter geometry.

Comparison with standard methods, flash-photography and impres
sion technique, showed a satisfactory agreement of mean diameters,
but wider size distributions were found by standard techniques.

A new data processing procedure was applied to the visibility
measurements made in a furnace under burning conditions and without
the double probe (Ref.13). This procedure has the capability of determining size distributions also with non monotonic V(d) curves.
As a consequence, larger β values, i.e. reduced probe volume dimensions and higher particle concentrations, can be allowed.

In general, the following relationship exists among f(d), g(V)
and V(d):

(4) $$\overline{V} = \int_{0}^{d_{Max}} f(d)\ V(d)\ \mathrm{d}d\ = \int_{0}^{1} g(V)\ V\ \mathrm{d}V$$

where $\overline{V}$ is the mean value of the visibility. The f(d) function cannot be directly deduced by eq.(4), also because the relationship
is not univocal. An iterative procedure, in connection with some
forecast of the form of the f(d) curve, can be used to determine
the size distribution function. The main idea is to test different
f(d) functions until a distribution is found able to lead to a g(V)
distribution very close to the measured one.

In Figure 4 is shown the experimental apparatus used in the
furnace. The burner maximum load was about 400,000 Kcal/hr, the
furnace inner diameter 0.8 m and the optical throw length 1.5 m.
The 488 nm line (200 mW) of an Argon Ion laser was used with an in-
terferential filter in front of the photomultiplier to reduce the
background radiation of the flame. Calibration check with alumina
particles of known sizes were performed, but were not necessary.
The effective probe volume was about 10 mm^3, due to the small diameters of the furnace windows, but single scattering conditions
were easily achieved indicating a particle concentration of $10^8 \mathrm{m}^{-3}$.
Without optical access limitations the maximum allowed concentration could rise up to 10^{10} m^{-3}, by using larger β .

In Figure 5 three distribution curves of droplet diameters a-
re reported, relative to three axial positions into the furnace.
It is demonstrated the possibility of resolving the size variation
along the axis of the flame.

In conclusion, droplet size measurements in the range 10 to
100 μm can be performed, also in hostile environments, from the
visibility of individual scattered signal. Advantages of this method are: simultaneous measurement of particle size, concentration
and velocity; no calibration is necessary; good spatial resolution
up to less than 1 mm^{-3}; the visibility is independent on particle
trajectory. Limitations are: individual scattered signal can be
obtained only with moderate particle concentration; it is difficult
to automatically process scattered signals to extract the visibility value and to check validation conditions; it seems very difficult to extend the technique to cover the entire spray distribution;
the lower limit in the small particle end of the distribution curve depends upon experimental sensitivities and V(d) curve flatness

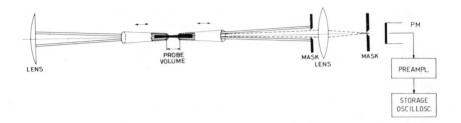

Figure 3. The double probe used in water sprays

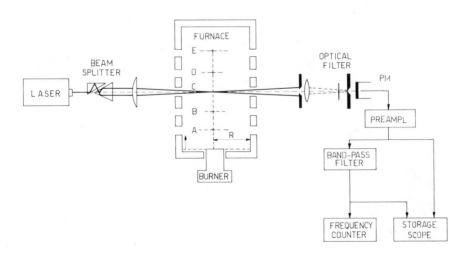

Figure 4. The optical arrangement for droplet-size measurement in an industrial furnace: R = .4 m, X_A = .26 m, X_C = .42 m, X_E = .68 m.

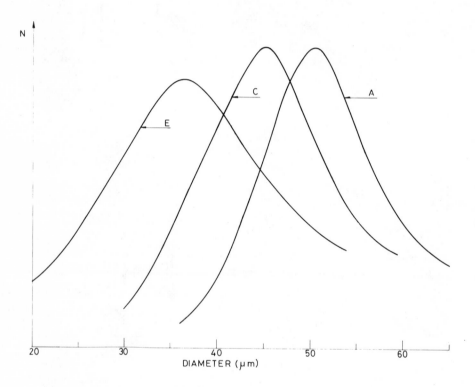

Figure 5. Distribution curves of droplet diameters relative to three axial positions into the furnace (A, C, and E)

and can be estimated 10 μm in practical situations. By increasing the β angle this limit can be lowered, but the full range will be also reduced, due to the V(d) shape and the reduction of the probe volume. In fact the technique requires probe volume dimensions larger than the maximum particle size.

Finally, under burning conditions, the refractive index variations produced by the gas temperature gradients cause deflections of the beams, hence drop-outs of the photodetector signal, but visibility measurements are still possible.

Abstract

This paper describes the development of a particle sizing technique utilizing the visibility parameter and the Mie scattering theory for spherical particles. The technique allows the measurement of size and velocity of individual particles and was tested in fuel sprays under both burning and non burning conditions.

Literature Cited

1. Farmer,W.M., Applied Optics, 11, 11, pp.2603-2612 (1972).
2. Robinson,D.M.,and Chu,W.P., Applied Optics, 14, 9,pp.2177-2183 (1975).
3. Hirleman,E.D.,Wittig,S.L., presented at Laser'77 Opto-Elektronik Conference, Munich, West Germany, June 20-24, 1977.
4. Chigier,N.A., Ungut,A. and Yule,A.J., Seventeenth Symposium (International) on Combustion, Leeds, U.K., 1978.
5. Bachalo,W.D., Third International Workshop on Laser Velocimetry, Purdue University, U.S.A., 1978.
6. Holve,D. and Self,S., Applied Optics, 18, 10, pp.1632-1645 (1979).
7. Holve,D., 18th AIAA Aerospace Sciences Meeting, Pasadena, U.S.A., January 1980.
8. Chu,W.P. and Robinson,D.M., Applied Optics, 16, 3, pp.619-626 (1977).
9. Adrian,R.J. and Orloff,K.L., Applied Optics, 16, 3, pp.677-684 (1977).
10. Coghe,A. and Ghezzi,U., Proceedings of Dynamic Flow Conference, DK-2740 Skovlunde, Denmark, pp.825-849, 1978.
11. Anglesio,P., Coghe,A. and Ghezzi,U., Meeting of the Aerodynamic and Oil Panel of the I.F.R.F., Lyon, France, 1978.
12. Ghezzi,U. and Coghe,A., Proceedings of the 4th International Symposium on Air Breathing Engines, Florida, U.S.A., 1979.
13. Ghezzi,U., Coghe,A. and Gamma,F., Symposium on Laser Probes for Combustion Chemistry, 178th ACS National Meeting, Washington, U.S.A., 1979.

RECEIVED March 18, 1980.

Continuous-Wave Intracavity Dye Laser Spectroscopy: Dependence of Enhancement on Pumping Power

STEPHEN J. HARRIS

Physical Chemistry Department, General Motors Research Laboratories, Warren, MI 48090

Intracavity dye laser spectroscopy (IDLS) can be a powerful technique for detecting trace species important in combustion. The technique is based on the phenomenal sensitivity of a laser to small optical losses within the laser cavity. Since molecular absorptions represent wavelength-dependent optical losses, the technique allows detection of minute quantities of free radicals by placing them inside the laser cavity and monitoring their effect on the spectral output of the laser.

IDLS was discovered nearly a decade ago (1,2), and, although there have been many demonstrations of the technique (3-6) and several theories proposed to explain it (4,7-10) there are few reports (11,12) of the technique actually being used to gain new chemical information. Hardly any work has been reported (13,14) in quantifying the experimental parameters affecting IDLS. The present work represents the first quantitative comparisons between theory and experiment for cw IDLS.

..The experimental arrangement is basically similar to that of Hänsch et al. (4). A Spectra Physics Ar$^+$ laser operating at 514.5 nm pumps a Rhodamine 6G dye laser tuned with a birefringent filter. The linewidth is 25 to 30 GHz, and the wavelength is tuned between 585.0 nm and 585.2 nm. The output mirror has a 1 meter radius of curvature and a reflectivity of 98% at 585.0 nm. The dye laser cavity is 74 cm long, and the laser is always run TEM$_{00}$ (this sometimes necessitates the use of an intracavity aperture).

I_2 is degassed and then distilled into a previously evacuated 23 cm long quartz cell with wedged (1.5°) anti-reflection coated windows epoxied on the ends. The cell is mounted in the laser cavity on X-Y-Z translation stages, and the I_2 is frozen into the sidearm by dipping it in a cold bath. The dye lasing threshold is measured, and the Ar$^+$ laser is then set to the desired power. For one set of experiments, threshold pump power is near (±11%) 550 mW, while for a second set the threshold is near 790 mW.

0-8412-0570-1/80/47-134-451$05.00/0
© 1980 American Chemical Society

The dye laser irradiates an external cell which contains 40 Pa (0.3 torr) I_2 vapor. A 1P28 photomultiplier whose face is covered by a 610 nm long pass filter measures the I_2 fluorescence, while a photodiode monitors a reflected spot of the dye laser. The ratio of fluorescence to dye laser power is displayed on a strip chart recorder. The sidearm temperature is gradually (1-2 hours) raised from about 210 K or 220 K, where fluorescence in the external cell is strong, to whatever temperature is required to reduce the signal by about 60%.

The extracavity absorption coefficient of I_2 at the laser wavelength is measured by detecting laser power with a thermopile before and after an I_2 cell.

Intracavity enhancement, relative to conventional single pass absorption spectroscopy, is due to mode competition and to threshold effects. A simple calculation of the latter for a single mode laser, starting with

$$I = I_0 \left(\frac{\alpha_0}{L} - 1\right)$$

gives

$$\xi^{(1)} \equiv \frac{d \ln I}{dL} = \frac{\alpha_0/L}{\alpha_0 - L} \tag{1}$$

where I is the dye laser intensity, I_0 is the saturation intensity, α_0 and L are the unsaturated single pass gain and loss, and $\xi^{(1)}$ is the enhancement of a single mode laser with an absorber dL inside the cavity. Equation (1) says, basically, that the dye laser output becomes very sensitive to additional intracavity loss when the laser is run near threshold (gain $\approx$ loss).

The effect of mode competition, which is generally the dominant effect, is more subtle since, to a first approximation, a cw dye laser has only one mode. Theories of IDLS which account for mode competition have been put forward by Hänsch, Schlawlow, and Toschek (HST) and by Brunner and Paul (BP). HST start with a realistic set of laser rate equations but use substantial approximations to solve them. BP use an approximate and very empirical set of rate equations which they solve analytically. Each theory yields a prediction for the dependence of enhancement on pumping power P relative to the threshold pumping power P^{th}.

Experimentally, fluorescence is measured as a function of I_2 pressure. Since IDLS is an absorption technique, and since fluorescence is proportional to the light "transmitted" by the intracavity cell at I_2 wavelengths, it makes sense to plot the logarithm of the fluorescence against pressure. We find a linear relationship, and the slope is then the intracavity absorption coefficient, ε_{int}. Enhancement is defined experimentally as $\xi = \varepsilon_{int}/\varepsilon_{ext}$, where ε_{ext} is the conventional single pass absorption coefficient. The results are compared with predictions of HST and BP in Figures 1 and 2, respectively. In

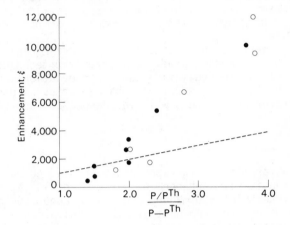

Figure 1. *A comparison of the theory of HST (– – –) with the data. It is assumed that there are 50 longitudinal modes. (●), A threshold of 790 mW; (○), a threshold of 550 mW (15).*

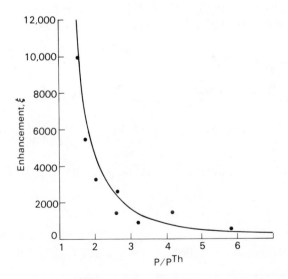

Figure 2. *A comparison of the data with the theory of BP (———). Threshold is 550 mW. M = 50; v = .050 ± .01 (15).*

judging these comparisons, it must be kept in mind that the theory of HST has no free parameters, while that of BP has a free empirical parameter whose physical significance is at best unclear. However, it is clear that HST's prediction of a large enhancement at high power (lim $(P\to\infty)$ $\xi \gtrsim 10^3$) is not consistent with these data.

Much more work, both theoretical and experimental, needs to be done for IDLS to become a well characterized technique. Numerical solutions to realistic laser rate equations, for example, as well as measurement of enhancement as a function of various laboratory parameters will increase IDLS' usefulness as an analytical technique.

Abstract

Intracavity absorption by I_2 vapor has been studied for a cw dye laser. The sensitivity enhancement varies from 10^4 at pump powers near threshold (550 mW and 790 mW) to about 500 at the highest pump powers (near 5 watts). The results can be interpreted quantitatively in terms of a previously proposed theory.

Literature Cited

1. Pakhomycheva, L. A.; Sviridenkov, E. A.; Suchkov, A. F.; Titova, L. V.; Churilov, S. S., JETP Letters, 1970, 12, 43.

2. Peterson, N. C.; Kurylo, M. J.; Braun, W.; Bass, A. M.; Keller, R. A., J. Opt. Soc. Am., 1971, 61, 746.

3. Thrash, R. J.; Weyssenhoff, H.; Shirk, J. S., J. Chem. Phys., 1971, 55, 4659.

4. Hänsch, T. W.; Schlawlow, A. L.; Toschek, P. E., IEEE J. Quantum Electron, 1972, QE-6, 802.

5. Schroder, H.; Neusser, H. J.; Schlag, E. W., Opt. Commun., 1975, 14, 395.

6. Atkinson, G. H.; Lavfer, A. H.; Kurylo, M. J., J. Chem. Phys., 1973, 59, 350.

7. Keller, R. A.; Zalewski, E. F.; Peterson, N. C., J. Opt. Soc. Am., 1972, 62, 319.

8. Brunner, W.; Paul, H., Opt. Commun., 1974, 12, 252.

9. Holt, H. K., Phys. Rev., 1976, A 14, 1901.

10. Tohma, K., J. Appl. Phys., 1976, 47, 1422.

11. Bray, R. G.; Henke, W.; Liv, S. K.; Reddy, K. V.; Berry, M. J., Chem. Phys. Letters, 1977, 47, 213.

12. Reilly, J. P.; Clark, J. H.; Moore, C. B.; Pimentel, G. C., J. Chem. Phys., 1978, 69, 4381.

13. Keller, R. A.; Simmons, J. D.; Jennings, D. A., J. Opt. Soc. Am., 1973, 63, 1552.

14. Childs, W. J.; Fred, M. S.; Goodman, L. S., Appl. Opt., 1974, 13, 2297.

15. Harris, Stephen J., J. Chem. Phys., 1979, 71, 4001.

RECEIVED February 1, 1980.

The Use of Photoacoustic Spectroscopy to Characterize and Monitor Soot in Combustion Processes

D. K. KILLINGER, J. MOORE, and S. M. JAPAR

Engineering and Research Staff, Ford Motor Company, Dearborn, MI 48121

Optical measurements of airborne combustion aerosols have been carried out for a number of years, usually with light scattering techniques. However, due to the particle size dependence of light scattering and the variable particle size distributions of smokes, it is extremely difficult to relate light scattering properties to particulate mass concentrations. The measurement of light absorption by particles can be directly related to particle mass if two conditions are met:

. Optical absorption per particle mass is independent of particle size; this holds for particle diameter (assuming spherical particles) $D << \lambda$, where λ is the incident wavelength ([1], [2])

. Ability to measure low absolute light absorption, < 1 part in 10^3, possibly in the presence of a comparable amount of light scattering.

For the measurement of light absorption by airborne carbonaceous particulate (soot), the conventional light absorption techniques fail due, primarily, to the second condition. However, photoacoustic spectroscopy has the necessary sensitivity ([3-6]) and is not subject to major interferences from light scattering. For these reasons photoacoustic spectroscopy was first used by Terhune and Anderson, in this laboratory, to study airborne soots produced by a number of combustion processes. ([4], [5], [6])

Experimental

The Photoacoustic Effect ([7]). The modulated absorption of light by material in a cell leads to the production of a sound wave at the modulation frequency. The sound wave is due to modulated pressure pulses in the cell arising from the liberation, as heat, of a portion of the absorbed light. The sound wave thus produced can be detected with a sensitive microphone and associated electronics, i.e., a spectrophone.

0-8412-0570-1/80/47-134-457$05.00/0
© 1980 American Chemical Society

The spectrophone response, S, is proportional to the amount of light absorbed at a given microphone sensitivity (assuming condition (1) above):

$$S = R(1-e^{-kx})W \tag{1}$$

where R is the cell response factor which is dependent on cell design and modulation frequency, W is the incident optical power, x is the cell length in meters and k is the absorption coefficient in meter^{-1}. For low light absorptions, $\leq 5\%$, equation (1) reduces to

$$S = R(kx)W \tag{2}$$

If the mass concentration, C, in gram m^{-3}, of particulate is known, then the specific absorption coefficient, a, in m^{2}g^{-1}, of the particulate can be determined from the relation,

$$k = aC \tag{3}$$

Instrumentation. The spectrophone has been described in detail elsewhere (4-6, 8). The output of a laser (either Ar ion, 1.2 watts at 514.5nm; or a dye laser, 600 mw at 600.0 nm, with a spectral range of 590nm to 625nm) is mechanically chopped at frequencies near 4 kHz, and the light is passed into an acoustically isolated cell containing a cylindrical brass cavity resonant near 4 kHz at atmospheric pressure. The magnitude of the sound waves produced are detected with a B & K model 4144 1" diameter condensor microphone, while signal processing is accomplished with lock-in analyzers and ratiometers.

Soot from an O$_2$-propane flame was produced in a 5 cm diameter 1-meter flow tube, with a total flow rate (N$_2$ carrier gas) of 3.5 liter min.$^{-1}$ 0.2 µm pore size Fluoropore filters (Millipore Corp.) were used for mass evaluation.

Automobile exhaust was sampled by passing it into a dilution tube where the flows were typically 300-500 ft^3 min.$^{-1}$ with dilution ratios of 5-10 to 1. Samples were removed from the tube at a flow rate of ∿1 liter min.$^{-1}$.

Results

Validation of the Photoacoustic Effect as a Soot Monitor (8). The photoacoustic spectrum and the absorption spectrum of airborne propane-generated soot have been simultaneously measured in the 590nm-625nm region using a tunable dye laser. The photoacoustic spectrum of the propane-generated soot is shown in Figure 1. The spectrophone response decreases 20% over the wavelength region investigated, while the structure shown in the Figure is not reproducible within the ±5% uncertainty.

The normalized light extinction spectrum is identical to the photoacoustic spectrum shown in Figure 1. At 600.0nm the extinction of the laser radiation was 2% for the 9.5cm spectrophone cavity optical path. Thus, from the Beer-Lambert Law,

$$I = I_o e^{-kx} \tag{4}$$

one can deduce an airborne soot extinction coefficient, k, at 600.0nm, of 2.1×10^{-1} m^{-1}. Since scattering at 600.0nm was found to be less than 0.1% of the total light extinction for the undiluted soot stream, the absorption coefficient for the airborne soot is identical to the extinction coefficient deduced above. Using eq. (3), and the measured soot concentration of 0.14 g m^{-3}, the specific absorption coefficient, at 600.0nm, for airborne propane-generated soot is calculated to be 1.5 m^{2}g^{-1}, with an overall uncertainty of $\pm 20\%$. This result is in agreement with those obtained (0.6-4.5 m^{2}g^{-1}) for bulk graphitic samples (9).

Engine Exhaust Measurements. Automobile exhaust, sampled from the dilution tube, was studied using the spectrophone and the 514.5nm line of an Ar ion laser.

The performance of the spectrophone on the dilution tube was checked in two ways. First, the dependence of the spectrophone signal on laser power was investigated. From eq. (1) and (2), the dependence should be linear. Figure 2 shows that this is the case for a 10-fold variation in laser output power, using a 2.4 liter Mercedes Diesel cruising at 55 mph. In addition, the effect of dilution tube flow rate on the spectrophone signal was studied. Since exhaust soot concentration is inversely proportional to the dilution tube flow rate, from eq. (2) and (3) a plot of such data should again be linear. This has been shown to be the case for dilution tube flow rates between 250 and 500 cfm. (Above 550 cfm, the pressure drop in the dilution tube adversely affected the spectrophone sampling system.)

The coupled spectrophone-dilution tube was first used to determine the absorption coefficient of airborne exhaust soot from a 2.4 liter Mercedes Diesel vs. NO$_2$ as a standard (8). The value was found to be (5.5 ± 1.0)m^{2}g^{-1} at 514.5nm. (The automobile was run on a vehicle dynamometer at a cruise speed of 55 mph, with an average particulate concentration of 0.050 g m^{-3}.)

The instrumentation has also been used to monitor exhaust soot concentrations from vehicles as a function of engine operation mode, using a vehicle dynamometer. Results showing spectrophone response for two vehicles (a Mercedes Diesel and a gasoline-powered Mercury Cougar PROCO) run through a portion of the Federal Test Procedure are presented in Figure 3. Note the 50-fold difference in the ordinate scale for the two vehicles. Integration of the spectrophone signals over the 8.5 minute test indicates that the Mercedes produced about 48 times more soot than did the Cougar. However, the two response profiles are very similar, with peaks in the spectrophone response correlating quite well with the acceleration modes in the FTP. The 3.5-minutes peak for the Diesel response (38.5mv) would correspond to a total particulate mass concentration of about 0.4 g m^{-3} in the diluted exhaust. The time resolution of the instrumentation in this mode was one second.

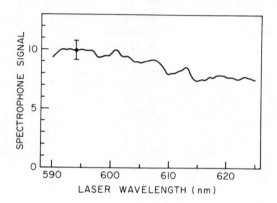

Figure 1.   The photoacoustic spectrum of airborne soot generated from a propane–
O₂ flame using a tunable dye laser light source.   This figure is adapted from Figure
2 of Ref. (8).

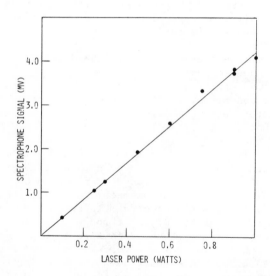

Figure 2.   Spectrophone response as a function of laser output power for exhaust
from a 2.4-L Mercedes diesel taken from a dilution tube

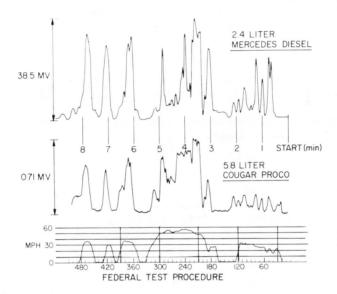

Figure 3. Spectrophone response to exhaust particulate from a 5.8-L Cougar PROCO and a 2.4-L Mercedes diesel (with a 55 mph cruise) run through a portion of the Federal Test procedure on a vehicle dynamometer. Dilution tube—300 CFM, 35°C; modulation frequency—3985 Hz; lazer—Ar+ (514.5 nm).

Conclusions

The data presented demonstrate that:
1) the photoacoustic effect can be used to measure the light absorption characteristics of airborne soot generated in combustion processes, and
2) a spectrophone can be used to monitor the concentration of carbonaceous particulate in automobile exhaust gas.

Acknowledgments

The authors would like to thank R. W. Terhune and W. R. Pierson for their continued interest and many helpful suggestions during the course of this investigation.

Literature Cited

1. Faxvog, F. R.; Roessler, D. M., Appl. Opt., 1978, 17, 2612.
2. Roessler, D. M.; Faxvog, F. R., Appl. Opt., 1979, 18, 1399.
3. Bruce, C. W.; Pinnick, R. G., Appl. Opt., 1977, 16, 1762.
4. Terhune, R. W.; Anderson, J. E., Opt. Lett., 1977, 1, 70.
5. Truex, T. J.; Anderson, J. E., "Proceedings of the Conference on Carbonaceous Particles in the Atmosphere," Lawrence Berkeley Laboratory, March 20-22, 1978; LBL-9037, CONF-7803101, UC-11, June, 1979.
6. Truex, T. J.; Anderson, J. E., Atm. Environ., 1979, 13, 507.
7. Pao, Y.-H., Ed., "Optoacoustic Spectroscopy and Detection," Academic Press: New York, 1977.
8. Killinger, D. K.; Japar, S. M., Chem. Phys. Lett., 1979, 66, 207.
9. Ergun, S.; McCarthy, J. T. ; Walline, R. E., Fuel, 1961, 40, 109.

RECEIVED February 1, 1980.

INDEX

A

A-state fluorescence, time-resolved 391
($A^2\Delta \rightarrow X^2\pi$), CH chemiluminescence 397
$a^3\Pi_u$
    produced by multiphoton UV
      excimer laser photolysis 381–387
    reactions with CO_2, $C_2X^1\Sigma_g^+$ vs. 387
    reactions with hydrocarbons and
      hydrogen, $C_2\ X^1\Sigma_g^+$ vs. 385
    reactions with O_2,$C_2X^1\Sigma_g^+$ vs. 387
$A^3\Pi_i$ state long-decay component,
    Stern–Volmer plot for quenching 394f
Absorption
    Bouger–Lambert law of 415
    CO laser resonance
      apparatus 404f
    studies of CH_2 radical
      reactions 403–409
    studies of $O(^3P)$ 1-alkynes
      reactions 403–409
    coefficient
      of I_2, extracavity 452
      for an individual line 75
      linear 427
      for soot 459
    cross section 109
    laser 7
    spectroscopy of combustion gases
      using a tunable IR diode..413–425
    spectroscopy, optical arrange-
      ment for tunable diode 414f
    line profile
      for CO in CH4–air flat flame,
        Voigt fit to 422f
      for CO in room temperature cell
        experiment, Voigt fit to 419f
      Lorentzian broadened 196
    measurements, comparison of
      concentration 55f
    measurements in a methane
      diffusion jet, multiangular ..427–433
    photon 62, 175
    in a hydrogen–oxygen–argon
      flame, one- 183–187
    in a hydrogen–oxygen–argon
      flame, two- 183–187
    profile of atoms in an atmospheric
      combustion flame 196
    system, SO $B^3\Sigma^- - X^3\Sigma^-$ 119

Absorption (*continued*)
    theory 415
    values for a six-angle case,
      reconstructed linear 430
Acetylene–air flame, LIF of the C_2
    Swan band system in 43f
Acetylene, photodissociation of
    hexafluorobutyne-2 and 383
Adaptive gridding 342
    intermittent 344
    method 343f, 347f
Air at one atmosphere pressure,
    rotational CARS spectrum of 42f
Airborne propane-generated soot 459
Algorithm based on the physics of
    the problem, choosing 340
Alkali metals in flames, ionization of .. 175
Alkali metals in flames, saturated
    fluorescence of 189–194
Alkyne reactions, mechanism 405
1-Alkyne reactions with $O(^3P)$ atoms,
    RCH diradicals formed in
      series of 405
Anti-stokes
    band 214
    Raman scattering, coherent 7, 20
    Raman spectroscopy (CARS),
      coherent 7, 20
    -to-Stokes intensity ratio from
      nitrogen, analysis of 260
    -to-Stokes vibrational Raman
      scattering, intensity ratio of 262f
Ar^+ beam, fluorescence downstream
    from 169
Ar^+ laser 168
$Ar–O_2–H_2$ flames, theoretical and
    experimental values of the
    fluorescence excitation profile
    halfwidth for 197
Asymptotic approach to modelling 340
Asymptotic integration method,
    selected 341
Atmospheric
    combustion flame, absorption
      profile of atoms in 196
    pressure cell, fluorescence spectra .. 162f
    -pressure hydrocarbon–air flames,
      primary reaction zone of 85
Atom(s)
    in an atmospheric combustion
      flame, absorption profile–Voigt 196

Atom(s) (*continued*)
　in flames, electronically excited ..175–181
　two-level 201
Atomic
　to molecular systems, extension of
　　saturation LIF spectroscopy
　　from 40
　oxygen profiles for unity initial
　　ozone mole fraction 371*f*
　species, spatial density profiles of .. 201
　species as three-level systems 65
　state by LIFS, measuring the
　　population of an excited 81
　systems, two-level model for 40, 47
Average CO product vibrational
　energies for CH_2 and O_2 average　408
Axial profiles 306*f*
Axial variation of temperature in a
　sooting flame 38*f*

B

Band strength at 273.2 K,
　CO fundamental 420*f*
Bandpass detection 14
Bandwidth, decreasing the detection .. 107
Bandwidth via different methods,
　comparison between the values
　of the laser spectral 198
Beams, CARS results with colinear 23
Beer–Lambert Law 458
Bench-scale laboratory flames 220
Benzene, microdensitometer traces of
　the OHD–RIKES single-pulse
　spectra of solutions of 323*f*
Beta (β) angles, curves for 445*f*
Boltzmann distribution in upper levels　147
Bouguer–Lambert–Beer law 427
Bouguer–Lambert law of absorption .. 415
BOXCARS 274, 311
　crossed-beam phase matching 273
　experimental arrangement 275*f*
　mode, folded 311
　spectrum(a)
　　experimental arrangement used
　　to generate 35*f*
　　of N_2 .... 277*f*
　　in a sooting flame 37*f*
　　room air CO_2 .... 313*f*
　　with background cancellation.. 314*f*
BP (*see* Brunner and Paul) 452
Branching chain reactions, Semenov's
　model and 357
Branching reactions of the hydrogen
　atom 358
Broad-band CARS 27
　signals, generation of 31*f*
　spectrum of CH_4 using OMA detec-
　　tion and a single-laser pulse 32*f*

Broad-band CARS (*continued*)
　spectrum of a gas mixture using an
　　OMA detector and a single-
　　laser pulse 34*f*
Broadening species, collision-
　broadened linewidth of 417
Brunner and Paul (BP) 452
　theory of IDLS 452
Bulk gas temperature, modeled CARS
　spectra for N_2 in a flame as a
　function of 27
Bulk gas temperature, plot of
　calculated CARS spectra of N_2
　as a function of 29*f*
Burning velocities 366
　over a range of ozone mole
　　fractions, comparison of 369*f*
(B–X), visible fluorescence from I_2 ..　169

C

C_2
　fluorescence emission signal vs. 49*f*
　fluorescence signal vs. I/I, plot of .. 51*f*
　fragments, preparation of 383
　partial excitation spectrum of the
　　Swan band system in 41
　radicals generation 381
　rate of depletion of 383
　Swan band system in an acetylene–
　　air flame, LIF of 43*f*
$C_2(a^3\Pi_u)$, rate constants for the
　disappearance of 386*t*
$C_2a^3\Pi_u$ reaction vs. quenching 385
CH
　chemiluminescence ($A^2\Delta \rightarrow X^2\pi$) ..　397
　CN investigations 293
　and CN, summary of saturated
　　fluorescence measurements for　295
　flame emission and laser-excited
　　fluorescence spectrum in
　　oxy-acetylene slot torch,
　　comparison of 294*f*
　kinetics 397
　productions 397
　radical(s)
　　apparatus for the productions
　　and detection of CH 398*f*
　　reactions important to hydro-
　　carbon combustion systems,
　　kinetics of 397–401
　　reactions at room temperature,
　　rate constants for 399
　studies on the reactions of 399
CH + N_2 reaction 399
　reaction at room temperature, pres-
　　sure effect on the rate of 400*f*
CH_2
　with CO_2 average CO product
　　vibrational energies for 408

CH_2 (*continued*)
with CO_2 reactions of 406
with O_2 average CO product
  vibrational energies for 408
with $O(^3P)$, average CO product
  vibrational energies for 408
with $O(^3P)$ reactions of 406
radical reactions, CO laser reson-
  ance absorption studies of ..403–409
$CH_2 + CO_2$, vibrational energy distributions of the CO formed in 407*f*
C_2H_2–air aspirating slot burner, plot
  of fluorescence signal vs. Na
  concentration in 53*f*
C_2H_2–O_2–N_2 flame, comparison of
  SH $A^2\Sigma^+ - X^2\Pi$ fluorescence and
  synthetic emission spectra for 118*f*
C_2H_2–O_2N_2 flame, laser excitation
  spectra for OH $A^2\Sigma^+ - X^2\Pi$ in 112*f*
CN
  flame emission and laser-excited
    fluorescence spectrum in a
    nitrous oxide-acetylene slot
    torch, comparison of 296*f*
  fluorescence intensity with laser
    spectral irradiance in a nitrous
    oxide-acetylene slot torch,
    variation of laser-excited 297*f*
  investigations, CH 293
  summary of saturated fluorescence
    measurements for CH and 295
C_2O
  fluorescence decay for 392*f*
  lifetime and quenching data 395
  pseudo-first-order disappearance
    rate constants for ground-state 394*f*
$C_2 + O_2$, product channels for 386*t*
$C_2O(A^3\Pi_i)$
  fluorescence lifetimes and quenching
    rate constants 389
  quenching rate constants for 395
  radiative lifetimes for 395
  $(X^3\Sigma^-)$ pulsed-laser studies of
    the kinetics of 389–396
$C_2O(A^3\Pi_i - X^3\Sigma^-)$, laser-induced
  fluorescence excitation spectrum
  of 392*f*
$C_2O(X^3\Sigma^-)$ reaction kinetics 393
$C_2O(X^3\Sigma^-)$ reaction rate constants
  at 298 K 395
$C_2(X\Sigma_g^+) + H_2$, plot of K^1 vs.
  pressure for 384*f*
$C_2(X^1\Sigma_g^+)$
  fragments, first-order decay for 383
  populations probing of 383
  produced by multiphoton UV
    excimer laser photolysis 381–387
  psuedo-first-order decay in the
    presence of excess H_2 .... 384*f*

$C_2(X^1\Sigma_g^+)$ (*continued*)
  rate constants for the disap-
    pearance of 386
  vs. $a^3\Pi_u$ reactions
    with CO_2 387
    with hydrocarbons and hydrogen 385
    with O_2 387
Carbon tetrachloride (CCl_4) 322
  microdensitometer traces of RIKES
    spectra of solutions of
    cyclohexane in 321*f*
  microdensitometer traces of spectra
    from solutions of cyclohexane
    in 324*f*
CARS (*see* Coherent anti-Stokes
    Raman scattering)
Cell experiments, room-temperature .. 417
Chemical lasers, flowing 167
Chemical reactivity differences in
    states 40
Chemistry–hydrodynamics coupling
    and feedback 338
Chromatix CMX-4, (flashlamp-
    pumped tunable dye laser) 104
Circuits, sample-and-hold (S/H) 242
Circuitry, electronic signal
    conditioning 242
Closed form similarity solution 345
Coefficient(s)
  expressions for the rate 372*f*
  kinetic 367
  ratio of the values of input 371*f*
  transport 368
Coherene anti-Stokes Raman scatter-
    ing (CARS) 7, 19, 20, 235, 272, 320
  advantages of 23
  broad-band 27
  characteristics 22
  coherent anti-Stokes Raman
    spectroscopy 272
  for combustion diagnostic 303
  combustion system of 315
  detection sensitivity of 312
  developments, future 36
  diagnostics of reactive media
    at ONERA 311–318
  experimental setups for 23
  experiments, signal intensity in 20
  flame spectrum, CO 36
  generation of signals in 20
  measurement(s)
    accuracy of 312
    performed in a sooting flame 303
  in simulated practical combustion
    environments 303–310
  in sooting flames, capability of .. 276
  of temperature from H_2 and O_2 .. 280
  for N_2 in a flame as a function of the
    bulk gas temperature, modelled 27

Coherene anti-Stoke Raman scattering (CARS) (*continued*)
N_2 thermometry, accuracy of 276
polarization-sensitive 288
and radiation-corrected thermocouple-derived temperature profiles, comparison of 304
research effort at ONERA 311
resonance 315
results with colinear beams 23
signal(s)
for D_2 gas within an electric discharge lamp 26*f*
generation
of broad-band 31*f*
in gases 34*f*
phase-matching diagrams for .. 21*f*
H_2 concentration in N_2 gas plot of 25*f*
levels 272
nonresonant and resonant 7
resonant 7
spatial resolution of 311
spectral resolution 312
spectrometer, development of 311
spectrometer in combustor facility, quantel 316*f*
spectrum(a)
of air at one atmosphere pressure, rotational 42*f*
of CH_4 using OMA detection and a single-laser pulse, broadband 32*f*
of CO_2 in the postflame region of a $CO-O_2$ flame, scanned 291*f*
comparison of averaged and single-pulse N_2 .... 282*f*
concentration measurements and 288
of D_2 gas 23
experimental approach to 274
of a gas mixture using on OMA detector and a single-laser pulse, broad-band 34*f*
of H_2 at 1% concentration, computed temperature sensitivity of 284*f*
of H_2 in a flat H_2-air diffusion flame 285*f*
H_2O
in a methane–air flame 287*f*
in a premixed methane–air flame 30*f*
measuring 24*f*
of N_2
comparison of a single-pulse and averaged 278*f*
as a function of bulk gas temperature, plot of calculated 29*f*
gas in the combustion zone of a homogeneous flat flame burner 28*f*

Coherene anti-Stoke Raman scattering (CARS) (*continued*)
spectrum(a) (*continued*)
of N_2 (*continued*)
probed within the homogeneous region of a flat flame burner 27
spatial variation of temperature from averaged 281*f*
of O_2
in H_2-air diffusion flame 290*f*
at 2000 K 289*f*
at one atmosphere pressure, rotational 39*f*
in the region of the CO Q-branch from a methane–air flat flame 39*f*
single-pulse 304
measurements 27
using one atmosphere of air 36
vibration–rotation H_2 280
temperature and average thermocouple temperature comparison of 309*f*
temperature profile 283*f*
theory and application of 272
vs. saturated-laser fluorescence 271
CO
CARS flame spectrum 36
in CH_4-air flat flame, Voigt fit to absorption line profile for 422*f*
collision halfwidth in CH_4-air flat flame as a function of rotational quantum number 424*f*
collision halfwidth in combustion gases on the vibrational and rotational quantum numbers .. 421
concentration, discrepancy between predicted and measured 97
formed in $O(^3P) + C_2H_2$ reactions, vibrational energy distributions of 404*f*
formed in $O(^3P) + C_4H_9C_2H$ reactions, vibrational energy distributions of 404*f*
in fuel-rich flames 421
fundamental band strength at 273.2 K 420*f*
laser resonance absorption apparatus 404*f*
studies of CH_2 radical reactions 403–409
studies of $O(^3P) + 1-$alkynes reactions 403–409
and NO, flat flame burner experiments measuring 418
partial pressure in CH_4-air flat flame as a function of fuel-air equivalence ratio 423*f*

CO (*continued*)
product vibrational energies,
for CH₂
with CO₂ average 408
with O₂ average 408
with O(³P) average 408
production at 300 K for O(³P)
+ 1-alkynes, absolute rate
constants for 408
productions at 300 K for O(³P)
+ 1-alkynes, average CO
vibrational energies 408
profiles for a CH₄–air flame 96*f*
Q-branch from a methane–air
flat flame, CARS spectra in
the region of 39*f*
in room temperature cell experi-
ment, Voigt fit to absorption
line profile for 419*f*
vibrational energies for CO pro-
duction at 300 K for O(³P)
+ 1-alkynes 408
CO–O₂ flame, scanned CARS
spectra of CO₂ in the postflame
region of 291*f*
CO₂
BOXCARS spectra room air 313*f*
with background cancellation 314*f*
in the postflame region of a
CO–O₂ flame, scanned
CARS spectra of 291*f*
profiles for a CH₄–air flame 94*f*
reactions of CH₂ with 406
Collision-broadened linewidth of
the broadening species 417
Collision halfwidth(s) 417
temperature dependence of 421
Collisional
de-excitation 63
ionization 175
of sodium atoms 183–187
quenching
of the laser-excited state 89
rates 6
by rotational relaxation 107
by vibrational relaxation 107
redistribution 77
of excited-state population fol-
lowing excitation
of OH 13
Combusting stratified charge engine,
nitrogen density in 259
Combusting stratified charge engine,
temperature in 259
Combustion
chamber, nitrogen density vs.
crank angle of 264
of cyclic nitramines, gas-phase
kinetic mechanisms in 365
detailed modelling of 331–354

Combustion (*continued*)
development of kinetic models of
hydrocarbon 85
diagnostic(s) 3
CARS for 303
preliminary assessment of the
topographic reconstruction
approach for 430
environments, CARS measure-
ments in simulated
practical 303–310
fuel spray 443
fundamental processes in 335
gases using a tunable IR diode
laser, absorption spectro-
scopy of 413–425
gases on the vibrational and rotational quantum numbers,
dependence of the CO colli-
sion halfwidth in 421
hydrocarbon-fueled 288
hydrogen 280
intermediates detected by LIF 12*t*
laser chemistry, and 3–17
modelling with measurement
capabilities, ordering of
predictive needs for 213
probes, laser 4
energy-level diagrams for
spectroscopic 5*f*
processes, photoacoustic spectroscopy to characterize and
monitor soot in 457–462
properties, Raman scattering
measurements of 207–228
single-pulse N₂ spectrum recorded
on OMA during 305*f*
system(s) 3
of CARS 315
kinetics of CH radical reactions
important to hydro-
carbon 397–401
modelling 332
NO formation in high tempera-
ture hydrocarbon 399
PCAH in 159
thermometry, H₂ and 280
tunnel, fan-induced 222*f*
co-flowing turbulent jet 220
zone of a homogeneous flat flame
burner, CARS spectrum of
N₂ gas in 28*f*
Combustor
facility, quantel CARS spectro-
meter for 316*f*
-optical layout 217
probing, spatially precise laser
diagnostics for practical 275–299
time-averaged spectrum of N₂ on
exit plane of 317*f*

Complicated reactions and flow 339
Computer-generated spectrum of N_2
    at 1700 K onto measured
    spectra, overlay of 305*f*
Computerized tomography 427
Concentration(s)
    and absorption measurements,
        comparison of 55*f*
    measurements and the CARS
        spectrum 288
    measurements, LIFS and 80
    profiles, coupled radical reactions
        that account for 124
    profiles in a H_2–O_2–N_2 flame with
        1% H_2S 123*f*
    and temperature, comparison of
        CH_4–O_2 flame species 93*f*
Continuous wave (CW) 274
    laser
        fast turbulent mixing in gases
            using 247–253
        scattering for pdf, basic quanti-
            ties in analyses of 248
        technique for pdf measure-
            ments, pulsed-laser Raman
            technique 253
    intracavity dye laser spectroscopy
        (CWIDLS) 451–454
Cougar PROCO and Mercedes
    diesel, spectrophone response to
    exhaust particulate from 461*f*
Coupled radical reactions that
    account for concentration
    profiles 124
Coupling and feedback, chemistry–
    hydrodynamics 338
Crank angle of the combustion
    chamber, nitrogen density vs. 264
Crossed-beam phase matching
    BOXCARS 273
CW (*see* Continuous-wave)
CWIDLS (*see* Continuous-wave
    intracavity dye laser spectro-
    scopy)
Cyclic nitramines, gas-phase mech-
    anisms in the combustion of 365
Cyclohexane solutions
    in CCl_4, microdensitometer
        traces of spectra from 324*f*
        RIKES 321*f*
        IRS-flashed 322
    microdensitometer trace(s) of
        spectra from 325*f*
        IRS and OHD–RIKES 326*f*

D

D_2 gas, CARS spectrum of 23
De-excitation, collisional 63

Density
    measurement, vibrational Raman
        scattering methods for 209
    profiles laser-excited fluorescence,
        flame temperatures 199–203
    profiles of laser-excited fluores-
        cence, plasma tempera-
        tures 199–203
    and temperature with velocity,
        correlations of 220
Depletion of the pair population 139
Detailed modelling
    of combustion 331–354
    physical complexity 335
    purpose of 333
Detectability
    interferences, LIFS 72
    limits for flame conditions 74*f*
    limits of LIFS 71
Detection bandwidth, decreasing 107
Detection sensitivity of CARS 312
Diagnostic conditions for making
    routine measurements on flame
    series, development of 109
Diatomic larger molecule spectrum 81
Diatomic molecules, vibrational
    Raman scattering from 235
Diffusion
    flame
        CARS spectra of H_2 in a flat
            H_2–air 285*f*
        CARS spectra of O_2 in H_2–air 290*f*
        radial temperature profiles in
            laminar propane 279*f*
        temperature measurements in
            H_2–air 286*f*
        jet, methane–argon asymetrical
            diffusion jet, methane–argon .. 428
        velocity, Stefan–Maxwell relation .. 366
Dilution tube, spectrophone– 459
Dirty flames 36
Disappearance rate constants for
    ground-state C_2O, pseudo-
    first-order 394*f*
Dispersion function, Lorentzian 195
Distribution curves of droplet
    diameters 466, 448*f*
Doppler-broadened linewidth 417
Droplet
    diameters, distribution curves 446, 448*f*
Doppler velocimeter, dual-beam
    laser 443
Double-probe used in water sprays 447*f*
    size measurement in an industrial
        furnace, optical arrange-
        ment for 447*f*
    size measurements in reacting
        flows by laser inter-
        ferometry 443–449

Dual-beam laser Doppler
 velocimeter 443
 optics, geometry of 445*f*
Dummy-level population 144
 as a function of transfer cross
  section 143*f*
Dye laser(s) 176
 laser-pumped 293
 Nd:YAG pumped 41
 system for nitric oxide fluorescence measurements in
  $CH_4 + O_2 + N_2$ flame 154*f*, 155*f*
 tunable 103, 160
  flashlamp-pumped 183
  Chromatix CMX-4 104
  phase-R corporation
   DL-1400 160
Dynamic excitation–physical
 quenching cycle 190

E

Elastic molecular light scattering,
 information from 208
Electronic
 ground-state rotational temperature measurement 89
 polarization of a medium 319
 quenching of the excited-state
  level 40
 signal conditional circuitry 242
 and vibrational structure, OH
  energy level diagram— 68*f*
Electronically excited atoms in flames
Electronically excited atoms
 in flames 175–181
Emission
 to lower levels, spontaneous 62
 in OH in a CH_4–air flame,
  fluorescence scans of 16*f*
 process, induced 62
Energy
 density
  required to saturate the
   excited transition 69
  saturation 63, 196
  for the two-level model saturation..
  for the two-level model
   saturation 69
 flux, saturation 71
 of initial rotational state, plot of
  LIF intensity per transition
  strength vs. 91*f*
 laser pulse 160
 -level diagram(s) 66*f*
  –electronic and vibrational
   structure, OH 68*f*
  for laser probe methods 4

Energy (*continued*)
 -level diagram(s) (*continued*)
  for MgO 41
  for low-lying electronic
   single states of 44*f*
  —rotational structure, OH 68*f*
  sodium 66*f*
  for spectroscopic laser
   combustion probes 5*f*
  for the OH molecule,
   diagram of 90*f*
 -state diagram for lithium 193*f*
Engine
 exhaust measurements 459
 nitrogen density in a combusting
  stratified charge 259
 temperature in a combusting
  stratified charge 259
 time-resolved Raman spectroscopy
  in a stratified charge 259–267
Equations describing premixed,
 laminar, unbounded flame for
 a multicomponent ideal gas
 mixture 365
Equilibration test in fuel-rich
 H_2–O_2–N_2 flames with 1% H_2S
 of $H_2 + SO_2 = SO + H_2O$ 127*f*
 of $S + H = SH + SH$ 126*f*
 of $SH + OH = SO + H_2$ 128*f*
 of $SO_2 + 2H_2 = SH + OH + H_2O$ .. 129*f*
Equilibrium test of $H_2 + OH = H_2O$
 in H in fuel-rich H_2–O_2–N_2
 flames 115*f*
Ethylene–air flame, fluorescence
 spectra: sample injected in 164*f*
Excitation
 dynamic, LIFS and 80
 fluorescense spectrum from
  near-resonant 78*f*
 from Level 1 to 2 steady-state rate
  equations 65
 in NO, fluorescence 298
 of the $Na(3^2P_{3/2})$ level, saturated .. 189
 profile(s), fluorescence 184
 saturation broadening in
  flames 195–198
 saturation broadening in
  plasmas 195–198
  halfwidth(s) 196
  for the Ar–O_2–H_2 flames,
   theoretical and experimental values of 197
 results, comparison of 178*f*
 scan of NH in a NH_3–O_2 flame 10*f*
 short-pluse 145
 of sodium in a H_2N–air flame,
  opto-galvanic signal for 177*f*
 of sodium, LIFS for near-resonant 77
 source, lasers as 62

Excitation (*continued*)
  spectrum(s)
    for NO 153
    for nitric oxide in $CH_4 + O_2 + N_2$
      flame laser 157*f*
    for nitric oxide in N_2, laser 156*f*
    of the Swan band system in C_2,
      partial 41
Excited
  atomic state by LIFS, measuring
    the population of 81
  level, fluorescence spectrum at
    the directly 151
  -state level, electronic quenching of 40
  -state population distribution for
    OH 70*f*
  transition, energy density required
    to saturate 69
  vibrational level population 131
Experimental approach to the
  CARS spectrum 274
Extracavity absorption coefficient
  of I_2 452

F

Fabry–Pedro etalon 418
Fan-induced combustion tunnel 222*f*
  co-flowing turbulent jet 220
Far field isolator and 532 nm pulsed
  multipass cell experiment 256*f*
Fast turbulent mixing in gases using
  a CW laser 247–253
Feedback, chemistry–hydrodynamics
  coupling and 338
Flame(s)
  arrival, pdf's of nitrogen number
    density near time of 266*f*
  bench–scale laboratory 220
  burner experiments measuring CO
    and NO, flat 418
  CH_4–air
    concentration profile(s) 378*f*
    experimental results for a
      fuel-lean 92
    hydroxyl, for a stoichiometric 98
    laser probes of premixed
      laminar 85–101
    NO 376*f*
    postulated mechanism for 87*t*
    radiative trapping of sodium in .. 76*f*
    rate constant measurement in 375
    species profiles for premixed
      laminar 86
    temperature profiles for premixed,
      laminar 86
    theoretical results for a fuel-lean 92
  $CH_4 + O_2 + N_2$ 153

Flame(s) (*continued*)
  chemistry, LIF as tool for the
    study of 103–129
  chemistry of the sodium system 50
  CO in fuel-rich 421
  comparison of SH $A^2\Sigma^+ - X^2\Pi$
    fluorescence and synthetic
    emission spectra for
    $C_2H_2 - O_2 - N_2$ 118*f*
  conditions, detectability limits for .. 74*f*
  detailed modeling of premixed,
    laminar, steady-state 365–372
  determination of temperatures in
    premixed laboratory 231
  diagnostics, applications of laser-
    enhanced ionization in 187
  diagnostic methods, Raman scatter-
    ing flame 239
  of different stoichiometries,
    quenching rate along 135*f*
  dirty 36
  distribution of lithium 189
  electronically excited atoms in 175–181
  fluorescence excitation profiles,
    saturation broadening in 195–198
  fluorescence measurements, experi-
    mental system for the NO 153
  as a function of the bulk gas tem-
    perature, modeled CARS
    spectra for N_2 in 27
  fluorescence spectra: sample in-
    jected in an ethylene–air 164*f*
  gas
    properties, laser light-scattering
      techniques for measuring 207
    temperature determination 239
    translational temperature of 17
  Gedanken
    calculation, space and time
      scales in 337*f*
    experiment 336
    important scales in 338
  geometry of 86
  with 1% H_2S, SH $A^2\Sigma^+ - X^2\Pi$
    fluorescence spectra in 118*f*
  with 1% H_2S, SH $A^2\Sigma^+ - X^2\Pi$
    laser excitation spectra in 117*f*
  hydrogen–oxygen–nitrogen
    $(H_2 - O_2 - N_2)$ 103
  fluorescence profile(s) in
    with added H_2S, SO_2 $^1B_2 - ^1A_1$ 122*f*
    SH $A^2\Sigma^+ - X^2\Pi$ 120*f*
    with added H_2S, SO $B^3\Sigma^- - X^3\Sigma^-$ 121*f*
    OH $A^2\Sigma^+ - X^2\Pi$ 113*f*
    for SH in a rich 119
  fuel-rich
    chemical model of the surfur
      chemistry 119

Flame(s) (*continued*)
  hydrogen–oxygen–nitrogen
    (H_2–O_2–N_2) (*continued*)
    fuel-rich (*continued*)
      equilibrium of H_2 + OH =
        H_2O + H in 115*f*
      with 1% H_2S, equilibration test
        of H_2 + SO_2 = SO + H_2O
          in 127*f*
        of S_2 + H_2 = SH + SH in .. 126*f*
        of SH + OH = SO + H_2 in 128*f*
        of SO_2 + $2H_2$ = SH + OH
          + H_2O in 129*f*
    Padley–Sugden burner,
      premixed 103
    with 1% H_2S
      added, fluorescence spectrum
        for SO_2 and SO in 120*f*
      added, SO $B^3\Sigma^-$–$X^3\Sigma^-$
        fluorescence spectrum for 121*f*
      concentration profiles in 123*f*
    laser excitation scan of 114
  hydrogen-rich 190
  ionization of alkali metals in 175
  laser
    excitation spectra for
      OH $A^2\Sigma^+$–$X^2\Pi$ in a
        C_2H_2–O_2–N_2 .... 112*f*
    fluorescence, nitric oxide
      detection in 153
    measurements, species amenable
      to 56*f*
    LIFS in 61–82
      applied to OH 131–136
      of PCAH 159–164
    luminescence 438
    model, assumptions made in 97*t*
    modelling for the data formations
      in 176
    for a multicomponent ideal gas mix-
      ture, equations describing pre-
      mixed, laminar,
      unbounded 365
    OH profile through the low-
      pressure flat 135*f*
    one-photon absorption in a
      hydrogen–oxygen–argon 183–187
    optical system for laser fluorescence
      measurement in 110*f*
    pdf for temperature × velocity for
      turbulent diffusion 219*f*
    photoacoustic spectrum of airborne
      soot generated from
      propane–O_2 .... 460*f*
    pressure, quenching rate vs. 134*f*
    probability density functions of
      temperature for H_2–air
      turbulent diffusion 219*f*

Flame(s) (*continued*)
  laser (*continued*)
    propagation problem using compu-
      tational expense of performing 344
    quenching studies in low-pressure .. 133
    recombination zone, measurement
      of OH concentration and
      temperature in 98
    RMS temperature profile in
      turbulent jet diffusion 440*f*
    on saturated fluorescence of alkali
      metals in 189–194
    series, development of diagnostic
      conditions for making routine
      measurements on 109
    and short-duration laser pulses,
      low-pressure 131
    sooting 159
      capability of CARS for measure-
        ments in 276
      CARS measurements performed
        in 303
    species
      experimental arrangement used
        to measure LIF signals from 42*f*
      and temperature of H_2–air flame,
        plots of 222*f*
    spectrum, CO CARS 36
    statistical nature of the turbulent .. 240
    studies of LIF on OH in 13
    temperature(s)
      density profiles laser-excited
        fluorescence 199–203
      measured 202
      measurement methods 200
      profile in turbulent jet diffusion .. 440*f*
    two-photon absorption in a
      hydrogen–oxygen–argon 183–187
    via Rayleigh scattering, temperature
      measurement in turbulent ..435–441
Flashlamp-pumped tunable dye laser.. 183
  Chromatix CMX-4 104
Flashlamp-pumped systems 41
Flashed cyclohexane solutions, IRS .. 322
Flat flame
  burner, CARS spectrum of N_2 gas
    in the combustion zone of
    a homogeneous 28*f*
  burner experiments measuring CO
    and NO 418
  CARS spectra in the region of the
    CO Q-branch from a
    methane–air 39*f*
  OH profile through the low-pressure 135*f*
Flow(s)
  approximation, slow 345
  calculations, reactive 340
  complicated reactions and 339

Flow(s) (continued)
  equations, laminar macroscopic 339
  equations, nonlinear time-dependent
    slow 345
  modelling, problems in reactive 333
  nonuniformity 169
  Stokes nitrogen density measure-
    ments in a turbulent 259
  visualization in supersonic 167–172
Flowing chemical lasers 167
Fluctuation measurement capabilities
  for laser source characteristics,
  comparison vibrational Raman
  scattering 211
Fluid dynamics equations 11
Fluid layer, Rayleigh–Taylor unstable 352
Fluoranthene, fluorescence spectra for 160
  of alkali metals in flames,
    on saturated 189–194
  decay for C_2O 392f
  detection system 160
  downstream from the Ar^+ beam 169
  efficiency expression 106
  efficiency, quenching term in 106
  emission in OH, in CH_4–air flame,
    rotationally resolved 15f
  emission signal vs. C_2 .... 49f
  excitation
    in NO 298
    profile(s)
      halfwidth of 196
      flames, theoretical and ex-
        perimental values of .. 197
      saturation broadening
        in flames 195–198
      saturation broadening
        in plasmas 195–198
    spectrum of the MgO $B^1\Sigma^+X^1\Sigma^+$
      transition in an C_2H_2–air
      aspirating slot burner 45f
    spectra and reaction rate con-
      stants, experimental system
      used for measuring 390f
  I_2 168
    (B–X) 169
  induction 148
  intensity(ies) 137, 190
    (I_t) 106
  S_2 114
  laser
    detection system, multiphoton
      UV-photolysis 382f
    excitation in OH, multilevel
      model of response to 137–144
    -excited
      flame temperatures density
        profiles of 199–203
      plasma temperatures density
        profiles of 199–203

Fluoranthene, fluorescence spectra for
  (continued)
  laser (continued)
    -excited (continued)
      spectrum in a nitrous oxide–
        acetylene slot torch, com-
        parison of CN flame
        emission and 296f
      spectrum in an oxy-acetylene
        slot torch, comparison of
        CH flame emission and .. 294f
    exciting and detecting wave-
      lengths 108t
    -induced 389
      excitation spectrum of
        $C_2O(A^3\Pi_i–X^3\Sigma^-)$ 392f
      spectroscopy, saturated 19, 36
      spectrum of $A^2\Delta–X^2\pi$ transi-
        tion of the ground-state
        CH radicals 400f
    nitric oxide detection in flames by 153
    saturated 189
      advantages of 292
      CARS vs. 271
      studies of sodium in fuel-rich
        $H_2–O_2–N_2$ flames 189–194
      theory of 292
    techniques for NO 153
  lifetimes and quenching rate con-
    stants, $C_2O(A^3\Pi_i)$ 389
  measurements 106
    for CH and CN, summary of
      saturated 295
    in $CH_4 + O_2 + N_2$ flame, dye
      laser system for nitric
      oxide 154f, 155f
    experimental system for the
      NO flame 153
    of the hydroxyl radical,
      saturated 145–151
  multiphoton excitation of 9
  photons 132
  power, collected 61, 64
  profiles
    for $H_2–O_2–N_2$ flames
      with added H_2S, SO_2 $^1B_2–^1A_1$ 122f
      with added H_2S, SO $B^3\Sigma^-–X^3\Sigma^-$ 121f
      with H_2S, SH $A^2\Sigma^+–X^2\Pi$ 120f
      OH $A^2\Sigma–X^2\Pi$ 113f
      S_2 $B^3\Sigma^-_u–X^3\Sigma^-_g$ 166f
    for SH in a rich $H_2–O_2–N_2$ flame 119
    for SO 119
  pulse(d) 80
  shape, SO_2 119
  radiance, evaluation of 195
  scans of emission in OH in a
    CH_4–air flame 16f
  signal 61
    detection 133
    intensity 47

Fluoranthene, fluorescence spectra for
(*continued*)
signal (*continued*)
for PCAH 163
and saturation 151
vs. I/I, a plot of the C_2 .... 51*f*
vs. laser power 150*f*
vs. Na concentration in a C_2H_2–
air aspirating slot burner,
plot of 53*f*
vs. time, normalized laser pulse .. 147*f*
spectrum (a)
atmospheric pressure cell 162*f*
in flames with 1% H_2S, SH
$A^2\Sigma$–$X^2\Pi$ 118*f*
for fluoranthene 160
for H_2–O_2–N_2 flame with 1% H_2S
added, SO $B^3\Sigma^-$–$X^3\Sigma^-$ 121*f*
from near-resonant excitation 78*f*
for pyrene 160
sample injected in an ethylene–
air flame 164*f*
spectrum
of the MgO $B^1\Sigma^+A^1\Pi$ transition 46*f*
for $S_2B^3\Sigma^-_u$–$X^3\Sigma^-_g$ in a
H_2–O_2–N_2 flame containing
1% H_2S 115*f*
of SH excited at 323.76 mn 114
for SO_2 and SO in H_2–O_2–N_2
flame with 1% H_2S added .. 120*f*
studies of lithium in fuel-rich
H_2–O_2–N_2 flames 189–194
and synthetic emission spectra for a
C_2H_2–O_2–N_2 flame, compari-
son of SH $A^2\Sigma^+$–$X^2\Pi$ 118*f*
time-resolved A-state 391
trapping in sodium, comparison of
Rayleigh and 79*f*
two-line 80
Formations in flames, modelling
the data 176
Forward scatter geometry 445*f*
Four-level molecular model ..145, 146*f*, 147
Fourier transforms of the function in
the two-coordinate systems 427
Fractional
distribution of sodium and lithium
over states in a H_2–O_2–N_2
flame 193*f*
populations as a function of rota-
tional quantum number 140*f*, 141*f*
uncertainty in photon flux 72
Free radicals, rate of methane oxidation controlled by 357–363
Frozen excitation model 145
Fuel
-air equivalence ratio, CO partial
pressure in CH_4–air flat flame
as a function of 423*f*

Fuel (*continued*)
consumption, rate equation for 358
-lean CH_4–air flame, experimental
results for 92
-lean CH_4–air flame, theoretical
results for 92
-oxidizer mixture, ignition of 344
-rich flames, CO in 421
-rich H_2–O_2–N_2 flames
chemical model of the sulfur
chemistry in 119
equilibrium of H_2 + OH =
H_2O + H 115*f*
with 1% H_2S, equilibration test
of H_2 + SO_2 = SO + H_2O in .. 127*f*
of S_2 + H_2 = SH + SH in 126*f*
of SH + OH = SO + H_2 in 128*f*
of SO_2 + $2H_2$ =
SH + OH + H_2O in 129*f*
premixed 103
saturated-laser fluorescence
studies of lithium in 189–194
saturated-laser fluorescence
studies of sodium in 189–194
spray combustion 443
Fundamental processes in combustion 335

G

Gas(es)
CARS signal generation in 34*f*
evolution 322
kinetic quench rate 107
mixture, equations describing pre-
mixed, laminar, unbounded
flame for a multicomponent
ideal 365
mixture using an OMA detector and
a single-laser pulse, broadband
CARS spectrum of 34*f*
-phase kinetic mechanisms in the
combustion of cyclic nitra-
mines 365
Raman scattering from 247
temperature, plot of calculated
CARS spectra of N_2 as a
function of bulk 29*f*
temperature and species concentration, spontaneous Raman
scattering for measuring 255
using a CW laser, fast turbulent
mixing in 247–253
Gasoline powered mercury cougar
PROCO 459
Gaussian laser profile–Voigt atom
profile 196
Gedanken flame calculation, space
and time scales in 337*f*
Gedanken flame experiment 336

Geometric complexity associated with
real systems 334
Global
implicit differencing 340
quenching ionization rates for
n-manifold states 179
quenching ionization rates for Na .. 179
rate constants for $l \geq 2$ states 180t
Grid, Lagrangian 351
Grid, triangular 351
Gridding, adaptive 342
intermittent 344
method 343f
injected 347f
Ground-state C_2O, pseudo-first-order
disappearance rate constants for 394f
Ground-state rotational temperature
measurement, electronic 89

H

HCN + N, formation of 401
HO + $O_3 \rightarrow HO_2 + O_2$, rate constant
for 337f
H_2N + air flame, opto-galvanic signal
for excitation of sodium 177f
H_2–O_2 premixed flame, temperature
analysis plot for rotational
Raman scattering from O_2 in 234f
H_2–O_2–AR (see Hydrogen–oxygen–
argon)
H_2 + SO_2 = SO + H_2O in fuel-rich
H_2–O_2–N_2 flames with 1% H_2S,
equilibration test of 127f
Half-cylinder
instantaneous pressure distributions
on 353f
pressure variations over 350f
triangular grid early in a calculation
of wave flow over 350f
Halfwidth(s)
for the Ar–O_2–H_2 flames, theoretical and experimental values of
the fluorescence excitation
profile 197
collision 417
of the fluorescence excitation
profile 196
temperature dependence 421
Heat release during methane oxidation, processes governing the
rate of 357
Hexafluorobutyne-2 and acetylene,
photodissociation of 383
High-intensity limit (saturation) 64
High-principal quantum numbers
ionization for 179
HST theory of IDLS 452

Homogeneous flat flame burner,
CARS spectrum of N_2 gas in the
combustion zone of 28f
Homogeneously broadened Raman
transition, resonant susceptibility
associated with 20
Hydrocarbon(s)
–air flames, primary reaction zone
of atmospheric-pressure 85
combustion
development of kinetic models of   85
systems, kinetics of CH radical
reactions important to 397–401
systems, NO formation in high
temperature 399
-fueled 288
and hydrogen, $C_2 X^1\Sigma^+_g$ vs. $a^3\Pi_u$
reactions with 385
Hydrodynamics coupling and feed-
back, chemistry 338
Hydrodynamics framework,
Lagrangian 339
Hydrogen
–air
diffusion flame, temperature
measurements in 286f
flame, plots of flame species and
temperature for 222f
turbulent diffusion flame, probability density functions of
temperature for 219f
atom, branching reactions of 358
$C_2 X^1\Sigma_g^+$ pseudo-first-order decay in
the presence of excess 384f
$C_2 X^1\Sigma_g^+$ vs. $a^3\Pi_u$ reactions with
hydrocarbons and 385
CARS spectrum(a)
vibration–rotation 280
at 1% concentration, computed
temperature sensitivity of
the CARS spectrum of 284f
air diffusion flame, O_2 in 290f
in a flat H_2–air diffusion flame .. 285f
combustion 280
concentration, discrepancy between
predicted and measured 97
concentration in N_2 gas, plot of
CARS signal 25f
–oxygen–argon (H_2–O_2–AR) 183
flame, one-photon absorption
in 183–187
flame, two-photon absorption
in 183–187
and O_2, CARS measurements of
temperature from 280
+ OH = H_2O + H in fuel-rich
H_2–O_2–N_2 flames, equilibrium
test of 115f

Hydrogen (*continued*)
  –oxygen–nitrogen (H$_2$–O$_2$–N$_2$)
    flame(s) 103
  containing 1% H$_2$S, fluorescence
    spectrum for S$_2$B$_u^3$–X$^3\Sigma_g^-$ .. 115*f*
  fluorescence profiles in
    with added H$_2$S, SO$_2$ ^{1}B$_2$–^{1}A$_1$.. 122*f*
    with added H$_2$S, SO
      B$^3\Sigma^-$–X$^3\Sigma^-$ 121*f*
    with H$_2$S, SH A$^2\Sigma^-$–X$^2\Pi$ 120*f*
    OH A$^2\Sigma^+$–X$^2\Pi$ 113*f*
  fractional distribution of sodium
    and lithium over states in 193*f*
  in fuel rich
    chemical model of the sulfur
      chemistry 119
    equilibrium test of H$_2$ + OH =
      H$_2$O + H 115*f*
    fluorescence profiles for SH 119
  with 1% H$_2$S, equilibrium test
    of H$_2$ + SO$_2$ = SO + H$_2$O .. 127*f*
    of S$_2$ + H$_2$ = SH + SH 126*f*
    SH + OH = SO + H$_2$ 128*f*
    of SO$_2$ = 2H$_2$ =
      SH + OH + H$_2$O 129*f*
  premixed 103
  with 1% H$_2$S, added, SO
    B$^3\Sigma^-$–X$^3\Sigma^-$ fluorescence
      spectrum for 121*f*
  with 1% H$_2$S, concentration
    profiles in 123*f*
  laser excitation scan of 114
  saturated laser fluorescence
    studies of lithium 189–194
  saturated laser fluorescence
    studies of sodium 189–194
  sodium line reversal temperature
    profiles in 110*f*
  profiles for a CH$_4$–air flame 96*f*
  sodium line-reversal temperature
    profile measurements were for   109
Hydroxy
  concentrations
    LIF and 89
    as measured by LIF 99*f*
  profile for a stoichiometric
    CH$_4$–air flame 98
  radical (OH$^-$) 12, 67, 137, 145–151
    A$^2\Sigma^+$, quenching rates of 111
    A$^2\Sigma^+$–X$^2\Pi$ in a C$_2$H$_2$–O$_2$N$_2$ flame,
      laser excitation spectra for.. 112*f*
    A$^2\Sigma^+$–X$^2\Pi$ fluorescence profiles
      in a H$_2$–O$_2$–N$_2$ flame 113*f*
    in a CH$_4$–air flame, fluorescence
      scans of emission in 16*f*
    in CH$_4$–air flame, rotationally re-
      solved fluorescent emission
      in 15*f*

Hydroxy (*continued*)
  radical (OH$^-$) (*continued*)
    collisional redistribution of
      excited-state population fol-
      lowing laser excitation of   13
    concentration measurement 133
    concentration and temperature in
      the flame recombination
      zone, measurement of   98
    energy-level diagram—electronic
      and vibrational structure 68*f*
    energy-level diagram—rotational
      structure 68*f*
    excited-state population dis-
      tribution for 70*f*
    in flames, LIFS applied to 131–136
    in flames, studies of LIF on   13
    LIF in   12
    measurements 111
    molecule, diagram of the energy
      levels for 90*f*
    profile through the low-pressure
      flat flame 135*f*
    radical balance reaction 111
    temperatures, LIF and 89

I

I$_2$
  extracavity absorption coefficient of 452
  fluorescence 168
  for LIF studies, advantages in using 168
  pressure, fluorescence as a function
    of 452
  vapor, He gas with 169
  I$_2$ (B–X), visible fluorescence from .. 171*f*
IDLS (*see* Intracavity dye laser
    spectroscopy)
Ignition of a fuel–oxidizer mixture 344
Induced emission process 62
Induction parameter 346
Induction period of methane oxidation 357
Industrial furnace, optical arrange-
    ment for droplet size measure-
    ment in 447*f*
Inelastic light scattering
  molecular information from 208
  processes 209
  space-resolved measurements from.. 210
  time-resolved measurements from .. 210
Initial ozone mole fraction(s) 366
  calculated atomic oxygen profiles
    for unity 371*f*
  calculated temperature profiles
    for unity 372*f*
Input coefficients, ratio of the
  values of 371*f*
Input coefficients, Warnatz' 366

Instantaneous pressure distributions
 on a half-cyclinder 353*f*
Instantaneous time concentration
 profiles 433
Integration method, selected
 asymptotic 341
Intensity
 fluorescence 190
 limit, low- 64
 limit (saturation), high- 64
 ratio of anti-Stokes to Stokes
  vibrational Raman scattering .. 262*f*
Internal quantum states, measure-
 ments on 4
Intracavity
 enhancement 452
 laser 438
 dye laser spectroscopy (IDLS) 451
  BP theory of 452
  continuous-wave (CWIDLS)..451–454
  experimental arrangement 451
  HST theory of 452
Inverse laser intensity, ratio of populations as a function of 142*f*
Inverse Raman spectroscopy (IRS) .. 319
 flashed-cyclohexane solutions 322
 and RIKES spectra, apparatus used
  to obtain 321*f*
Ionization
 of alkali metals in flames 175
 collisional 175
  on of sodium atoms 183–187
 for high-principal quantum
  numbers 179
 laser-enhanced 187
  in flame diagnostics, applications
   of 187
 laser-induced 192
 rate(s)
  for *n*-manifold states 179
  constant, state-specific 176
  for *n*-manifold states, global
   quenching 179
  for Na, global quenching 179
 signals of sodium atoms 183
 signal vs. laser power, log–log
  plot of 187
Irradiation, laser 147
IRS (*see* Inverse Raman spectroscopy)
Isothermal Rayleigh scattering 435

J

Jet diffusion flame, temperature profile
 in turbulent 440*f*
 RMS 440*f*

K

K_I vs. pressure for $C_2(X^1\Sigma_g^+ + H_2$,
 plot of 384*f*
 coefficients 367
 mechanisms in the combustion of
  cyclic nitramines, gas-phase 365
 model, reaction trajectories calculated using a detailed 362*f*
 models of hydrocarbon combustion,
  development of 85
Kinetics of CH radical reactions important to Hydrocarbon com-
 bustion systems 397–401
 modelling and 9

L

Lagrangian
 grid 351
 hydrodynamics framework 339
 pathlines at various stages of a
  Rayleigh–Taylor collapse 353*f*
Laminar
 microscopic flow equations 339
 premixed
  CH_4–air flames, species profiles
   for 86
  CH_4–air flames, laser probes
   of 85–101
  CH_4–flames, temperature profiles
   for 86
  steady-state flames, detailed
   modelling of 365–372
  unbounded flame for a multicomponent ideal gas mixture,
   equations describing 365
 propane diffusion flame, radial
  temperature profiles in 279*f*
Laser(s)
 absorption 7
  spectroscopy of combustion gases
   using a tunable IR 413–425
  spectroscopy, optical arrangement for tunable diode 414*f*
 applications as a combustion
  diagnostic 19
 Ar^+ 168
 chemistry and combustion 3–17
 combustion probes 4
  energy-level diagrams for
   spectroscopic 5*f*
 and computer system control flow
  chart 241*f*
 as diagnostic instruments 19
 diagnostics for practical combustor
  probing spatially precise 275–299
 Doppler velocimeter, dual-beam 443

Laser(s) (*continued*)
Doppler velocimetry 207
dye 176
flashlamp-pumped tunable
(Chromatrix CMX-4) 104
laser-pumped 293
Nd:YAG pumped 41, 145
molectron 148
system for nitric oxide fluoresc-
ence measurements in
$CH_4 + O_2 + N_2$ flame..154*f*, 155*f*
tunable 103, 160
-enhanced ionization 187
in flame diagnostics, applications
of 187
excitation
scan of a H_2–O_2–N_2 flame 114
spectrum(a)
in flames with 1% H_2S,
SH $A^2\Sigma^+$–$X^2\Pi$ 117*f*
for nitric oxide in
$CH_4 + O_2 + N_2$ flame 157*f*
for nitric oxide in N_2 .... 156*f*
for OH $A^2\Sigma^+$–$X^2\Pi$ in a
C_2H_2–O_2–N_2 flame 112*f*
source 62
to the $Na(^2P_{3/2})$, quenching
processes for saturated 191*f*
to the $Na(^2P_{3/2})$, radiative
processes for saturated 191*f*
of OH, collisional redistribution
of excited-state population
following 13
excited
CN fluorescence intensity with
laser spectral irradiance in
a nitrous oxide–acetylene
slot torch 297*f*
fluorescence
flame temperatures density
profiles of 199
plasma temperatures density
profiles of 199–203
spectrum in a nitrous oxide–
acetylene slot torch, comparison of CN flame
emission and 296*f*
spectrum in an oxy-acetylene
slot torch, comparison of
CH flame emission and .. 294*f*
level, steady-state rate equation
for 69
NO fluorescence spectrum in an
NO-doped methane–air
premixed flat flame 297*f*
state, collisional quenching of 89
flame measurements, species
amendable to 56*f*

Laser(s) (*continued*)
flashlamp-pumped tunable dye 183
flowing chemical 167
fluorescence
excitation in OH, multilevel
model of response to 137–144
exciting and detecting wave-
lengths 108*t*
measurements in flames, optical
system for 110*f*
saturated 189
advantages of 292
CARS vs. 271
studies of lithium in fuel-rich
H_2–O_2–N_2 flames 189–194
studies of sodium in fuel-rich in
H_2–O_2–N_2 flames 189–194
theory of 292
techniques for NO 153
-induced fluorescence (LIF) ..6, 167, 389
combustion intermediates
detected by 12*t*
excitation spectrum of the C_2
Swan band system in an
acetylene–air flame 43*f*
excitation spectrum of
$C_2O(\tilde{A}^3\Pi_r$–$X^3\Sigma^-)$ 392*f*
and hydroxyl concentration(s) .. 89
as measured by 99*f*
and hydroxyl temperatures 89
instrumentation 161*f*
intensity per transition strength
vs. energy of initial rota-
tional state, plot of 91*f*
measurements, problems 131
in OH 12
on OH in flames, studies of 13
optical saturation in 137
of PCAH in a flame 159–164
signals from flame species,
experimental arrangement
used to measure 42*f*
signals, quantitative interpreta-
tion of 6
spectroscopy (LIFS) 61
from atomic to molecular sys-
tems, extension of
saturation 40
and concentration measure-
ments 80
detectability interferences 72
detectability limits of 71
and excitation dynamics 80
in flames 61–80
applied to OH 131–136
measuring the population of an
excited atomic state by .. 81
and molecular systems 67

Laser(s) (*continued*)
  -induced fluorescence (LIF)
    (*continued*)
    spectroscopy (*continued*)
      for near-resonant excitation
        of sodium 77
      saturated 19, 36
      signal, theoretical considera-
        tions of 62
      and temperature measure-
        ments 80
      spectrum of the $A^2\Delta \longleftrightarrow X^2\pi$
        transition of the ground-
        state CH radicals 400*f*
      studies, advantages in using I_2 168
      for studying the chemistry of
        sulfur 103
      as tool for the study of flame
        chemistry 103–129
  -induced ionization 192
  intensity 137
    becomes large, populations as 139
    ratio of populations as a function
      of inverse 142*f*
  interferometry, droplet-size
    measurements in reacting
    flows 443–449
  intracavity 438
  irradiation 147
  light-scattering techniques for
    measuring flame gas properties   207
  multipass cell for pulsed Raman
    scattering diagnostics,
    Nd:YAG 255–258
  output power, spectrophone
    response as a function of 460*f*
  phase-R corporation DL-1400
    flashlamp-pumped tunable 160
  power, fluorescence signal vs. 150*f*
  power, log–log plot of ionization
    signal vs. 187
  probe(s)
    for combustion applications 19
    methods, energy-level diagrams
      for 4
    of premixed laminar CH_4–
      air flames 85–101
    purpose of 3
    results 11
  pulse(s)
    energy 160
    fluorescence signal vs. time,
      normalized 147*f*
    low-pressure flames and short-
      duration 131
    pumped dye lasers 293
    Q-switched neodymium 295
  Raman spectroscopy, species pro-
    files obtained by 89

Laser(s) (*continued*)
  Raman technique vs. CW laser tech-
    nique for pdf measurements,
    pulsed 253
  resonance absorption
    apparatus, CO 404*f*
    studies of CH_2 radical reactions,
      CO 403–409
    studies of $O(^3P)$ 1-alkynes
      reactions, CO 403–409
  scattering apparatus 437*f*
  spectral bandwith via different
    methods, comparison between
    the values of 198
  spectral irradiance in a nitrous
    oxide–acetylene slot torch 297*f*
  spectroscopic probe methods 4
    experimental setups for 8*f*
  source characteristics, comparison
    of vibrational Raman scatter-
    ing fluctuation measurement
    capabilities for 211
  studies of the kinetics of,
    $C_2O(A^3\Pi_i$ and $X^3\Sigma^-)$,
    pulsed 389–396
  technique for pdf measurements,
    pulsed laser Raman technique
    vs. CW 253
  velocimetry and Raman scattering
    diagnostics optical layout for.. 245*f*
LDV optics, geometry of the dual-
  beam 445*f*
LED (*see* Light emitting diode)
Level populations of a four-level
  molecular model, for 147
Level 1 to 2 steady-state rate equations, for excitation from 65
LIF (*see* Laser-induced fluorescence)
LIFS (*see* Laser-induced
  fluorescence spectroscopy)
Lifetimes for $C_2O(A^3\Pi_i)$, radiative .. 395
Lifetime and quenching data, C_2O 395
Light
  absorption by particles, measure-
    ment of 457
  absorption by soot, measurement of   457
  emitting diode (LED) 250, 251*f*
    output, quasi-sinusoidal 250
    source, experimental results from 252*f*
  extinction spectrum, normalized 458
  -scattering
    inelastic 209
    information from molecular 208
    space-resolved measurements
      from 210
    time-resolved measurements
      from 210
    information from elastic
      molecular 208

Light (*continued*)
-scattering (*continued*)
probes 207
properties and particulate mass
concentrations 457
Line
analysis, Voigt 176
intensity 415
reversal temperature profiles in
H_2–O_2–N_2 flame, sodium 110*f*
reversal temperature profile
measurements were made for
H_2–O_2–N_2, sodium 109
source, Lorentzian atom profile 195
strength 415
correction factors for pure rota-
tional Raman scattering,
rotational–vibrational 233*f*
Linear absorption
coefficient 427
values for a six-angle case,
reconstructed 430
values for a twelve-angle case,
reconstructed 430
Linear two-line methods 199
Lineshape function 416
Linewidth, Doppler-broadened 417
Liquid photolytic reaction, application of single-pulse nonlinear
Raman techniques to 319–327
Lithium
energy-state diagram for 193*f*
flame distribution 189
in fuel-rich H_2–O_2–N_2 flames, saturated laser fluorescene studies
of 189–194
over states in a H_2–O_2–N_2 flame,
fractional distribution of
sodium and 193*f*
-scattering techniques for measur-
ing flame gas properties, laser 207
Lorentzian
atom profile, line source 195
broadened absorption line profile .. 196
dispersion function 195
Low-intensity limit 64
Low-pressure flame(s)
quenching studies in 133
and short-duration laser pulses 131
flat, OH profile through 135*f*
LV data, acquisition, optical
layout for 220
LV–Raman scattering timing
sequence 243*f*

M

Macroscopic flow equations, laminar .. 339
Magnesium oxide
B^1 + A^1II transition, fluorescence
excitation spectrum of 46*f*

Magnesium oxide (*continued*)
B^1 + $-X^1$ + transition in an
C_2H_2–air aspirating slot
burner, fluorescence excitation
spectrum of 45*f*
energy level diagram for 41
n-Manifold states 176
global quenching ionization rates
for 179
overall ionization rates 179
Mean quenching rate 132
Mercedes diesel 459
spectrophone response to exhaust
particulate from a cougar
PROCO and 461*f*
Mercury cougar PROCO, gasoline
powered 459
Metals in flames, on saturated
fluorescence of alkali 189–194
Methane
–air
flame(s)
CARS spectrum of H_2O in 287*f*
premixed 30f
CH_4 profiles for 94*f*
CO profiles for 96*f*
CO_2 profiles for 94*f*
concentration profiles from 387*f*
experimental results for a
fuel-lean 92
flat
CARS spectra in the region
of the CO Q-branch
from 39*f*
as a function of fuel–air
equivalence ratio, CO
partial pressure in 423*f*
as a function of rotational
quantum number CO
collision halfwidth in .. 424*f*
premixed, laser-excited NO
fluorescence spectrum
in 297*f*
Voigt fit to absorption line
profile for XO in 422*f*
fluorescence scans of emission
in OH in 16*f*
H_2 profiles for 96*f*
H_2O profiles for 95*f*
H_2O spectra for a premixed 27
hydroxyl concentration profile
for a stoichiometric 98
laser probes of premixed
laminar 85–101
NO concentration profiles 376*f*
O_2 profiles for 95*f*
OH in 13
postulated mechanism for 87*t*
radiative trapping of sodium in 76*f*

Methane (*continued*)

  –air (*continued*)

    rate constant measurement in .. 375

    rotationally resolved fluores-
      cent emission in OH, in .. 15f

    species profiles for premixed,
      laminar 86

    temperature profiles for 94f, 95f, 96f
      premixed, laminar 86

    theoretical results for a fuel-
      lean 92

  –argon asymmetrical diffusion jet .. 428

  jet

    cross section for six-angle case,
      reconstruction of 432f

    cross section for twelve-angle
      case, reconstruction of 431f

    diffusion, multiangular absorp-
      tion measurements in 427–433

    experiment, apparatus for 429f

    in a mixed gas system, single-pulse
      measurements made for 27

  –O_2 flame species concentrations
    and temperature, comparison
    of 93f

  oxidation

    controlled by free radicals,
      rate of 357–363

    induction period of 357

    processes governing the rate of
      heat release during 357

  profiles for CH_4–air flame 94f

  single-pulse measurements for 27

  using OMA detection and a single-
    laser pulse, broad-band CARS
    spectrum of 32f

  $CH_4 + O_2 + N_2$ flame 153

  dye laser system for nitric oxide
    fluorescence measurements
    in 154f, 155f

  laser excitation spectrum for nitric
    oxide 157f

Microdensitometer trace(s)

  of IRS and OHD–RIKES spectra
    of cyclohexane solutions 326f

  of the OHD–RIKES single-pulse
    spectra of solutions of benzene 323f

  of RIKES spectra of solutions of
    cyclohexane in CCl_4 .... 321f

  of spectra from solutions of
    cyclohexane 325f

  of spectra from solutions of cyclo-
    hexane in CCl_4 .... 324f

Mie scattering 73

  from particles 438

Mixing in gases using a CW laser,
  fast turbulent 247–253

Mode competition, effect of 452

Model(s)

  for atomic systems, two-level 40, 47

  phenomenological 331

  three-dimensional 334

  two-dimensional 334

  verification 11

Modeled CARS spectra for N_2 in a
  flame as a function of the bulk
  gas temperature 27

Modelling

  asymptotic approach to 340

  combustion systems 332

  the data formations in flames 176

  detailed

    of combustion 331–354

    purpose of 333

    physical complexity of 335

  and kinetics 9

  onset and other transient turbulence
    phenomena 339

  problems in reactive flows 333

Molectron Nd:YAG pumped dye laser 148

Molecular

  input parameters 3

  light scattering, information from
    elastic 208

  light scattering, information from
    inelastic 208

  model 145, 146f

    for the level populations of 147

  scattering processes, primary 214

  species, saturation in 77

  species, spatial density profiles
    of 201

  systems, extension of saturation
    spectroscopy from atomic to .. 40

  systems, LIFS and 67

Molecule concentration, radiation
  intensity and 4

Molecule spectrum, diatomic larger 81

Mollow's theory 77

Multiangular absorption measurements in a methane diffusion
  jet 427–433

Multifluid conservation equations 331

Multilevel model of response to laser
  fluorescence excitation in OH 137–144

Multipass cell(s) 438

  for enhancement of CW Raman
    scattering, optical 255

  experiment, far field isolator and
    pulsed 256f

  nitrogen Stokes vibrational Raman
    signal from pulsed 257f

  performance of pulsed 258

Multiphoton

  excitation of fluorescence 9

  UV excimer laser photolysis, ($a^3\Pi_u$)
    produced by 381–387

Multiphon (*continued*)
  UV excimer laser photolysis,
    $C_2(X^1\Sigma_g^+)$ produced by 381–387
  UV–photolysis–laser fluorescence
    detection system 382*f*
Multiple space scales 334
Multiple time scales 333

N

N_{2-1} at 2000 K, calculated band profiles of Stokes vibrational Raman
  scattering from 236*f*
NH in a NH_3–O_2 flame, excitation
  scan of 10*f*
NH_3–O_2 flame, excitation scan of
  NH in 10*f*
Nd:YAG (*see* Neodymium-doped
  yttrium aluminum garnet)
Near-resonant
  excitation, fluorescence spectrum
    from 78*f*
  excitation of sodium, LIFS for 77
  Rayleigh scattering 75
Neodymium-doped yttrium aluminum
  garnet (Nd:YAG) 86, 295
  laser multipass cell for pulsed
    Raman scattering
    diagnostics 255–258
  pumped dye laser(s) 41, 145
  molectron 148
Neodymium laser, Q-switched 295
Nitramines, gas-phase kinetic mechanisms in the combustion of cyclic   365
Nitric oxide (NO) 153–158
  -acetylene slot torch, comparison of
    CN flame emission and laser-
    excited fluorescence spectrum
    in 296*f*
  concentration profiles 375
    from CH_4–air flame 376*f*
  detection in flames by laser
    fluorescence 153
  -doped CH_4–air premixed flat flame,
    laser-excited NO fluorescence
    spectrum in 297*f*
  excitation spectra for 153
  flame fluorescence measurements,
    experimental system for 153
  flat flame burner experiments meas-
    uring CO and 418
  fluorescence excitation in 298
  fluorescence spectrum in an NO-
    doped CH_4–air premixed flat
    flame, laser-excited 297*f*
  formation 377
    in high temperature hydrocarbon
      combustion systems 399
  laser-fluorescence techniques for 153
  in N_2, laser excitation spectrum for 156*f*

Nitrogen (N_2)
  analysis of the anti-Stokes-to-Stokes
    intensity ratio from 260
  BOXCARS spectrum of 277*f*
  in a sooting flame 37*f*
  CARS spectrum(a)
    comparison of single-pulse and
      averaged 278*f*, 282*f*
    in a flame as a function of the
      bulk gas temperature,
      modelled 27
    as a function of bulk gas tem-
      perature plot of calculated .. 29*f*
    gas in the combustion zone of a
      homogeneous flat flame
      burner 28*f*
    probed within the homogeneous
      region of a flat flame
      burner 27
    spatial variation of temperature
      from averaged 281*f*
  CARS, time averaged 304
  concentration vs. temperature in
    H_2–air turbulent diffusion
    flame 224*f*, 225*f*, 226*f*, 227*f*
  density
    in a combusting stratified charge
      engine 259
    fluctuations in 264
    measurements in a turbulent
      flow, Stokes 259
    vs. crank angle of the combus-
      tion chamber 264
  on exit plane of combustor, time-
    averaged spectrum of 317*f*
  at 1700 K onto measured spectra,
    overlay of computer-generated
    spectrum of 305*f*
  laser excitation spectrum for nitric
    oxide in 156*f*
  number density
    one standard deviation of
      fluctuations in 265*f*
    near time of flame arrival, pdf's
      of 266*f*
    vs. crank angle, relative mean 265*f*
  Q-branch spectra of 304
  Raman and Rayleigh scattering
    from 216*f*
  relative Stokes vibrational Raman
    intensity for 261*f*
  rotational Raman spectra 232
  spectrum recorded on OMA during
    combustion, single-pulse 305*f*
  Stokes–anti-Stokes intensity ratio 220
  Stokes vibrational Raman signal
    from pulsed multipass cell 257*f*
  temperature measurements from 276

Nitrogen (N_2) (*continued*)
temperatures in primary reaction
zone and recombination
regions, comparison of OH and 100*f*
thermometry 276
accuracy of CARS 276
NO (*see* Nitric oxide)
Normalized laser pulse, fluorescence
signal vs. time 147*f*
Normalized light extinction spectrum 458
Non-adiabatic quenching process,
physical 190
Noncolinear beam phase-matching
techniques 33
Nonlinear Raman techniques to a
liquid photolytic reaction, application of single-pulse 319–327
Nonlinear time-dependent slow flow
equations 345
Nonresonant CARS signals 7
Number density of any rotational
state, steady-state rate equation
for 67

O

O + C_2H_2 reaction 405
O + N_2 reaction rate 375–380
constant 375
O + $N_2 \rightarrow$ NO + N, rate constants
for 379*f*
OH⁻ (*see* Hydroxyl radical)
O(^{3}P)
1-alkynes
absolute rate constants for CO
production at 300 K for 408
average CO vibrational energies
for CO production at 300 K
for 408
reactions, CO laser resonance
absorptions studies of 403–409
atoms, RCH diradicals formed in
series of 1-alkyne reactions
with 405
reactions of CH_2 with 406
O(^{3}P) + CH_2, vibrational energy
distributions of the CO formed in 407*f*
O(^{3}P) + C_2H_2 reactions, vibrational
energy distributions of the CO
formed in 404*f*
O(^{3}P) + $C_4H_9C_2H$ reactions, vibrational energy distributions of the
CO formed in 404*f*
Office National d'Etudes et de
Recherches Aerospatiales
(ONERA) 311–318
CARS diagnostics of reactive
media at 311–318
CARS research effort at 311

OHD–RIKES (*see* Optical heterodyne
detection of the RIKES signal)
OMA (*see* Optical multichannel
analyzer)
One-photon absorption in a hydrogen–
oxygen–argon flame 183–187
One-photon transition(s) 187
of sodium 184
ONERA (*see* Office National d'Etudes
et de Recherches Aerospatiales)
Optical
arrangement for droplet-size measurement in an industrial
furnace 447*f*
arrangement for tunable diode laser
absorption spectroscopy 414*f*
data acquisition system for vibrational Raman scattering temperature measurement 218*f*
heterodyne detection of the RIKES
signal (OHD–RIKES) 322
single-pulse spectra of solutions
of benzene, microdensi-
tometer traces of 323*f*
spectra of cyclohexane solutions,
microdensitometer traces of
IRS and 326*f*
layout for laser velocimetry data
acquisition 220
layout for laser velocimetry and
Raman scattering diagnostics .. 245*f*
multichannel analyzer (OMA) 276, 304
detector and a single-laser pulse,
broad-band CARS spectrum
of a gas mixture using 34*f*
during combustion, single-pulse
N_2 spectrum recorded on 305*f*
multipass cells for enhancement of
CW Raman scattering 255
saturation in LIF 137
system
fluorescence power of 61
for laser fluorescence measurements in flames 110*f*
single-beam 413
transition rates calculation of 179
Optogalvanic signal for excitation of
sodium in a H_2–air flame 177*f*
Optogalvanic spectroscopy 7, 175
Overall ionization rates for *n*-mani-
fold states 179
Oxide–acetylene slot torch, compari-
son of CN flame emission and
laser-excited fluorescence
spectrum in a nitrous 296*f*
Oxide–acetylene slot torch, variation
of laser-excited CN fluorescence
intensity with laser spectral
irradiance in a nitrous 297*f*

Oxy-acetylene slot torch, comparison
of CH flame emission and laser-
excited fluorescence spectrum
in 294*f*
Oxygen
$C_2X^1\Sigma_g^+$ vs. $a^3\Pi_u$ reactions with 387
CARS measurements of tempera-
ture from H_2 and 280
CARS spectrum(s) of
at one atmosphere pressure, a
rotational 39*f*
in H_2–air diffusion flame 290*f*
at 2000 K 289*f*
in an H_2–O_2 premixed flame, temperature analysis plot for rotational Raman scattering from .. 234*f*
jet, Raman scattering from 250
profiles for a CH_4–air flame 95*f*
profiles for unity initial ozone mole
fraction, calculated atomic 371*f*
Q-branch spectra of 304
reactions of CH_2 with 406
rotational Raman spectra 232
Ozone
decomposition reaction, tempera-
ture data for 370
mole fraction
calculated atomic oxygen profiles
for unity initial 371*f*
calculated temperature profiles
for unity initial 372*f*
comparison of burning velocities
over a range of 369*f*
computed profiles for unity initial 369*f*
initial 366
results for 365–372

P

Padley–Sugden burner 103
sectional sketch of 105*f*
Pair population, depletion of 139
Pair population, repletion of 139
Particle(s)
concentration, maximum 444
mass concentrations, light-scattering properties and 457
measurement of light absorption by 457
number concentration 444
sizing measurements in water sprays 444
PCAH (*see* Polycyclic aromatic
hydrocarbons)
PDECOL 366
Pdf's (*see* Probability density
functions)
Peak detection, saturation method
with 200

Phase
-matching
diagrams for CARS signal
generation 21*f*
diagram for three-wave mixing .. 273
techniques, noncolinear beam 33
-R corporation DL-1400 flashlamp-
pumped tunable dye laser 160
Phenomenological models 331
Photoacoustic
effect 457
as a soot monitor, validation of .. 458
spectroscopy to characterize and
monitor soot in combustion
processes 457–462
spectrum of airborne soot generated
from a propane–O_2 flame 460*f*
spectrum of the propane-generated
soot 458
Photodissociation of hexafluoro-
butyne-2 and acetylene 383
Photolysis, $(a^3\Pi_u)$ produced by multiphoton UV excimer laser 381–387
Photolysis, $C_2(X^1\Sigma_g^+)$ produced by
multiphoton UV excimer laser 381–387
Photolytic reaction, application of
single-pulse nonlinear Raman
techniques to a liquid 319–327
Photomultiplier tube (PMT) 259
Photon(s)
absorption 62, 175
count 72
distribution(s) 247
using a LED source, detection
of 251*f*
fluorescence 132
flux (P_F) 71
fractional uncertainty 72
Physical
complexity of detailed modelling 335
non-adiabatic quenching process 190
quenching cycle, dynamic
excitation– 190
Plasmas fluorescence excitation profiles, saturation broadening in 195–198
Plasma temperatures density profiles
laser-excited fluorescence 199–203
PMT (*see* Photomultiplier tube)
Poisson statistics 72
Polar coordinates, projection 429*f*
Polarization of a medium 319
Polarization-sensitive CARS 288
Polycyclic aromatic hydrocarbons
(PCAH) 159–164
in combustion systems 159
in a flame, LIF 159–164
fluorescence signal for 163

Population(s)
  depletion of the pair 139
  distribution(s) 4
    for OH, excited-state 70f
    rotational 139
  dummy-level 144
  of an excited atomic state by
    LIFS, measuring 81
  excited vibrational level 131
  as laser intensity becomes large 139
  as a function
    of inverse laser intensity, ratio of 142f
    of rotational quantum number,
      fractional 140f, 141f
    of transfer cross section, dummy-
      level 143f
  ratio, quasi-equilibrium 65
  repletion of 139
Premixed flames
  laboratory, determination of
    temperatures in 231
  full-rich H_2–O_2–N_2 103
  laminar
    CH_4–air, laser probes of 85–101
    CH_4–air species profiles for 86
    CH_4–air, temperature profiles for 86
  steady-state, detailed
    modelling of 365–372
  unbounded for multicomponent
    ideal gas mixture, equations
    describing 365
Pressure
  distributions of a half-cylinder
    instantaneous 353f
  effect on the rate of the CH + N_2
    reaction at room temperature .. 400f
  forces on a half-cylinder, model
    calculation of wave-induced 348
  -jet nozzle, experimental results with 444
  regions of the long components 391
  variations over a half-cylinder 350f
Primary
  molecular scattering processes 214
  reaction zone of atmospheric-pres-
    sure hydrocarbon–air flames .. 85
  reaction zone and recombination
    regions, comparison of OH
    and N_2 temperatures in 100f
Probability density function(s) (pdf) .. 209,
    221f, 242
  basic quantities in analyses of CW
    laser scattering for 248
  calculation 247, 249
  data obtained with the Raman
    Stokes–anti-Stokes technique,
    accuracy of temperature 220
  data, recent results for 217

Probability density function(s) (pdf)
    (continued)
  measurements, pulsed-laser Raman
    technique vs. CW laser tech-
    nique for 253
  of nitrogen number density near
    time of flame arrival 266f
  for temperature × velocity for
    turbulent diffusion flame ..221f, 243f
Product channels for C_2 + O_2 386
Profiles
  of laser-excited fluorescence, flame
    temperatures density 199–203
  of laser-excited fluorescence, plasma
    temperatures density 199–203
  for SO, fluorescence 119
Projection in polar coordinates 429f
Propane
  diffusion flame, radial temperature
    profiles in a laminar 279f
  -generated soot, airborne 459
  -generated soot, photoacoustic
    spectrum of 458
  -O_2 flame, photoacoustic spectrum
    of airborne soot generated
    from 460f
Pseudo-first-order disappearance rate
    constants for ground-state C_2O .. 394f
Pulse count distribution 250
Pulse shape, SO_2 fluorescence 119
Pulsed
  fluorescence 80
  -laser Raman technique vs. CW
    laser technique for pdf
    measurements 253
  -laser studies of the kinetics of
    $C_2O(A^3\Pi_i$ and $X^3\Sigma^-)$ 389–396
  multipass cell
    experiment, far field isolator and
      532 nm 256f
    nitrogen Stokes vibrational
      Raman signal from 257f
    performance of 258
  Raman-scattering diagnostics, Nd:
    YAG laser multipass cell
    for 255–258
Pumped tunable dye laser flashlamp
    (Chromatix CMX-4) 104
Pumping pulse in electronic states,
    rotational relaxation during 40
Pumping, saturation method for
    sequential 200
Pyrene, fluorescence spectra for 160

Q

Q-branch(es) 214
  from a CH_4–air flat flame CARS
    spectra in the region of CO 39f

Q-branch(es) (*continued*)
spectra of N_2 304
spectra of O_2 304
transitions 235
Q-switched neodymium laser 295
Quantel CARS spectrometer in combustor facility 316*f*
Quantum dependence of the CO collision halfwidth in combustion gases on the vibrational and rotational 421
Quantum number(s)
CO collision halfwidth in CH_4–air flat flame as a function of rotational 424*f*
fractional populations as a function of rotational 140*f*, 141*f*
ionization for high-principal 179
Quasi-equilibrium ratio 147
population 65
Quasi-equilibrium
radical concentration 359
Quasi-sinusoidal LED output 250
Quench
rate, kinetic 107
summation term 107
term expanded 107
Quenching 63, 176
$A^3\Pi_i$ state long-decay component, Stern–Volmer plot for 394*f*
$C_2a^3\Pi_u$ reaction vs. 385
data, C_2O lifetime and 395
collisional
of the laser-excited state 89
rates 6
by rotational relaxation 107
by vibrational relaxation 107
of the excited-state level, electronic 40
ionization rates for Na, global 179
process for saturated laser excitation to the $Na(^2P_{3/2})$ 191*f*
rate
constants
for $C_2O(A^3\Pi_i)$ 395
fluorescence lifetimes 389
long-component 391
mean 132
of OH $A^2\Sigma^+$ 111
vs. flame pressure 134*f*
studies in low-pressure flames 133
term in the fluorescence efficiency .. 106

R

Raman
effect, fundamentals of 212
experiment, experimental arrangements of time-resolved 263*f*

Raman (*continued*)
-induced Kerr effect spectroscopy
(RIKES) 319
spectra 320
apparatus used to obtain IRS and 321*f*
of solutions of cyclohexane in CCl_4, microdensitometer traces of 321*f*
intensity for nitrogen, relative Stokes 261*f*
and Rayleigh scattering from N_2 .... 216*f*
scattering 6
diagnostics 212
Nd:YAG multipass cell for pulsed 255–258
optical layout for laser velocimetry 245*f*
flame diagnostic methods 239
from the gas 247
measurements of combustion properties 207–228
measurement, geometry for typical 216*f*
optical multipass for enhancement of CW 255
from oxygen jet 250
properties of 214
rotational 231
rotational–vibrational line strength correction factors for pure 233*f*
spontaneous (SRS) 6
disadvantage of 271
for measuring gas temperature and species concentration 255
timing sequence 243*f*
vibrational
data, temperature–velocity correlation measurements for turbulent diffusion flames from 239–246
from diatomic molecules 235
fluctuation measurement capabilities for laser source characteristics, comparison of 211
intensity ratio of anti-Stokes-to-Stokes 262*f*
methods for density measurement 209
from N_{2-1} at 2000 K, calculated band profiles of Stokes 236*f*
spontaneous 259
temperature from rotational and 231–237

Raman (continued)
  scattering (continued)
    vibrational (continued)
      signal from pulsed multipass cell,
        nitrogen Stokes vibrational .. 257f
      spectra, N_2 rotational 232
    spectroscopy
      anti-Stokes 19
      coherent anti-Stokes (CARS) 20
      species profiles obtained by laser 89
      in a stratified charge engine,
        time-resolved 259–267
    Stokes–anti-Stokes technique,
      accuracy of the temperature
      pdf data obtained with 220
    technique vs. CW laser technique
      for pdf measurements, pulsed-
      laser 253
    techniques to a liquid photolytic
      reaction, application of single-
      pulse nonlinear 319–327
Radiance, evaluation of the
  fluorescence 195
Radial
  profiles, X 308f
  profiles, Y 307f
  temperature profiles in a laminar
    propane diffusion flame 279f
Radiation-corrected thermocouple-
  derived temperature profiles,
  comparison of the CARS and 304
Radiation intensity and molecule
  concentration 4
Radiative
  lifetimes for $C_2O(A^3\Pi_i)$ 395
  processes for saturated-laser excita-
    tion to the $Na(^2P_{3/2})$ 191f
  trapping 73, 74f
    of sodium in CH_4–air flame 76f
Radical
  balance process, equilibrium of 124
  chain branching reactions 124
  concentration, quasi-equilibrium 359
Radioactive trapping effect for sodium 75
Rate
  coefficient, expressions for 372f
  constant(s)
    for $C_2O(A^3\Pi_i)$, quenching 395
    fluorescence lifetimes and 389
    for CH radical reactions at room
      temperatures 399
    for CO production at 300 K for
      $O(^3P)$ 1-alkynes 408
    for the disappearance of $C_2(a^3\Pi_u)$ 386t
    for the disappearance of
      $C_2(X^1\Sigma^+_g)$ 386t
    experimental system used for
      measuring fluorescence exci-
      citation spectra and reaction 390f

Rate (continued)
  constant(s) (continued)
    for ground-state C_2O pseudo-
      first-order disappearance
      rate 394f
    for $HO + O_3 \rightarrow HO_2 + O_2$ 337f
    at 298 K, $C_2O(X^3\Sigma^-)$ reaction 395t
    long-component quenching 391
    $O + N_2$ reaction 375
    for $O + N_2 \rightarrow NO + N$ 379f
    measurement in a CH_4–air flame 375
    state-specific ionization 176
  equation(s) 62
    for fuel consumption 358
    for the level populations of a
      four-level molecular level .. 147
    steady-state
      for laser-excited level 69
      for the number density of any
        rotational state 67
    for two-level system 63
Rayleigh
  cross section, constant 436
  and fluorescence trapping in
    sodium, comparison of 79f
  to resonance fluorescence signal,
    ratio of 78f
  scattered light, intensity of 435
  scattering 72
    experiments 435
    from N_2, Raman and 216f
    near-resonant 75
    temperature measurement in tur-
      bulent flames via 435–441
    temperature and species variation 436
  –Taylor collapse, Lagrangian path-
    lines at various stages of 353f
  –Taylor unstable fluid layer 352
Reaction(s)
  and flow, complicated 339
  rate constants, experimental system
    used for measuring fluores-
    cence excitation spectra and 390f
  trajectories calculated with Sene-
    nov's model 361f
  trajectories calculated using a
    detailed kinetics model 362f
  zones 133
Reactive flow calculations 340
Reactive flows modelling, problems in 333
Recombination regions, comparison of
  OH and N_2 temperatures in pri-
  mary reaction zone and 100f
Recombination zone, measurement of
  OH concentration and tempera-
  ture in the flame 98
Reconstructed linear absorption
  values for a six-angle case 430

Reconstructed linear absorption
  values for a twelve-angle case 430
Reconstruction approach for combus-
  tion diagnostics, preliminary
  assessment of the tomographic 430
Redistribution, collisional 77
Relative Stokes vibrational Raman
  intensity for nitrogen 261f
Relaxation, collisional quenching by
  rotational 107
Relaxation, collisional quenching by
  vibrational 107
Relaxation, rotational 69
Repletion of the pair population 139
Resolution of CARS, spatial 311
Resolution of CARS, spectral 312
Resonance CARS 315
Resonance fluorescence signal, ratio
  of Rayleigh-to- 78f
Resonant CARS signals 7
Resonant susceptibility associated
  with a homogeneously broadened
  Raman transition 20
RIKES (see Raman-induced Kerr
  effect spectroscopy)
RMS temperature profile in turbulent
  jet diffusion flame 440f
Room-temperature
  cell experiments 417
    Voigt fit to absorption line profile
      for CO in 419f
  pressure effect on the rate of the
    CH + N$_2$ reaction at 400f
  rate constants for CH radical
    reactions at 399
Rosin–Rammler mean diameter 444
Rotational
  CARS spectrum of air at one
    atmosphere pressure 42f
  CARS spectrum of O$_2$ at one
    atmosphere pressure 39f
  distribution(s) 13
  population 139
  interactions, effects of
    vibrational– 231–237
  Quantum number(s)
    CO collision halfwidth in CH$_4$–
      air flat flame as a function
      of 424f
    dependence of the CO collision
      halfwidth in combustion
      gases on the vibrational and 421
    fractional populations as a
      function of 140f, 141f
  Raman scattering 231
  from O$_2$ in an H$_2$–O$_2$ premixed
    flame, temperature analysis
    plot for 234f
  rotational–vibrational linestrength
    correction factors for pure 233f

Rotational (continued)
  from O$_2$ in an H$_2$–O$_2$ premixed flame,
    temperature analysis plot for
    (continued)
    and vibrational, temperature
      from 231–237
    –vibrational linestrength correc-
      tion factors for pure 233f
  Raman spectra, N$_2$ 232
  Raman spectra, O$_2$ 232
  relaxation 69
  during the pumping pulse in
    electronic states 40
  state, steady-state rate equations for
    the number density of any 67
  structure, OH energy-level
    diagram— 68f
  temperature of determination 14
  temperature measurement, elec-
    tronic ground-state 89

S

Sample–and–hold (S/H) circuits 242
Saturated
  excitation of the Na(3^{2}P$_{3/2}$) level 189
  fluorescence
    of alkali metals in flames on ..189–194
    measurements for CH and CN,
      summary of 295
    measurements of the hydroxyl
      radical 145–151
  laser
    excitation to the Na(^{2}P$_{3/2}$),
      quenching processes for 191f
    excitation to the Na(^{2}P$_{3/2}$),
      radiative processes for 191f
    fluorescence 189
      advantages of 292
      CARS vs. 271
      studies of lithium in fuel-rich
        H$_2$–O$_2$–N$_2$ flames 189–194
      studies of sodium in fuel-rich
        H$_2$–O$_2$–N$_2$ flames 189–194
      theory of 292
    -induced fluorescence
      spectroscopy 19, 36
  LIF spectroscopy, experimental
    setup for 41
  LIF spectroscopy, historical
    development of 36
Saturation 69
  broadening in flames fluorescence
    excitation profiles 195–198
  broadening in plasmas fluorescence
    excitation profiles 195–198
  energy density 63, 196
  for the two-level model 69
  energy flux 71
  fluorescence signal and 151
  high-intensity limit 64

Saturation (*continued*)
   in LIF, optical 137
   LIF spectroscopy from atomic to
     molecular systems, extension
     of 40
   method with peak detection 200
   method for sequential pumping 200
   in molecular species 77
   spectroscopy data two-level model
     used for 48f
   use of 77
Sautes mean diameter (SMD) 444
Scaling for turbulence modelling 339
Scattergram of temperature and
   velocity in turbulent diffusion
   flame 244f
Scattered light, intensity of Rayleigh .. 435
Scattering
   diagnostics, optical layout for laser
     velocimetry and Raman 245f
   Mie 73
   for pdf, basic quantities in analyses
     of CW laser 248
   processes, primary molecular 214
   processes, typical cross section
     values for 215
   Raman 6
     timing sequence, LV– 243f
   Rayleigh 72
     from N_2 .... 216f
     near-resonant 75
Selected asymptotic integration
   method 341
Semenov's model
   and branching chain reactions 357
   calculations using 360f
   reaction trajectories calculated with 361f
Sensitivity of CARS, detection 312
Sequential pumping, saturation
   method for 200
SH
   excited at 323.76 nm, fluorescence
     spectrum of 114
   measurements 114
   + OH = SO + H_2 in fuel-rich
     H_2–O_2–N_2 flames with 1%
     H_2S, equilibration test of 128f
   in a rich H_2–O_2–N_2 flame, fluores-
     cence profiles for 119
SH $A^2\Sigma + X^2\Pi$
   fluorescence profiles for H_2–O_2–N_2
     flames with H_2S 120f
   fluorescence spectra in flames with
     1% H_2S 118f
   fluorescence and synthetic emission
     spectra for C_2H_2–O_2–N_2 flame,
     comparison of 118f
   laser excitation spectra in flames
     with 1% H_2S 117f
S/H circuits (sample–and–hold) 242

Short-duration laser pulses, low-pressure flames and 131
Short-pulse excitation 145
Similarity
   and detailed model solutions,
     comparisons of 349f
   solution, closed form 345
   solution plotted as a function of
     time 347f
Signal
   averaging, effect of 225
   in CARS, generation of 20
   conditioning circuitry, electronic 242
   intensity in CARS experiments 20
   intensity, fluorescence 47
   levels, CARS 272
   –to–noise ratio (SIN) 255, 322
SIN (*see* Signal–to–noise ratio)
Single
   -beam optical system 413
   -laser pulse, broad-band CARS
     spectrum of CH_4 using OMA
     detection and 32f
   -laser pulse, broad-band CARS
     spectrum of a gas mixture
     using an OMA detector and 34f
   -pulse
     and averaged CARS signatures,
      comparison of 276
     and averaged CARS spectra of
      N_2, comparison of 278f
     CARS for combustion work 320
     CARS spectrum(a) 304
     measurements 27
     measurements for methane 27
     in a mixed gas system 27
     N_2 CARS spectra, comparison of
      averaged and 282f
     N_2 spectrum recorded on OMA
      during combustion 305f
     nonlinear Raman techniques to a
      liquid photolytic reaction,
      application of 319–327
     spectra of, microdensitometer
      traces of RIKES 323f
Six-angle case, reconstructed linear
   absorption values for 430
Six-angle case, reconstruction of
   methane jet cross
     section for 432f
Slot burner 86
   diagram of 88f
   fluorescence excitation spectrum of
     the MgO $B^1\Sigma^+ \leftarrow X^1\Sigma^+$ transi-
     tion in an C_2H_2–air aspirating 45f
   plot of fluorescence signal vs. Na
     concentration in a C_2H_2–air
     aspirating 53f
Slow flow
   approximation 345

Slow flow (*continued*)
equations, nonlinear time-
dependent 345
technique 343*f*
SO
fluorescence profiles for 119
in a H_2–O_2–N_2 flame with 1% H_2S
added, fluorescence spectrum
for SO_2 and 120*f*
and SO_2 measurements 119
SO $B^3\Sigma^-$–$X^3\Sigma^-$
absorption system 119
fluorescence profiles in H_2–O_2–N_2
flames with added H_2S 121*f*
fluorescence profiles in H_2–O_2–N_2
flames with 1% H_2S added 121*f*
SO_2
fluorescence pulse shape 119
+ 2H = SH + OH + H_2O in
fuel-rich H_2–O_2–N_2 flames
with 1% H_2S, equilibration
test of 129*f*
measurements, SO and 119
and SO in a H_2–O_2–N_2 flame with
1% H_2S added, fluorescence
spectrum for 120*f*
SO_2 $^1B_2$–$^1A_1$ fluorescence profiles in
H_2–O_2–N_2 flames with added
H_2S 122*f*
Sodium (Na)
atoms, collisional ionization 183–187
atoms, ionization signal of 183
in CH_4–air flame, radiative trapping
of 76*f*
comparison of Rayleigh and
fluorescence trapping 79*f*
concentration in C_2H_2–air aspirating slot burner, plot of fluores-
cence signal vs. 53*f*
emission signal vs. I/I, plot of 52*f*
energy-level diagram 66*f*
in fuel-rich H_2–O_2–N_2 flames, saturated-laser fluorescence studies
of 189–194
global quenching ionization rates
for 179
in a H_2–air flame, optogalvanic
signal for excitation of 177*f*
LIFS for near-resonant excitation
of 77
line reversal temperature profiles in
H_2–O_2–N_2 flame 110*f*
line-reversal temperature profile
measurements were for
H_2–O_2–N_2 109
and lithium over states in a
H_2–O_2–N_2 flame, fractional
distribution of 193*f*

Sodium (Na) (*continued*)
one-photon transition of 184
($^2P_{3/2}$), quenching processes for
saturated-laser excitation to 191*f*
($^2P_{3/2}$) radiative processes for saturated-laser excitation to 191*f*
$3^2P_{3/2}$) level saturated excitation of 189
radioactive trapping effect for 75
system, flame chemistry of 50
two-photon transitions of 175, 184
Soot
absorption coefficient for 459
airborne propane-generated 459
in combustion processes, photoacoustic spectroscopy to characterize and monitor 457–462
measurement of light absorption by 457
monitor, validation of the photoacoustic effect as 458
photoacoustic spectrum of the
propane-generated 458
Sooting flame(s) 159
axial variation of temperature 38*f*
BOXCARS spectrum of N_2 in 37*f*
capability of CARS for measure-
ments in 276
CARS measurements performed in 303
Space
scales, multiple 334
-resolved measurements from
inelastic scattering 210
and time scales in the Gedanken
flame calculation 337*f*
Spatial
density profiles of atomic species 201
density profiles of molecular species 201
resolution of CARS 311
variation of temperature from
averaged CARS spectra of N_2 281*f*
Spatially precise laser diagnostics for
practical combustor
probing 275–299
Species
amenable to laser flame measure-
ments 56*f*
concentration, spontaneous Raman
scattering for measuring gas
temperature and 255
profiles obtained by laser Raman
spectroscopy 89
profiles for premixed, laminar
CH_4–air flames 86
variation Rayleigh scattering,
temperature and 436
Spectral resolution of CARS 312
Spectrophone–dilution tube 459
performance of 459

Spectrophone response 457
    to exhaust particulate from a cougar
       PROCO and mercedes diesel .. 461f
    as a function of laser output power 460f
Spectroscopic
    laser combustion probes, energy-
       level diagrams for 5f
    probe methods, laser 4, 8f
Spectroscopy, optogalvanic 7, 175
Spectroscopy, staurated laser-induced
    fluorescence 36
Spin component conservation 13
Spontaneous emission to lower levels .. 62
Spontaneous Raman scattering (SRS)   6
    disadvantage of 271
    measurements 19
    for measuring gas temperature and
       species concentration 255
    vibrational 259
SRS (see Spontaneous Raman
    scattering) . . .
State-specific cross sections and global
    rate constants for $l \geq 2$ states 180
State-specific ionization rate constant 176
$l \geq 2$ States, state-specific cross sections
    and global rate constants for 180
Steady-state flames, detailed modelling
    of premixed laminar 365–372
Steady-state rate equations
    for excitation from level 1 to 2 65
    for laser-excited level 69
    for the number density of any
       rotational state 67
Stefan–Maxwell relation for diffusion
    velocity 366
Stern–Volmer plot for quenching $A^3\Pi_i$
    state long decay component 394f
Stiff equations 334
Stiff phenomenon 341
Stilbene dye laser 295
Stokes
    –anti-Stokes technique, accuracy of
       the temperature obtained with 220
    band 214
    intensity ratio from nitrogen,
       analysis to the anti-Stokes to .. 260
    nitrogen density measurements in a
       turbulent flow 259
    vibrational Raman
       intensity for nitrogen, relative 261f
    scatering, intensity ratio of anti-
       Stokes to 262f
    scattering from N_{2-1} at 2000 K,
       calculated band profiles of .. 236f
    signal from pulsed multipass cell,
       nitrogen 257f
Stratified charge engine
    nitrogen density in a combusting 259

Stratified charge engine (continued)
    temperature in a combusting 259
    time-resolved Raman spectroscopy
       in 259–267
Sulfur (S)
    B–X bands 114
    $B^3\Sigma^-_u$–$X^3\Sigma^-_g$ . . .
       fluorescence profiles in a
       H_2–O_2–N_2 flame 116f
       in a H_2–O_2–N_2 flame containing
       1% H_2S, fluorescence
       spectrum for 115f
       system 111
    chemistry in fuel-rich H_2–O_2–N_2
       flames, chemical model of 119
    fluorescence intensities 114
    + H_2 = SH + SH in fuel-rich
       H_2–O_2–N_2 flame with 1% H_2S,
       equilibration test of 126f
    LIF for studying the chemistry of .. 103
    measurements 111
Supersonic flows, flow visualization
    in 167–172
Susceptibility associated with a homo-
    geneously broadening Raman
    transition, resnant 20
Susceptibility, third-order 20
Swan band system in C_2, partial
    excitation spectrum of 41
Synthetic emission spectra for a
    C_2H_2–O_2–N_2 flame, comparison
    of SH $A^2\Sigma^+$–$X^2\Pi$ fluorescence
    and 118f

T

Temperature
    analysis plot for rotational Raman
       scattering from O_2 in an H_2–O_2
       premixed flame 234f
    from averaged CARS spectra of N_2,
       spatial variation of 281f
    in a combusting stratified charge
       engine 259
    comparison of CH_4–O_2 flame
       species concentration and 93f
    corrections caused by higher-order
       effects 236t
    data for the ozone decomposition
       reaction 370
    dependence of collision halfwidths 421
    determination, flame gas 239
    of determination, rotational 14
    of flame gas, translational 17
    as a function of position 349f
    for H_2–air
       flame, plots of flame species and 222f

Temperature (*continued*)
for H$_2$–air (*continued*)
turbulent diffusion flame,
nitrogen concentration vs. .. 224*f*,
225*f*, 226*f*, 227*f*
turbulent diffusion flame, probability density functions of .. 219*f*
from H$_2$ and O$_2$, CARS measure-
ments of 280
at L/D = 65 R/D = 0.0 .... 439*f*
measured flame 202
measurement(s) 109
electronic ground-state rotational 89
in H$_2$–air diffusion flame 286*f*
LIFS and 80
from N$_2$ 276
in turbulent flames via Rayleigh
scattering 435–441
pdf data obtained with the Raman
Stokes–anti-Stokes technique .. 220
in premixed laboratory flames,
determination of 231
profile(s)
CARS 283*f*
for a CH$_4$–air flame 94*f*, 95*f*, 96*f*
comparison of the CARS and
radiation-corrected thermocouple-derived 304
in H$_2$–O$_2$–N$_2$ flames, sodium line
reversal 110*f*
in a laminar propane diffusion
flame, radial 279*f*
measurements were for
H$_2$–O$_2$–N$_2$, sodium line-
reversal 109
for premixed, laminar CH$_4$–air
flames 86
in turbulent jet diffusion flame 440*f*
RMS 440*f*
for unity initial ozone mole
fraction, calculated 372*f*
from rotational and vibrational
Raman scattering 231–237
sensitivity of the CARS spectrum of
H$_2$ at 1% concentration,
computed 284*f*
in a sooting flame, axial variation of 38*f*
spatially resolved average 304
and species variation Rayleigh
scattering 436
with velocity, correlations of
density and 220
–velocity correlation measurements
for turbulent diffusion flames
from vibrational Raman
scattering data 239–246
× velocity for turbulent diffusion
flame, pdf for 219*f*, 243*f*

Temperature (*continued*)
and velocity in the turbulent diffusion flame, scattergram of 244*f*
vibrational Raman scattering
methods for 209
Thermalization 13
Thermocouple-derived temperature
profiles, comparison of the CARS
and radiation-corrected 304
Thermocouple temperature, comparison of CARS temperature and
average 309*f*
Thermometry, H$_2$ and combustion 280
Thermometry, N$_2$ 276
Third-order susceptibility 20
Three
-body reactions 124
-dimensional models 334
-level systems, atomic species as 65
-wave mixing 272
phase-matching diagram for 272
Time
averaged N$_2$ CARS 304
-averaged spectrum of N$_2$ on exit
plane of combustor 317*f*
concentration profiles,
instantaneous 433
dependence 139
normalized laser pulse, fluorescence
signal vs. 147*f*
resolved
A-state fluorescence 391
measurements from inelastic light
scattering 210
Raman experiment, experimental
arrangements 263*f*
Raman spectroscopy in a strati-
fied charge engine 259–267
scales in a Gedanken flame calcula-
tion, space and 337*f*
scales, multiple 333
Timestep splitting 341
Topographic reconstruction approach
for combustion diagnostics,
preliminary assessment of 430
Transfer cross section, dummy-level
population as a function of 143*f*
Transient turbulence phenomena,
modelling onset and other 339
Transition(s)
energy density required to saturate
the excited 69
one-photon 184, 187
rates calculation of, optical 179
rovibronic 41
two-photon 175
3S–5S 187
of sodium 175, 184

Translational temperature of flame gas   17
Transmissivity   415
Transport coefficients   368
Trapping, radiative   73
Triangular grid   351
  early in a calculation of wave flow
    over a half-cylinder 350*f*
Tunable
  diode laser absorption spectroscopy,
    optical arrangement for 414*f*
  IR diode laser, absorption spectro-
    scopy of combustion gases using   413–
                                              425
  dye lasers 103, 160
    flashlamp-pumped   183
    (Chromatix CMX-4)   104
    phase-R corporation DL-1400   160
Turbulence   338
  modelling, efficiency of   339
  modelling, scaling for   339
  phenomena, modelling onset and
    other transient   339
  combustor geometry for vibrational
    Raman-scattering temperature
    measurement 218*f*
  diffusion flame
    nitrogen concentration vs. tem-
      perature in H_2–air 224*f*, 225*f*,
                                       226*f*, 227*f*
    pdf for temperature $\times$ velocity
      for 219*f*
    probability density functions of
      temperature for H_2–air 219*f*
    scattergram of temperature and
      velocity in 244*f*
  from vibrational Raman scatter-
    ing data, temperature–
    velocity correlation
    measurements for 239–246
  flame, statistical nature of   240
  flames via Rayleigh scattering, tem-
    perature measurement in 435–441
  flow, Stokes nitrogen density
    measurements in   259
  jet combustion tunnel, fan-induced
    co-flowing   220
  jet diffusion flame, temperature
    profile in 440*f*
  mixing in gases usin a CW laser,
    fast 247–253
Twelve-angle case, reconstructed
  linear absorption values for   430
Twelve-angle case, reconstruction of
  CH_4 jet cross 431*f*
Two
  -coordinate systems, Fourier trans-
    forms of the function in   427
  -dimensional models   334

Two (*continued*)
  -level
    atom   201
    model
      for atomic systems 40, 47
      saturation energy density   69
      used for saturation spectro-
        scopy data 48*f*
    system   137
    rate equation for   63
  -line fluorescence   80
  photon
    absorption in a hydrogen–
      oxygen–argon flame 183–187
    transitions   175
    in Na 175, 184
    3S–5S   187

U

Unity initial ozone mole fraction
  calculated atomic oxygen profies for   371*f*
  calculated temperature profiles for..   372*f*
  computed profiles for 69*f*
UV, multiphoton
  excimer laser photolysis ($a^3\Pi_u$)
    produced by 381–387
  excimer lased photolysis, $C_2(X^1\Sigma_g^+)$
    produced by 381–387
  photolysis-laser fluorescence detec-
    tion system, multiphone 382*f*

V

Velocimeter, dual-beam laser Doppler   443
Velocity(ies) burning   366
  correlations of density and
    temperature with   220
  correlation measurements for tur-
    bulent diffusion flames from
    vibrational Raman scattering
    data, temperature 239–246
  Stefan–Maxwell relation for   366
  for turbulent diffusion flame, pdf
    for temperature $\times$ 219*f*
  in the turbulent diffusion flame,
    scattergram of temperature and   244*f*
Vibration–rotation H_2 CARS
  spectrum   280
Vibrational
  distributions in the states   40
  energy distributions of the CO
    formed
      in $CH_2 + CO_2$ 407*f*
      in $CH_2 + O_2$ 407*f*
      in $O(^3P) + CH_2$ 407*f*
      in $O(^3P) + C_2H_2$ reactions 404*f*
      in $O(^3P) + C_4H_9C_2H$ reactions ..   404*f*
  -level population, excited   131

Vibrational (*continued*)
  Raman intensity for nitrogen,
    relative Stokes 261*f*
  Raman-scattering data
    temperature–velocity correlation
      measurements for turbulent
      diffusion flames from 239–246
    from diatomic molecules 235
    fluctuation measurement capa-
      bilities for laser source
      characteristics, comparison
      of 211
    intensity ratio of anti-Stokes-to-
      Stokes 262*f*
    methods for density measure-
      ments 209
    methods for temperature 209
    from N$_{2-1}$ 2000 K, calculated
      band profiles of Stokes 236*f*
    from pulsed multipass cell,
      nitrogen Stokes 257*f*
    spontaneous 259
  temperature
    measurement, optical data
      acquisition system for 218*f*
    measurement, turblent
      combustor geometry for .. 218*f*
    from rotational and 231–237
    relaxation in the states 40
    –rotational interactions, effects of .. 231–237
  and rotational quantum numbers,
    dependence of the CO collision
    halfwidth in combustion gases
    on 421
  structure, OH energy-level
    diagram—electronic and 68*f*
  transfer rates, X-state 139
Visibility in a functional form 443

Visible fluorescence from I$_2$(B–X) 169, 171*f*
Voigt
  atom profile, Gaussian laser profile–
    fit to absorption line profile for CO
    in CH$_4$–air flat flame 196
    fit to absorption line profile for CO
    in room temperature cell
    experiment 422*f*
  function 419*f*
  line analysis 417
                                                176

W

Warnatz' input coefficients 366
Water
  CARS spectrum 280
    in a CH$_4$–air flame 287*f*
    premixed 30*f*
  profiles for a CH$_4$–air flame 95*f*
  spectra for a premixed CH$_4$–air
    flame 27
  sprays, double-probe used in 447*f*
  sprays, particle sizing measure-
    ments in 444
Wave
  flow over a half-cylinder, triangular
    grid early in a calculation of .. 350*f*
  -induced pressure forces on a half-
    cylinder, model calculation 348
  tank experiments 351
Wavelengths, laser fluorescence
  exciting and detecting 108*t*

X

X radial profiles 308*f*

Y

Y radial profiles 307*f*